The Chemistry and Technology of Edible Oils and Fats and their High Fat Products

FOOD SCIENCE AND TECHNOLOGY

A SERIES OF MONOGRAPHS

Editor Emeritus

Bernard S. Schweigert
University of California, Davis

Series Editor

Steve L. Taylor
University of Nebraska, Lincoln

Food Science and Nutrition Editorial Advisory Board

Douglas Archer
Center for Food & Applied Nutrition
FDA, Washington

Jesse F. Gregory, III
University of Florida, Gainsville

Susan K. Harlander
University of Minnesota, St Paul

John E. Kinsella
Cornell University, Ithaca

Daryl B. Lund
State University of New Jersey,
New Brunswick

Barbara O. Schneeman
University of California, Davis

A complete list of the books in this series is available from the publishers on request

The Chemistry and Technology of Edible Oils and Fats and their High Fat Products

G. Hoffmann

Meeuwenlaan 26
3055 CM Rotterdam
The Netherlands

ACADEMIC PRESS
Harcourt Brace Jovanovich, Publishers
London   San Diego   New York   Berkeley
Boston   Sydney   Tokyo   Toronto

ACADEMIC PRESS LIMITED
24–28 Oval Road
LONDON NW1 7DX

United States Edition published by
ACADEMIC PRESS INC.
San Diego, CA 92101

Copyright © 1989, by
ACADEMIC PRESS LIMITED

All Rights Reserved
No part of this book may be reproduced in any form by photostat, microfilm, or any other means, without written permission from the publishers

British Library Cataloguing in Publication Data

Hoffmann, G.
 The chemistry and technology of edible oils
 and fats and their high fat products
 1. Fats & Oils
 I. Title
 664'.3

ISBN 0-12-352055-X

Typeset by Bath Typesetting Ltd, Bath
Printed in Great Britain by St Edmundsbury Press Ltd, Bury St Edmunds, Suffolk

Foreword

The chemistry and technology of edible oils and fats, as well as the products which contain these ingredients, constitutes an area in which many books deal both with the general and specific aspects of this subject. Consequently it requires courage and creativity to be able to add to the existing range of publications in the field. Géza Hoffmann, who can draw on more than thirty years of experience involving the many facets of the chemistry and technology of edible oils and fats during his work as a research chemist in Unilever Research, has filled the gap in the range of publications. He has produced a concise review which enables the reader to familiarize himself quickly and thoroughly with the subject while at the same time providing up-to-date information to the processors of edible oils and fats in a readily accessible form. I feel confident that this work will meet a long-felt need.

C. Okkerse,
R & D-Director Edible Fats,
Unilever Research Laboratorium Vlaardingen,
Vlaardingen, The Netherlands

Contents

Introduction ix

Acknowledgements xv

1 The Chemistry of Edible Fats
  I On oils and fats in general 1
  II Major lipid components 3
  III Minor accompanying (lipid) components 14
  IV Fat modifying reactions 22

2 Vegetable-oil Separation Technology
  I Raw-material handling 29
  II Preparatory treatments 45
  III Vegetable oil milling 68
  IV Main and side products of oil-milling operations 111

3 Water- and Heat-promoted Fat Separation from Animal and Plant "Fatty Tissue"
  I Theoretical background 121
  II Edible-fat processing (rendering) from terrestrial animals 124
  III Edible-fat processing from marine animals 131
  IV Edible-fat processing from oily fruit pulps and nuts 133
  V Special quality aspects 137
  VI Surface-water and air pollution 138

4 Refining
  I "Chemical" versus "physical" refining 140
  II Deacidification by neutralization 145

viii   CONTENTS

   III   Dry deacidification methods   162
   IV   Bleaching or decolorization   163
   V   Deodorization/deacidification by distillation/heat bleaching   182
   VI   Refinery effluent treatment/fat recovery   198
   VII   Quality control of the refining operation   200

5 Fat-modification Processes
   I   Hydrogenation of edible oils   203
   II   Fractional crystallization and winterization/dewaxing   245
   III   Interesterification: ester interchange/randomization   264

6 Production of Edible-fat Products of High Fat Content
   I   Margarines/minimum calory margarines   279
   II   The production of mayonnaise and related products   325
   III   Shortenings   327
   IV   Vanaspati   329
   V   Prepared mixes   330
   VI   Fundamentals of baking   330
   VII   Packaging and packaging materials   332
   VIII   Warehousing, distribution, storage life   336
   IX   Product-quality assessment   336
   X   Cleaning methods   338

7 Speciality fats   339

8 Miscellaneous subjects
   I   Storage and transport of oils and fats   343
   II   Energy demands of the oil-milling and edible-fat processing operations   354
   III   Automation   362

References   365

Index   371

Introduction

The use of animal and vegetable fats and oils for edible purposes has as long a history as mankind itself. Their use in cooking in the everyday life of ancient cultures, like the Chinese, Egyptian and Greco-Roman, was just as important as it is in ours, notwithstanding their much lower production and per capita use.

The oldest known fats are those found in pots in the tombs of pharaohs. Some written reports in 1400 B.C. described the use of, probably, animal fats when "moving large stones on wooden logs", the biggest engineering achievement of ancient Egypt. Residues of soap have been found in a soap boiler, which was used to produce soap from potash and fats, in the ruins of Pompei. Ancient night lights used animal fat with a sort of cane or reed as the wick. The Romans knew candles made from tallow, the better sorts containing a mixture of tallow and beeswax. Regarding the use of oils, olives of the Mediterranean, rape seed of Europe, sesame seed of India and soya beans of China are amongst the oldest oil plants described by the ancients.

The sources of fats and oils were then, as they are now, plants and animals, the latter being primarily land animals, the use of marine animals being secondary. Mainly under the pressure of an ever growing world population, new fatty raw material resources have been discovered. Agricultural techniques have also been improved by chemical, biochemical and mechanical technological research and development, all this leading to high yielding, climate and disease resistant crop varieties. In times of serious shortage (e.g. wars, crises) both the fatty acids and glycerol, the main building blocks, could be synthesized if not for edible then at least for technical purposes. The biological synthesis of fats by means of microorganisms is completely feasible, although this is not done because of the relatively low pressure on the part of the fat/oil industry.

As regards the required technology, the Chinese knew the use of wedges, whereas Plinius the Elder described the use of double-edged running grinding stones rotating in a stone pan for crushing olives. The heating of crushed masses of seeds before pounding the mass, e.g. by stones in sacks, was a well known part of early technology. The use of animal, wind and water energy instead of human labour when grinding and stamping or pounding the mass were medieval developments. The microbiological way of "rotting" the oil-bearing seeds was also well known, although only practised for technical purposes, e.g. for linseeds. However, the real breakthrough came in the preindustrial age with the application of roller-type grinders, introduced by Smeaton in 1752. The introduction of hydraulic presses for other purposes by the discovery of Brahma (1795) and Neubauer (1800) led to the application of plate presses by Montgolfier (1819). The dehulling and air aspiration of cotton seed was introduced by Waring (1826), whereas the power requirements of hydraulic pressing were reduced by the accumulators developed by Armstrong (1843). The first patents for solvent extraction were issued to Deiss (1855) and Seyferth (1858), proposing carbon disulphide as solvent. Irvine, Richardson and Lundy (1864) issued a patent for the use of hydrocarbon solvents, the basis of the present-day technology. Marticke (1877) proposed the use of screw presses, although the first true expeller was introduced by Anderson in 1906. Mege Mouries obtained the first patent on the production of margarines in 1869, thus providing the fundamental basis for a breakthrough in feeding in the industrializing Western world with a cheap butter-like high-calory spread. The invention of catalytic fat-hardening by Normann in 1902 gave the industry independence from hard animal fats, like lard and tallow, which thus made possible the use of cheap marine oils, such as fish and whale oils. It is a sad fact that this development has led to a near extinction of certain whale populations, but at least today the whales killed serve primarily as a protein source and are used mainly in the petfood and delicatessen industries and not in the edible fat industries. Among other things the continuous-chain and basket-type extractors invented by Bollmann in the 1920s gave a huge impetus to the development of a primarily animal-feed (protein) directed soya-bean industry, using many different types of extractors in order to produce enormous daily amounts of defatted meal and the "side product", the oil.

Having thus reviewed briefly the main steps in the technological development of the industry, it is fair to recall some major points of the scientific development, which have made it possible to reveal the fine structure of oils and fats. In 1741, Geoffroy observed that soap acidified by mineral acids is decomposed to a substance that is soluble in alcohol, this in contrast to the original raw material, the fats. Scheele, in 1779, when saponifying fats with lead oxide, produced a water miscible product, which he called "sweet oil". Today we call this product glycerol. Chevreul, over the years 1813–1825, identified the acid character of the substances we know as fatty acids, and established the alcohol character of glycerol and the role of water in saponifying fats. Because of this work, Chevreul is considered the founder of modern fat chemistry. Berthelot established the triglyceride (triacylglycerol) character of fats and hydrolysed and esterified them, respectively, in closed vessels under pressure. The findings of Berthelot over the years 1853–54 were the basis for the use of fatty acids for many technical purposes. The introduction of many useful group characteristics, known as "values", are connected with some personal names, e.g. the "acid-value" with Merz (1879), the "saponification value" with Koettstorfer (1879), and the first iodine value with Huebl (1879). The modern elucidation of the triglyceride composition was begun by Hilditch and his school in the 1930s by means of fractional crystallization. The introduction of chromatography, executed originally on columns by Tswett (1906) and developed in the gas phase by James and Martin (1952), for the separation of fatty acids and derivatives opened a previously unknown chapter in establishing the chemical composition of fats and oils and other (fatty) substances. Chromatography, in combination with different methods of instrumental physical analysis and the application of computers here and on the side of production, processing and design has completely changed the picture of the edible fat industry in the last 20 years.

One might wonder why, after the publication of many excellent books on the chemistry and technology of oils and fats, some of them quite recently, another book on the same subject was felt necessary for such a relatively limited field of intellectual and practical interest. Some of these books cover broadly all the theoretical and practical aspects associated with the chemistry and technology of oils and fats, and all the fields where they can be used, not only for edible, but also for technical purposes. Others are written in languages less mastered by a broad public than English. This book gives, as an introduction, a concise treatment of fat chemistry. Thereafter the most relevant fat-separation technologies for the production of the important types of crude fats and oils from seeds, fruits and animal tissues are described. The oil and fat resources are not enumerated, and the use of defatted meals as animal feed and oil-seed proteins for human consumption are not elaborated further. Following this the refining and the most important fat-modification technologies are discussed, as these lead to those edible oils and fats which are applied in different products of high fat content, e.g. margarines (except dairy products), shortenings, mayonnaise, etc., and the speciality fats.

This book is thus written on the one hand for the high-fat food manufacturer, who is often in control of only the finishing stage and not the raw-material production and processing stages. If the manufacturer is able to exert some influence on the selection and production of the raw or partly processed raw materials (intermediate/half products) used, he might do that on knowledge secured from books and other publications. On the other hand, this book might also be of interest to the oil miller or the refiner/hardener/fat fractionator, who wants some insight into the needs of the high-fat product manufacturer by taking a look at his specific products and methods. This aim will be achieved by disclosing many quality aspects of the half and final products to be controlled, besides some quality criteria of the important raw/intermediate materials involved in the processing.

The basics of the most important quality-control methods, as standardized by the International Union of Pure and Applied Chemistry (IUPAC), by the American Oil Chemists' Society (AOCS), by the German Society of Fat Science (Deutsche Gesellschaft fuer Fettwissenschaft, DGF), by some of those of the International Association of Seed Crushers (IASC) and by the British Standards Institution (BSI) were highlighted previously by the present author in *Quality Control in the Food Industry* (edited by S. M. Herschdoerfer, Vol. 2, 2nd edn., London: Academic Press, 1986), which can thus be used as a reference.

This book is dedicated to my wife Edith, whose solidarity, understanding and patience supported and encouraged me, as always.

Acknowledgements

The author wishes to express his sincere gratitude to all those friends and colleagues within the Unilever organization and also outside it, who have reviewed individual chapters in their own area of specialization, including those who have helped him in the improvement of the draft manuscript as regards his English. Their advice and comments have been invaluable. The permission to use the library, internal documentation, some illustration material and the word-processing facilities of the Unilever Research Laboratory, Vlaardingen, The Netherlands, the institution in which he served with pleasure for 33 years, are most gratefully acknowledged. In spite of this long service contact, all the views given by the author remain solely his own responsibility and are not necessarily those of Unilever.

Thanks are due to all those publishers, copyright owners, patented process inventors or equipment manufacturers whose published material has been used to complete this work. The description or mention of any process or firm under the illustrations or in the text does not imply any recommendation whatsoever of the author, the items serving only as illustrations for the subjects handled.

When describing the technology and processes it seemed to be of little value to collate all the items discussed with immediate literature references. All those publications (books, monographies, journals, etc.) which are considered to be relevant to the subjects discussed and which can be used as sources for further information on the matter, are fully cited in the general and the chapter-wise special literature references at the end of the book. As an exception, authors whose work contains information more specific to certain subjects, are individually cited by name and year of publication (in parentheses) in the text and are fully cited in the special reference list.

1

The Chemistry of Edible Fats

I. On Oils and Fats in General

Oils and fats used for edible purposes are of either vegetable or animal origin. Whilst most of these natural sources are grown primarily for their fat content, some are only side products of animal-feed manufacture or of animal-carcass working-up (meat packing) operations. Because of their favourable intrinsic (e.g. caloric, physiological) and extrinsic (flavour, palatability, appearance, consistency and texture) quality characteristics they are consumed as such and they are also very important ingredients in many home-made and industrial food preparations.

Chemically, oils and fats are esters of glycerol and fatty acids (longer chain aliphatic monocarboxylic acids). By a complete (biological) esterification process a maximum of three fatty acids (substituent acyl-groups) can be incorporated into one molecule of glycerol, a trihydric alcohol. Because of this fact, fats are (tri)acylglycerols (previously called triglycerides).

On the basis of their glycerol-esterified character we can discriminate between true (glycerol based) oils and two other groups of "oils": the mineral oils, which cannot be hydrolysed ("saponified") by water, catalysts and alkalis, and the ethereal oils, which may be partly hydrolysable (saponifiable) but do not give glycerol and fatty acids.

Acylglycerols and many other compounds which by origin are either fat-incorporated or fat-soluble substances are called lipids. Lipids which contain the glycerol residue are called glycero-lipids. Lipids which upon hydrolysis give only two cleavage products, such as glycerol and fatty acids or fatty alcohols and fatty acids (the latter are called "waxes"), are known as simple lipids.

Simple glycero-lipids, depending on their character, and the number and position of substituent fatty acids (acyl substituents) in the acylglycerol molecule, manifest themselves according to the temperature of the surroundings in different physical states and consistencies. Substances which are solid or semi-solid (ointment like) are called fats, and those which are liquid are called oils. Hereafter the two types will often be referred to as fats.

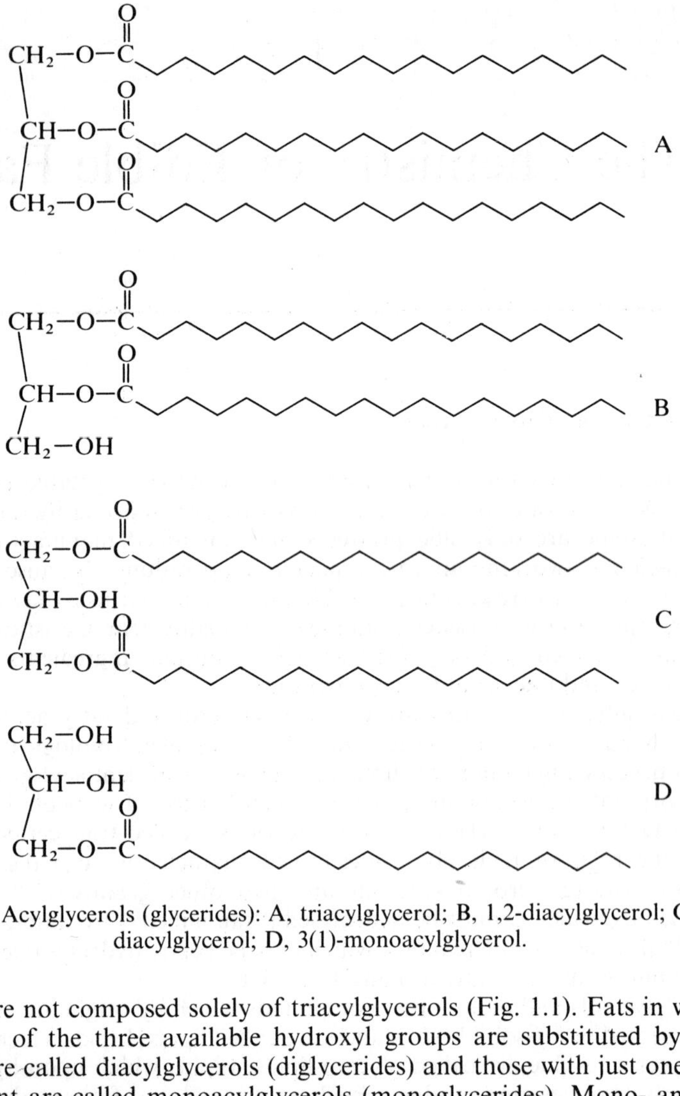

Fig. 1.1 Acylglycerols (glycerides): A, triacylglycerol; B, 1,2-diacylglycerol; C, 1,3-diacylglycerol; D, 3(1)-monoacylglycerol.

Fats are not composed solely of triacylglycerols (Fig. 1.1). Fats in which only two of the three available hydroxyl groups are substituted by acyl groups are called diacylglycerols (diglycerides) and those with just one acyl substituent are called monoacylglycerols (monoglycerides). Mono- and di-acylglycerols (glycerides) are also called partial acylglycerols or glycerides, according to the traditional denomination.

When separating (rendering) fats from their natural surroundings, the processes used rupture and destroy the cell structure of the housing tissues. For this reason fats in the crude state are always mixed physically (mechanically) or colloid chemically as micellar solutions or colloid dispersions with smaller or larger amounts of the constituents of their parent sources (the so-called accompanying components) some of which are also lipids. The parent sources are quite characteristic of the nature and source of the fats and, therefore, may give very important analytical clues regarding their origin and quality.

In this respect we should first mention the free (non-esterified) fatty acids (FFA). These, if present in higher amounts than is accepted for the given source, are the first and most important (necessary, although insufficient) measures for deciding quickly on the below-standard quality of a given fat sample.

Besides the full and partially substituted acylglycerols and the free fatty acids it is also necessary to enumerate as further accompanying lipids, the phosphoacylglycerols (phosphoglycerides or glycerophospholipids). The main building element of these compounds is an ester of glycerol with orthophosphoric acid, the common basis for fat and phospholipid biosynthesis. This ester is further esterified on the glycerol residue with fatty acids, and on the phosphoric acid residue with an organic base (e.g. an amino-alcohol) or a sugar-like compound (e.g. inositol). Non glycerol based orthophosphoric acid ester derivatives are also well known. Since the hydrolysis of these derivatives gives more than two types of cleavage products, they might also be called complex lipids.

All the lipids mentioned above occur in (crude) fats in quite distinct amounts and comprise up to 99% of the fat components. These lipids form the so-called major lipid components. Before describing the so-called minor accompanying components, some of which are also lipids, we shall discuss the major ones in some detail.

II. Major Lipid Components

A. Fatty Acids

Normal (straight chain) aliphatic carboxylic acids with 4 or more carbon atoms are called fatty acids. In nature they occur with an even number of carbon atoms, with few exceptions.

1. *Saturated Fatty Acids*

The term "saturated" means that all the carbon valencies (except in the carboxylic acid group) are satisfied independently.

(a) *Nomenclature.* The first real fatty acid is called by its systematic name butanoic acid. Its common (trivial) name is butyric acid. The IUPAC (International Union of Pure and Applied Chemistry) rule system, based on the Geneva nomenclature, states that the denomination is determined by the longest straight chain, as the parent compound, having the maximum occurrence of the principal functional group. The order of the hierarchy of principal functional groups is given by Chemical Abstracts Service rules, the carboxylic group being primordial. The proper chain

EDIBLE OILS AND FATS

Table 1.1 *Important fatty acids*

Systematic name (acid)	Trivial name (acid)	Chain length (ω notation)	m.p. (°C)	Triacylglycerol m.p.[a] (°C)
Decanoic	Capric	10:0	31.6	31.5
Dodecanoic	Lauric	12:0	44.4	46.4
Tetradecanoic	Myristic	14:0	54.3	57.0
Hexadecanoic	Palmitic	16:0	62.9	
Octadecanoic	Stearic	18:0	70.0	73.1
9-Octadecenoic	Oleic	18:1, ω-9	13.0	5.5
9-*trans*-Octadecenoic	Elaidic	18:1, ω-9	36.0	42.0
13-Docosenoic	Erucic	22:1, ω-9	33.5	30.0
9,12-Octadeca-dienoic	Linoleic	18:2, ω-6,9	−3	−13.1
9,12,15-Octa-decatrienoic	α-Linolenic	18:3, ω-3,6,9	−11.9	−24.2
4,8,12,15,19-Docosapentaenoic	Clupanodonic	22:5, ω-3,7,10,14,18		
12-Hydroxy-9-octadecenoic	Ricinoleic	18:1, ω-9(OH)	5.5	

[a] The melting points given are those of the highest melting modification with due allowance for variations found in the literature.

length of the fatty acid is expressed by the Greek name of the saturated hydrocarbon, with the suffix "-oic" and the separated word "acid". This completes the systematic name of the saturated fatty acid.

Another possibility is to name the longest straight-chain hydrocarbon part of the molecule separately and to consider the carboxylic group as a substituent. In this case butyric acid is called propane carboxylic acid, i.e. the words carboxylic acid are placed separately as a postfix. The "n" prefix for normality (straightness of the chain) is generally omitted.

The systematic and trivial names of frequently occurring fatty acids are given in Table 1.1.

(b) *Physical properties*. Going along the series of saturated fatty acids from those of short chain length to those of long chain length the physical state changes from liquid to a solid consistency, the members of longest chain lengths having, under ambient temperature in moderate climates, a wax-like appearance. The melting point (m.p.), boiling point (b.p.) and the refractive index (n_D) all increase with increasing chain length, whilst the density (d) decreases with increasing chain length. The shorter-chain members are easily distilled at atmospheric pressure, whereas the longer-chain ones, which have very low vapour pressures, should preferably be distilled under vacuum (because of the danger of decomposition). The shortest-chain fatty acids with up to 12 carbon atoms (lauric acid) can be distilled easily by steam distillation. Fatty acids with more than 12 carbon atoms are not distillable even by steam under practical (analytical) conditions. The solubility in water decreases as the carbon number increases. Decanoic acid (or capric acid) is only very slightly soluble even in boiling water. The solubility or practical miscibility of fatty acids with different organic solvents in the liquid state (above their melting points) depends on the affinity of the acid to the solvent; different solvents being more or less polar (apolar) in character. However, in practical terms, nearly all fatty acids are at least fairly soluble or readily miscible with polar (e.g. alcohols, ketones) and apolar (e.g. aliphatic, aromatic, branched or cyclic hydrocarbons) solvents.

(c) *Chemical properties.* In general fatty acids are easily transformed by alkalis or metallic oxides to their corresponding salts, called soaps. This neutralization reaction is easily performed even in heterogeneous phases (just-mixed aqueous alkalis, or metallic oxides heated with fatty acids). In spite of their saturated nature the saturated fatty acids are not quite inert. Free radicals initiate oxidation and halogenation processes which become very enhanced at higher pressures and elevated temperatures, and ionic mechanisms easily cause substitution of one or more hydrogen atoms in the chain. The cleavage of the chain (rupture) and the formation of lower molecular weight scission products at exaggerated reaction conditions (e.g. temperatures above 300°C) are quite feasible. Heating to high temperatures in an inert atmosphere may cause decarboxylation of fatty acids by loss of carbon dioxide, with the formation of hydrocarbons.

2. *Unsaturated Fatty Acids*

Fatty acids having hydrogen deficient carbon atoms bonded by double or triple valencies are called unsaturated. According to the number of double bonds present they are called monoethenoic (monoenoic), or di-, tri-, tetra-, penta-enoic, etc., or, in general, polyenoic fatty acids. Ethylene is the simplest "ethene". The less-common triply-bonded ones are called "ethynoic" acids, acetylene being the simplest "ethyn".

(a) *Nomenclature.* The nomenclature of unsaturated fatty acids is assigned by modifying the name of the saturated fatty acid of corresponding chain length: the "a" in the last syllable of the Greek name is changed to "e" for the ethenoic types and to "y" for the ethynoic types. Repetitive presence of the same unsaturation is built into the name by writing di-, tri-, tetra-, etc., according to the occurrence. The final part of the name is "enoic acid": e.g. octadecadienoic acid is double ethenoic, unsaturated, 18 carbon (C18) fatty acid.

The place (locant) of any change to or substituent on the carbon chain is given by an Arabic numeral, describing the (first) substituted carbon, which is counted from the carboxyl group as carbon number one (C1). The name of the substituent group (oxo-, hydroxy-, epoxy-, methyl-, ethyl-, methoxy-, cyclopropane-, cyclopentene-, chloro-, bromo, etc., except unsaturation!) is written as a prefix, the locant as an Arabic numeral before the prefix. Unsaturation as locant is indicated between the prefix and the modified Greek name by an Arabic numeral. E.g., 12-hydroxy-9-octadecenoic acid is C18 fatty acid at the 9 position mono-unsaturated, at the 12 position substituted by a hydroxyl group (trivial name, ricinoleic acid).

(b) *Isomerism*. Unsaturation can be found at different locations, which give rise to the phenomenon of positional isomerism (isomers are compounds which have the same molecular formula and molecular weight, but are different in structure). Another type of isomerism is geometric isomerism. For fatty acids this concerns the relative position of the two hydrogen atoms attached to the double-bonded carbon atoms against a symmetry plane, parallel to the longitudinal axis of the fatty acid molecule. If the two hydrogen atoms are located on the same side of the bisecting plane, then the double bond formed is called a *cis* (*c*) bond or, as agreed by IUPAC, a Z (German: Zusammen = together) configuration. If the hydrogen atoms are on opposite sides of the plane, the bond formed is called a *trans* (*t*) bond or an *E* (German: Entgegen = opposite) configuration. The geometrical isomerism of the unsaturated fatty acid structure is expressed by writing the type of configuration and the locant of the double (triple) bond as an Arabic numeral separately (with hyphens) between the prefix (if any) and the modified Greek generic name. No sign at all could mean uncertainty or just the natural *cis* (*c*) or Z configuration. E.g. *cis*-9, *cis*-12-octadecadienoic acid ≡ Z-9,Z-12-octadecadienoic acid ≡ (trivially) 9c-12c-linoleic acid.

(c) *Shorthand notations*. There are two shorthand notation systems for unsaturated fatty acids, the one used by organic and analytical chemists, and the one used by biochemists. The first system abbreviates fatty acids in the form $x:y, z\,c$ (or $z\,t$), where x stands for the number of carbon atoms in the fatty acid chain, y for the number of double bonds present, z for the position(s) of the double bond(s), all numbered from the carboxyl group as C1, plus the geometric configuration(s) of the double bonds: *cis* or *trans* (Z or E).

In the second system, which is the ω or end-of-carbon-chain (ECC) or (more-up-to-date) the *n-m* system, ($x:y, \omega, m$; or $x:y, n\text{-}m$), the meaning of x and y is the same as before. The terms ω and n stand for the fact that the terminal methyl group of the fatty acid chain is taken as C1 and m is now the position of the locant of the first double bond observed from the methyl end.

The shorthand and systematic notations are summarized in Table 1.2. In spite of the many, perhaps somewhat too artificial, rules for nomenclature and shorthand notations, all the trivial names of fatty acids which are generally used, are conserved and accepted.

(d) *Physical and chemical properties*. Double bonds in the *cis* configuration seen sterically give a bending in the straight-line profile of a fatty acid molecule. Actually, because of the valence angles (about 109 grades for a single bond) the profile is of zig-zag type. Double bonds in the *trans*

1. THE CHEMISTRY OF EDIBLE FATS

Table 1.2 *Shorthand notations for fatty acids*

Trivial name	Analytical shorthand
α-Linolenic acid	18 : 3, 9c, 12c, 15c

$$\overset{C15}{CH_3-CH_2-\overset{\cdot}{C}H}=CH-CH_2-\overset{C12}{CH}=CH-CH_2-\overset{C9}{CH}=CH-CH_2-(CH_2)_6-COOH$$

ECC shorthand	Systematic name
18 : 3, n-3	Z-9, Z-12, Z-15-Octadecatrienoic acid
18 : 3, ω3(6,9)	

configuration fit sterically more smoothly into the zig-zag line of the rest of the molecule, which calls our attention to the distinct differences between the physical properties of isomeric acids. For example, *trans* acids melt at higher temperatures than do *cis* ones, which explains the liquid state of the latter at room temperature. The *cis* isomers occur more frequently in nature than do the *trans* ones.

The presence of double (or triple) bonds between vicinal carbon atoms provides a special reactivity to the neighbouring, so-called α-methylenic carbon atoms. These bonds exert a very strong attraction on the electrons of the valences binding its hydrogen atoms. This interaction causes acids containing double bonds to be very reactive or sensitive towards certain chemical reagents and/or external physical conditions.

Unsaturated fatty acids may undergo reaction when (even unintentionally) confronted with oxygen, hydrogen, halogens, inorganic or organic acids, sulphur compounds, heavy metals, heat, radiation or generally free-radical or ion induced processes (see later). Fatty acids containing more than two so-called isolated (at least one methylene group interrupted) or conjugated (non methylene group interrupted) double bonds are very easily (auto)oxidized, isomerized, polymerized (see later). These reactions can occur even in the cold (for example, drying oils). Fatty acids can also accept the above reagents by addition or substitution reactions.

Most of the polyunsaturated fatty acids are to be found in the triacylglycerols of oils of vegetable origin and in the oils of aquatic animals. The latter contain fatty acids with 4, 5 and even 6, mainly isolated double bonds. Because of this high accumulation of unsaturation (many 1,4-pentadienoic groups in series) these fatty acids are even more prone to the alterations mentioned above.

The average degree of unsaturation of fatty acids can be expressed analytically by the iodine value. Differences in the stereoconfiguration can easily be detected spectroscopically in the infrared (IR) region by measuring the specific absorption of light (by the out-of-plane vibrating CH linkages) in the *trans* configuration. The refractive index (n_D) increases with the number of double bonds present and with increasing conjugated character of the double bonds. The density decreases with decreasing unsaturation. Therefore changes in these two criteria may serve as quick controls for saturation during the catalytic hydrogenation of fats.

3. Other Types of Fatty Acids

Besides saturated and unsaturated fatty acids, nature also produces oxygenated species, of which ricinoleic acid is a hydroxylated member, vernolic acid an epoxized member and licanic acid a keto group containing C18 representative, all of vegetable (plant) origin.

Exceptionally, some fatty acids, for example those present in microorganisms and some synthetic ones, are not built up from straight aliphatic chains, but may be substituted on one or more carbon atoms of the chain by one or more methyl (ethyl) groups. These fatty acids are called branched. Certain bacteria in the gastric tract of ruminants can decompose food to C3 units, more specifically to propionic acid. A resynthesis can then incorporate some propionyl coenzyme A building blocks into the chain, the coupling being performed again on the β-carbon atom of the coenzyme bound propionyl group. As the molecular weights of branched fatty acids are the same as those of the corresponding non-branched (straight chain) ones, they are called iso- or isomeric fatty acids. Branched fatty acids are to be found in small amounts in the milk and body fat of ruminants (tallows) and can be detected by using a suitable technique, such as gas chromatography (GC). Ingested linoleic and linolenic acids, e.g. ex grass, can be partially hydrogenated bacterially in the rumen leading to the formation of 5–10% natural "*trans*" isomers and some positional isomers of linoleic and oleic acids, again in the depot fats of ruminants.

Some fatty acids found in plants contain at the end of their chains a cyclopentenoic, unsaturated five-membered-ring. The existence of cyclopropane and cyclopropene, three-membered rings, in the middle of a fatty acid chain is an established fact, exemplified by the low-level presence of sterculic, malvalic and dihydrosterculic acids in cotton-seed and hemp oils.

4. The Biosynthesis of Fatty Acids

Both plant and animal cells are surrounded by non-rigid, semipermeable membranes, which allow the diffusion of small molecules into and out of the cells. Plant cells have a more definite shape, because of the cellulose content of their more rigid cell wall surrounding the membrane. Cells have a nucleus which contains chromosomes. Chromosomes are built of nucleic acids and carry the genetic information required for the maintenance of the cells. The cytoplasm inside the cell is an aqueous colloidal solution of proteins and other constituents. The endoplasmic reticulum is a complex network of filaments, which, together with the ribosomes, are the sites of protein build-up from amino acids. The green matter, called chloroplasts, is responsible for the photosynthesis of organic material in the green plants. More specifically chlorophyll catalyses the transformation of the sun's radiant energy to chemical energy, by which the photolysis (cleavage) of water to its elements and the binding of atmospheric carbon dioxide is made possible. And last but not least, the mitochondria, small rod-like structures in the cytoplasma, are the "power houses" of the cell, releasing the chemical energy required by the cell for its metabolic processes. The enzymes needed for fat synthesis are housed either inside or outside the mitochondria.

One of the most important pathways for producing fatty acids is the stepwise two carbon atom "*de novo*" synthesis of a saturated carbon chain up to hexadecanoic (palmitic) acid. Fatty acid synthesis is executed by two mechanisms: chain elongation and desaturation. Elongation is accomplished by the coupling (polycondensation) of malonyl-coenzyme A molecules with a "starter" acetyl-coenzyme A molecule to a butyryl group. All these acetyl and malonyl building blocks are attached to a so-called "acyl carrier protein" (ACP). Although the malonyl-coenzyme A constitutes a C3 unit, it ends up as a C2 unit, in each step losing its acquired carboxyl group by decarboxylation. (The carbon dioxide is originally delivered by the enzyme biotin, fixing carbon dioxide from the atmosphere in the "active" form.) This synthesis via further malonyl coenzyme A molecules is terminated with the hexadecanoic acid. The necessary condensation, reduction, dehydration and hydrogenation steps (four in total) are executed by the suitable interaction of different plant enzymes and coenzymes, all of which are proteinaceous and/or nucleic acid in nature. This so-called multienzyme complex is not present in the mitochondria, but is found in the cytoplasm of the cell.

The further elongation to longer chain lengths by two carbon atom (acetyl coenzyme A) units is regulated by enzymes present in the mitochondria, leading primarily to stearic acid. The elongation of some unsaturated fatty acids is accomplished in the (plant) cell by this mitochondrial mechanism, as for example the formation of 22:1, *n*-9 (erucic acid) in Cruciferae (rape or mustard seed) oils from 18:1, *n*-9 (oleic acid).

Unsaturation of the fatty acids in plants occurs by "oxidative desaturation". For example, by this enzymatic process the 18:0 saturated fatty acid (stearic acid) already synthesized in the plant organism, is first transformed to the monounsaturated 9-octadecenoic (oleic) acid.

Animal organisms can also desaturate stearic acid to oleic acid, but cannot desaturate the latter to linoleic or linolenic acids. These two acids belong to the umbrella group of the so-called polyunsaturated fatty acids (PUFA). There are some polyunsaturated fatty acids which are very important for animal health, but cannot be synthesized by the animal organism. These fatty acids must then be introduced ready made by ingesting food containing them. This group of unsaturated fatty acids is known as the "essential fatty acids (EFA)".

The starting point of this synthesis is again stearic acid. Suitable plant enzymes introduce the first double bond on the n-9 carbon, oleic acid having been formed. The second double bond is placed at the n-6 carbon, thus between the existing double bond and the terminal methyl group. The pattern is the n- 6,9 (lc-, 4c-pentadienoic or skipped) one:

$$\overset{cis}{-CH=CH}-CH_2-\overset{cis}{CH=CH}-$$

This n-6(9) pattern or configuration has been proven as the common denominator for essentiality (physiological activity), independent of the chain length of the acid.

Linoleic acid (18:2, 9c,12c or n-6,9) is the most important "essential" fatty acid, its absence in food causing growth retardation in young animals, skin permeation problems (dermatoses); the deficiency is also relevant to humans. Linoleic acid is well converted by the enzymes of the animal organism via chain elongation and desaturation to di-homo-γ-linolenic acid (20:3, 8c,11c,14c or n-6,9,12) and, consequently, to arachidonic (20:4, 5c,8c,11c,14c or n-6,9,13,15) acid, all of which are essential fatty acids. γ-Linolenic acid (C18:3, 6c,9c,12c or n-6,9,12) present in, for example, the evening primrose (Oenothera) seeds, is also considered as an essential fatty acid. On the other hand the polyunsaturated α-linolenic (18:3, 9c,12c,15c or n-3,6,9) acid is not absolutely required by animal organisms.

Some essential fatty acids form the building blocks of the physiologically important substances called prostaglandins. These are short-lived smooth muscle and vasodepressor regulators formed in animal organisms by enzymatic cyclization of the above-mentioned polyunsaturated fatty acids with the n-6 configuration, but with a chain length extended to C20.

Bio- and agro-techniques have provided the industry with plant hybrids, which have great uniformity in the quality and quantity of their oil production. This and their resistance to disease was mainly achieved by breeding techniques based on selection and crossing or tissue-culture methods, the latter being very successfully applied in the case of oil palm trees. Similarly, the very significant changes in the unsaturated fatty acid pattern of sunflower, safflower and rape seed oils has been achieved mainly by single intact seed selection and/or tissue-culture methods. On the other hand, more detailed studies of the biochemical pathways of elongation, and even more so of the desaturation of fatty acids, are also of great commercial interest. Therefore, there are enormous challenges in modifying the chain length and unsaturation of fatty acids by using the new genetic-engineering techniques.

B. Acylglycerols (Glycerides)

1. *Nomenclature*

In (tri)acylglycerols the component fatty acids esterified with the glycerol are named by adding the postfix -oyl or -yl (no "acid"!) to the trivial or systematic name of the substituting fatty acid and attaching it to the word glycerol: e.g. 1-stearoyl-, 2-oleyl-, 3-linoleyl-glycerol, or 1-octadecanoyl-, 2-octadecenoyl-, 3-octadecadienoyl-glycerol. The numbers indicate the place of substitution in the glycerol residue.

2. *Physical and Chemical Properties*

Triacylglycerols are synthesized in living cells by the interaction of an intermediate of carbohydrate metabolism called glycerol phosphate, formed by the reduction of dihydroxyacetone phosphate and the complete fatty acid residue coupled to coenzyme A. Direct coupling delivers first a monoacyl-, then a diacyl-glycerol phosphate, which after release of the phosphoric acid residue by a phosphatase enzyme is substituted by the third coenzyme A coupled fatty acid residue.

Tri-, di- and mono-acylglycerols, because of the trifunctionality of glycerol, can be composed of a wide variety of fatty acid (acyl-) substituents. As the triacylglycerol molecules in natural fats may contain more than one type of fatty acid, fats are composed of the so-called mixed triacylglycerols. Basically these fatty acids are those principal components which primarily determine the physical and chemical properties of a given fat. These properties manifest themselves in values which are often very close to the average value of the component fatty acids, if allowances are made for the higher molecular weight of the triacylglycerol molecule as compared with those of the component fatty acids.

Di- and mono-acylglycerols possess one or two free hydroxyl groups and can be considered as tracers of a not fully completed biological triacylglycerol synthesis (unripe fruits, seeds) or as the signs of an early hydrolysis caused by enzyme activity in deteriorating fatty commodities. Depending on the fatty acid composition they may be liquid or solid at ambient temperatures. Their main importance (besides being important quality criteria) is their strong affinity for both water and apolar substances, such as other lipids, hydrocarbons, etc., which explains their (decreasing) use as emulsifiers in many foods (e.g. margarine) and technical preparations.

The low vapour pressure of triacylglycerols makes it impossible to distil them under conventional conditions without decomposition of, primarily, the glycerol and, secondarily, of the fatty acid residues. The volatilities of di- and even more of mono-acylglycerols are increasingly somewhat higher, due to their lower molecular weights.

As some of the physical characteristics of triacylglycerols have a profound influence on their technical quality and applicability, they will not be discussed here in much detail. Instead the reader will find the relevant information in the Chapter 5 (Section II) on fractional crystallization and winterization/dewaxing, where the aspects of melting and solidification, crystallization and its mechanism are discussed. Fat consistency and the phenomenon of polymorphism are described in Chapter 6 (Section I) on margarine/minimum calory margarine. Only one example is given here to demonstrate the effect of the nature and position of the substituent fatty acids in the glycerol residue on the melting and solidification of triacylglycerols. As usual the two primary (1-, 3-) hydroxyl groups of glycerol are taken, for simplicity, as being equivalent as regards substitution; e.g. the "symmetric" triacylglycerols containing Stearic-Palmitic-Stearic acids (StPSt) and their "asymmetric" PStSt isomers. These two isomeric triacylglycerols have different "melting points", that of the former being 68.5°C and that of the latter 63.3°C. However, if both are hydrolysed to the component mixture of fatty acids (two parts of stearic acid and one part of palmitic acid), the "titre" (melting points) of the two acid mixtures is 63.2°C in each case. Triacylglycerol related positional isomerism leads sometimes to slight "optical activity", because of a possible "asymmetric carbon centre" at the 2 position of the glycerol.

The density of acylglycerols (fatty acids) is higher in the solid state than in the liquid state. For this reason they show a (specific) volume expansion, when being transformed from the solid to the liquid phase by heating. As fats are (varying) mixtures of solid and liquid acylglycerols at a given temperature, the measurement of this melting dilatation can be used to determine the solid fat index (SFI) of the given fat. The SFI of the U.S.A. practice is only proportional to, but not equal to the percentage solid fat content (SFC) of fats, and must be converted to a percentage by using suitable formulae. As we shall see, the SFC is a very important quality characteristic in edible fat processing.

The differences in structure related to the position of the fatty acids in the triacylglycerol molecule are the source of the tendency of these substituents in a natural fat to attain a so-called fully-random distribution during catalysed intra- and inter-molecular ester interchange. Chapter 5 (Section 3) on interesterification—ester interchange (randomization) deals with the fatty acid distribution in triacylglycerols in general, the fatty acid distribution in natural fats, and the "random" types of fatty acid distribution caused by chemical changes.

Other chemical reactions of triacylglycerols are generally comparable in effect, although not always in rate, with those carried out on mixtures of the component fatty acids and glycerol. Chapter 5 (Section II) on the hydrogenation of edible oils gives details of a special, but technologically most important, chemical reaction, the formal addition of hydrogen to the double bonds of fats containing many, liquid unsaturated fatty acids.

In the absence of the separation techniques required for a more individual analysis, fats in the past have very often been characterized on the basis of their average fatty acid composition. The similar reactivity of given samples of any fat towards different functional-group reagents has resulted in the so-called "values" or "numbers", which give the mass average of certain properties. We have already mentioned the FFA for the total non-esterified fatty acids (free carboxyl groups) and the "iodine value" for the average unsaturation in terms of the amount of iodine added on the double bonds. Now we might mention the "saponification value" for the average chain length in terms of the amount of alkali required to totally saponify the sample and the "titre" which is the average solidification temperature of the mass of fully hydrolysed, thus free fatty acids. These values still have some commercial importance, as the tests do not require complicated equipment or techniques and the values give a quick clue to the identity and quality of the fat. Modern analytical techniques tend to individualize the determinations, the more-or-less individual acylglycerol separation and analysis becoming as feasible as individual fatty acid (and accompanying components) analysis.

C. Phosphoacylglycerols (Phosphatides)

The cell wall of living animal and plant material and some structural elements of certain animal organs contain numerous derivatives of orthophosphoric acid. For our purposes the most important ones are the glycerol esterified derivatives, also called glycerophospholipids. Others, like the sphyngosine based sphyngophospholipids of the nerves and the carbohydrates containing glucosphyngolipids, such as the ceramids, cerebrosides, etc., of plant parts and animal organs, blood, brains, etc., are of great biological importance.

Of the phosphoacylglycerols at least one, and generally both, hydroxyl groups of glycerol are esterified with long-chain fatty acids. Actually at this stage of the biosynthesis it has already been decided whether a triacylglycerol or a phosphoacylglycerol will be formed. Sometimes the glycerol is etherified with a fatty alcohol, either saturated or unsaturated. On the other hand the phosphoric acid is esterified with, for example, an amino alcohol or inositol (a hexahydroxycyclohexene). All these substances, trivially called also phosphatides, are generally oil insoluble in the molecular sense, but they are co-extracted with the triacylglycerols during oil milling as colloidal aggregates called micelles. They are present in crude vegetable oils in amounts of 0.1–3%, and are generally easily removed during refining as a sludge, because of their high affinity to water. The Ca and Mg salts of phosphatidic acids and the free acids themselves, being less "hydrophilic", are less easily removed, if at all, and so they are called the "non-hydratable" phosphatides.

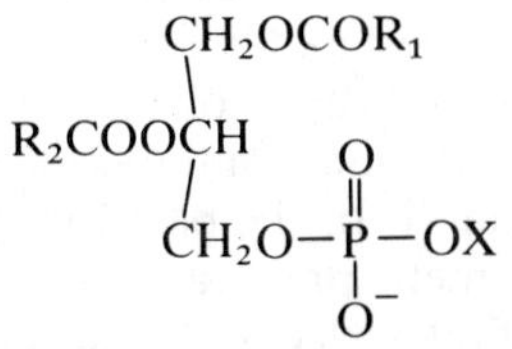

$$\begin{array}{l} CH_2OCOR_1 \\ | \\ R_2COOCH \\ | \\ CH_2O-P(=O)(O^-)-OX \end{array}$$

Fig. 1.2 The structure of phosphoacylglycerols. X = H, phosphatidic acid; X = $CH_2CH_2N(CH_3)_3$, phosphatidylcholine; X = $CH_2CH_2NH_2$, phosphatidylethanolamine; X = $CH_2CH(NH_2)COOH$, phosphatidylserine; X = inositol, phosphatidylinositol.

Glycerophospholipids are classified into five main groups (Fig. 1.2.) according to the nature of the substituents on the glycerophosphoric acid:

(i) Phosphatidic acids (no substitutents), PA; their dimers are called cardiolipins, CL;
(ii) Phosphatidyl ethanolamine (cephalin), PE;
(iii) Phosphatidyl choline (lecithin), PC;
(iv) Phosphatidyl serine, PS;
(v) Phosphatidyl inositol, PI.

If only one fatty acid (O-acyl residue) is present in the molecule, all the compounds above are considered to be partially hydrolysed members and so the prefix lyso- is put in front of the above-given names.

The presence or diminished presence of phosphoacylglycerols (commercially all called "lecithin" or by the misnomer "gums") is an important clue when assessing the quality of fats. The amount of these compounds can be measured by phosphorus or phosphate determination. Commercial "lecithin" is used as a dispersion stabilizer, an antispattering agent and a viscosity reducer in many foods, and in technical, cosmetic and pharmaceutical preparations.

III. Minor Accompanying (Lipid) Components

A. Alcohols

Glycerol or 1,2,3-trioxypropane, although the basic component of triacylglycerols, giving them their name, is not found in its free form in fats, but is produced if hydrolysed by aqueous alkalis, acids or enzyme type catalysts of plant, animal or microbial origin. Free glycerol is water soluble, in contrast to its state when esterified with fatty acids.

Primary long carbon chain alcohols can be derived from the fatty acids by converting the carboxyl group to a hydroxymethylenic ($-CH_2$-OH) one. Alcohols up to 22 carbon atoms are called fatty alcohols and those with longer chains are called wax alcohols. The C14–18 and the 18:1, 9c members are mainly present in the tissue fat of aquatic animals (mammals and fish), as palmitic, stearic, oleic and linoleic monoesters. They have a wax-like appearance. The C16 (cetyl) alcohol esterified with oleic acid is more liquid and is present in large amounts in the skull fat of the sperm whale and is called spermaceti or spermoil. In the past the spermaceti was hydrolysed and the alcohol used for cosmetic preparations and detergents (fatty alcohol sulphates). The discovery of fatty alcohols in the Jojoba (Simmondsia) shrub fat and catalytic reduction of fatty acids to the corresponding alcohols has at last relieved the need for whaling of sperm whales.

The C26–30 alcohols can be found as esters of fatty acids in mineral, plant and animal and insect waxes, described later separately (see Section III.D this chapter).

B. Sterols

The basic sterol structure is the so-called cyclopentanophenanthrene condensed four-ring system with 17 carbon atoms (Fig. 1.3, **A**). Sterols are alcohols. The substitution at the skeleton by one hydroxyl and by one or more methyl and side-chain alkyl groups, together with some unsaturation, causes the main differences between the different classes of these compounds. Powerful techniques are available in all areas of chromatography, which,

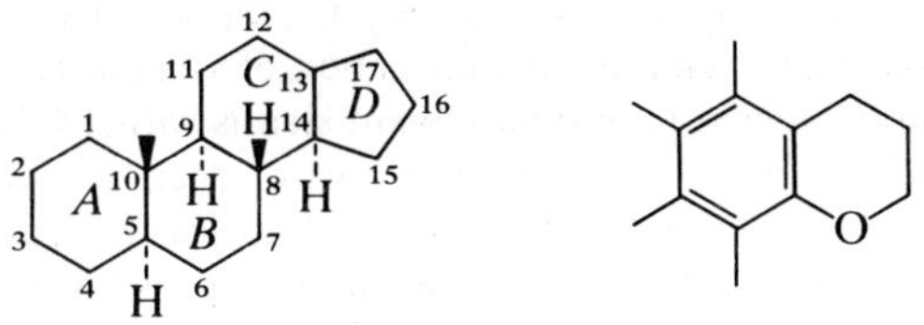

Fig. 1.3 Basic structures of: A, cyclopentanophenathrene for steroids; B, tocol for tocopherols and tocotrienols

if sensibly combined, can differentiate between them. After saponification of the fat to be analysed, the compounds are recovered in the so-called "unsaponifiable matter". Therefore, this previously routine measurement for detecting mineral oil contamination has become of increasing importance in modern quality assessment. Sterol levels vary in the range 0.05–0.6%, the extreme range being 0.6–1.7% in wheat and maize germ oils.

Sterols, sterol esters and sterol glucosides (steroids) are more-or-less fat soluble and completely water insoluble. Sterols dissolve easily in ether, pyridine, chloroform, acetone and also in warm alcohol or glacial acetic acid. Predominantly free sterols and sterol glucosides (see Section III. G on glucosides), having polar hydroxyl groups and apolar condensed ring elements, have a strong affinity both to water and to fats. Because of this they have the ability to reduce the interphase tension of water/oil mixtures. This makes them powerful emulsifiers for fats, when occurring, for example, in the blood plasma or in medicinal and cosmetic preparations.

Sterols can be classified by their origin as animal sterols (zoosterols) or as plant sterols which can be subdivided into higher-plant sterols (phytosterols) and sterols present in the fat of yeasts and fungi (mycosterols).

Instead of this classification of origin, which remains important for the food analyst when dealing with unknown fatty mixtures, we can classify them in a more up-to-date way into: (i) 4-demethylsterols, and (ii) 4-methylsterols and 4,4-dimethylsterols.

1. *Free sterols*

(a) *4-Demethylsterols.* Cholesterol, with a C27 carbon skeleton, is the typical animal sterol to be found in larger concentrations in organs such as the brain, the liver and the spleen (as sediment), in the adrenal gland, in white nerve tissue, in blood plasma and in egg yolk. Cholesterol is very useful as an analytical characteristic, in the identification of animal fats and their mixtures, adding extra clues in addition to the common GC determination of the fatty acid composition. Palm (kernel) oils contain a few percent of cholesterol among their other plant-type sterols, and this fact must be considered when analysing mixtures containing palm or palm kernel oils.

Ergosterol, a C28 sterol present in yeast, is one representative of the lower plant sterols. It is the starting material for the formation of vitamin D_2.

Stigmasterol and β-sitosterol are typical plant sterols with a C29 carbon skeleton and they can be found in the unsaponifiable part of nearly all vegetable oils. The same situation exists in campesterol, a C28 sterol. Brassicasterol, also a C28 sterol, is characteristic of Cruciferae oils such as rape and mustard-seed oils.

(b) *4-Methyl- and 4,4-Dimethylsterols.* As 4-methylsterols are only minor constituents of the unsaponifiable part, so vegetable oils can be better discriminated (additionally to the 4-demethylsterols) by analysing the 4,4-dimethylsterols, which are also called triterpenic alcohols. The most abundantly found representatives are cycloartenol (C30) and 24-methylenecycloartenol (C31), accompanied by small amounts of α- and β-amyrins, lupenol, butyrospermol and cycloartanol, all being C30 4,4-dimethylsterols. The varying amounts of these compounds present in different oils can help the analyst to arrive at a better assessment of the origin and composition of an unknown mixture of fats.

2. Sterol esters

In some vegetable oils a part of the sterols is esterified with fatty acids, because of which their fat solubility is increased and their emulsifier activity is decreased. Particularly in germ oils, such as maize- and wheat-germ oils, a quite significant percentage of the sterols present is in the esterified form.

C. Tocopherols

The tocopherols, widely distributed in nearly all vegetable fats but not in animal ones, are the most important natural antioxidants of phenolic nature. All tocopherols are fat soluble and generally occur in their free form, although some fats contain them in a partially esterified form, the phenolic hydroxyl being esterified by a fatty acid. Therefore, in esterified form their solubility in fats increases, but their antioxidant activity is totally removed (see Section IV. B, this chapter, on autoxidation and antioxidants). Their basic structure is the hypothetic tocol molecule, a chromanol structure containing a farnesyl side chain, substituted by one, two or three methyl groups (Fig. 1.3, **B**). A homologous series of tocotrienols is also known (e.g. in palm oil), which differs only by the presence of three double bonds in the farnesyl side chain. The standard (100%) vitamin E activity is ascribed by agreement with α-tocopherol. The amount of tocopherols varies between 30 and 700 mg kg^{-1}, the extremes being 1000–2800 mg kg^{-1} as measured in germ, linseed, soya bean and cotton seed oils.

The other tocopherols (β-, γ- and δ-) thus have less vitamin E activity than does the α compound. The different amounts and even more importantly, the different types and proportions of tocopherols present in different oils can additionally help the analyst when trying to determine the origin of unknown mixtures of fats. Because of their oxidizability, the proportions can change during storage and processing to such an extent that the clue may be rather uncertain. Tocopherols (and tocotrienols), while protecting the fat, are generally themselves first oxidized to quinones and tocopherol dimers.

Some oxidation products have distinct colours, like Tocored (5,6-chromanquinone) in heavily oxidized soya bean oil. The *in vitro* antioxidant activity of tocopherols in oils and fats is different according to the temperature and possible contaminants (heavy metals) present in the system. It is generally accepted that the least-substituted ones, which more easily form stabilized free-radicals, and thus lose their mobile hydrogen atoms in the hydroxyl group more readily, are the more potent ones, thus

$$\delta > \beta = \gamma > \alpha$$

D. Waxes

Waxes are, by definition, esters of long-chain aliphatic alcohols and fatty acids, partly of plant and partly of animal origin. The so-called mineral waxes are ultimately also of plant origin, representing the wax remnants of past vegetation in a progressively carbonified state. The waxes of vegetable oils, as minor constituents, are generally mere contaminations, originating from seed covers such as brans and husks, like waxes of rice-bran oil or sunflower-seed oil. Waxes derived from leaves, like Candelilla and Carnauba waxes, or from animal secretions, like beeswax or woolwax of sheep, also contain, among many others, aliphatic alcohols, free fatty acids, sterols, triterpenic alcohols and aliphatic hydrocarbons, which together form the substances popularly known as waxes. Some marine animals contain acylglyceride etherified fatty alcohols which are wax-like materials.

E. Pigments

We have seen above that the colour of different oils and fats is highly influenced by the presence in traces of some oil-soluble pigments, like the carotenoids, the chlorophylls and gossypol, the characteristic colouring pigments of, for example, red palm oil, green olive oil and yellow-amber cotton seed oil, respectively (Fig. 1.4).

1. *Carotenoids*

The carotenoids are oil-soluble polyisoprenoids of hydrocarbon, alcohol or carboxylic acid nature. They contain a considerable amount of conjugated double bonds, which explains their pronounced colour in the yellow, orange or red region of the visible spectrum. About 70 different carotenoids have been isolated, the most important being the α-, β- and γ-carotenes and lycopene, which all have the molecular formula $C_{40}H_{56}$.

Carotenoids are widely distributed in the plant and animal kingdoms,

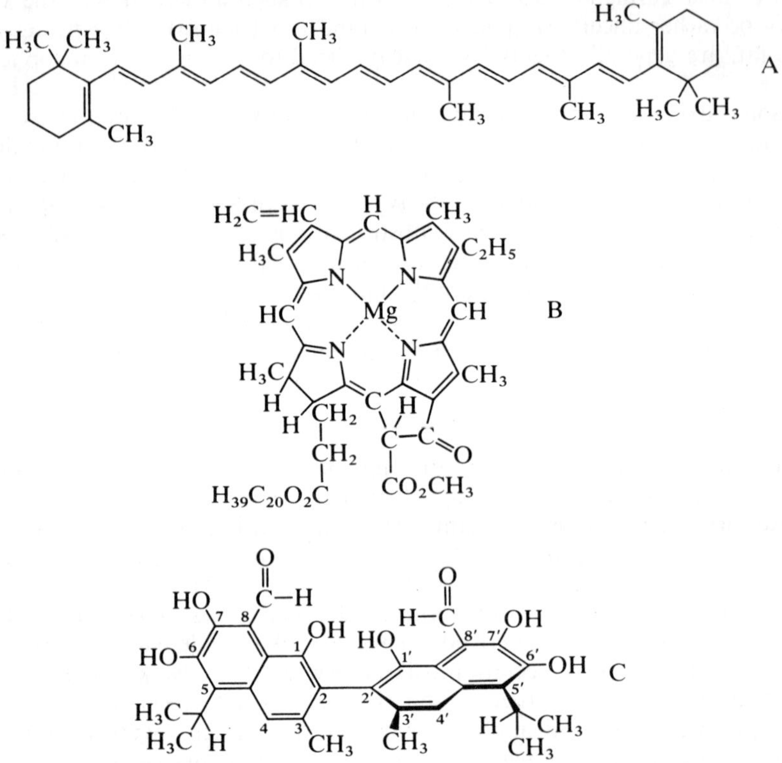

Fig. 1.4 Pigments: A, β-carotene; B, chlorophyll *a*; C, gossypol.

although their concentration in fats (except in palm oils, see later) is generally low. They are highly oxidizable, yet this generally provides a protective action against radiation (light) damage of the medium in which they are dissolved. The oxidation, a free-radical process, causes their ultimate destruction and in this way a bleaching of the colour.

Carotenoids are not sensitive towards dilute alkalis, but they suffer considerable damage under acidic conditions and even more by heating above 150°C. They can be removed by adsorbents like bleaching earth and activated carbon. Hydrogenation of the double bonds, if done selectively and without saturating too much of the unsaturated fatty acids of the medium, i.e. the oil as fat also causes a chemical type of hydro-bleaching. Bleaching of oils and fats by chemicals in the oxidative way, use of peroxides and generally all reagents producing active oxygen or other free radicals, are out of the question for use in the edible oil fat trade.

Some carotenoids are closely related to vitamin A. The α- and β-carotenes are precursors of vitamin A and are used as such (provitamin A). Carotenoids, produced from roots and seeds (carrots, anatto seeds, etc.) and also synthetically, are used in large amounts as safe colouring agents in foods and drinks.

2. Chlorophylls

The green-plant pigments chlorophyll *a* and *b* are found mainly in crude olive and grape seed oils, but also in vegetable oils harvested from unripe seeds, like rape seed and soya beans. The latter two both contain the so-called porphyrin ring, centred around a magnesium atom. The magnesium centre and the strongly conjugated structure of this ring are responsible for the green colour. If the ring contains iron as the anchoring metal ion in this so-called "chelate", the colour is brownish, and the compound is called haemin. When chemically bound to proteins, like haemoglobin in blood, myoglobin in meat and also in different enzymes and catalysts, like cytochromes, peroxidases, catalases, it becomes of the highest importance for animal metabolism, as an oxygen transporter or transformer. Some animal fats, mainly tallows, may have greenish tinges, most probably because of the decomposition of the brown haemoglobin and myoglobin of animal blood and meat tissues, which have been in contact with, for example, copper appliances, like tubes or valves. In plants the chlorophylls were the earliest discovered catalysts of photosynthesis, the radiation-mediated assimilation of water and carbon dioxide in plant material. Chlorophylls are not really oil soluble, and yet, if combined with oxidized fatty acids, it is very difficult to remove them by either chemical (acids, alkalis) or adsorptive means from the greenish coloured oils. The loss of magnesium turns chlorophylls a brownish colour and leads, in the first instance, to pheophytins.

3. Gossypol

Gossypol, an aldehydic polyphenol, is the precursor of the brownish discoloration of cotton-seed oils of less than average quality. It is originally of yellowish colour and only by oxidation, for example because of bad conditions during harvesting, delintering and seed pretreatment before storage, is it transformed to the brown-red coloured quinones or dimers of quinoid structure. Once dissolved in the oil and probably combined with oxidized fatty acids, its chemical or adsorptive removal during refining becomes a very troublesome operation. There are methods which can preferentially accumulate the pigment either in the extracted meal or in the oil. Gossypol causes male sterility if orally administered and has been used for this purpose.

F. Hydrocarbons

The biosynthesis of fatty acids, fatty alcohols, tocopherols, carotenoids and steroids is all based on the action of coenzyme A, the sulphhydril group containing protein, which has been mentioned before. When producing methyl group substituted repetitive linear or ring structures, the enzyme uses mevalonic acid as the main building block in order to synthesize them in consecutive anaerobic steps. One of the most imporant intermediates is squalene, a poly-isoprenoid C30 hydrocarbon present in higher amounts in the fatty tissue of a certain species of shark, living in the Pacific Ocean around Japan. However, in many vegetable oils (in olive oil up to 0.5%) and animal fats it can also be found as evidence of the intermediate of the biosynthesis of other compounds formed. Squalene contains six non-conjugated double bonds in its molecule. The rubber-like mass of about 10% karitene in the vegetable butter Shea-fat, pristane in the liver oil of fishes, and other saturated and unsaturated, branched and unbranched hydrocarbons, are examples of the existence of these unsaponifiable constituents in plant and animal fats.

G. Glucosides

Glucosides are ethers of mostly monosaccharide tetrosid, pentosid and hexosid sugars with a non-sugar component. The non-carbohydrate part, which can be liberated for example by acids or enzymes from the ether linked compound, is called the aglucon. This is generally the biologically active or technically important part, providing the specific character of the total oil. Many oil-bearing seeds and fruit parts contain glucosides, the best known being amygdalin, the glucoside of bitter almonds. The aglucon is benzaldehyde cyanohydrin, which after further decomposition liberates benzaldehyde having the characteristic bitter almond flavour. Sesamolin is the characteristic glucoside of sesame seeds and their oil, its aglucon: the phenolic substance sesamin has been used as a margarine tracer for the detection of the falsification of butter. It gives a characteristic red discolouration when reacted with concentrated hydrochloric acid and furfural (Villavecchia–Baudouin test).

Thioglucosinolates in Cruciferae oils are also important glucosides, e.g. sinigrin in mustard seed oil, their aglucon originally being a sulphate coupled mustard oil, allylisothiocyanate, which may form toxic (goitrogenic) cyclized sulphur containing derivatives, such as vinylthiooxazolidons.

Non sterol or diacylglycerol glucosides are generally oil insoluble, but if extracted with more-or-less colloidally soluble, micelle-forming lipid substances, they can be traced back in small amounts in some crude oils and fats of vegetable origin. They are generally easily removed by adding water to the fat, water causing their hydration and separation from the bulk oil.

1. *Sterol and Diacylglycerol Glucosides*

Small amounts of mainly β-sitosterol, etherified with tetrosid or pentosid sugars, all in the cyclized furanose or pyranose form, can be found in some vegetable oils. Besides these compounds, diacylglycerols are also often etherified with monosaccharides. Together with the phospholipids the above constitute the main building blocks of the cell-membrane bilayer material.

Saponins are sterol glucosides which possess surface-active, soil-removing, and haemolytic properties. The richest sources of saponins are saponaria roots and horse chestnuts, but they are also found in, for example, the vegetable butter Illipe fat and in smaller amounts in soya beans (soya sapogenins). The sapogenins are mainly 4,4-dimethylsterols, like the amyrins and lupenol etherified with monosaccharide sugars.

Some of these glucosides are also esterified by fatty acids via the hydroxyl groups, which explains their colloidal solubility in the relatively apolar triacylglycerols. All these glucolipids can often be found together with the phosphatides in the so called "lecithin sludges" of the oil degumming process.

H. Vitamins

Vitamins are organic substances which must be acquired by the human organism from outside sources, i.e. with its food. They regulate certain metabolic processes, being autonomic in their action or being involved in some enzymic systems as coenzymes. Some vitamins are fat soluble and others are water soluble. We mention here only the most important fat-soluble vitamins which occur in different parts of plant or animal organisms.

1. *Vitamin A or Retinol*

The principal sources of vitamin A (retinol) are fish-liver oils, eggs, liver, butter, cheese and milk. It has been totally synthesized. It is a polyisoprenoic primary alcohol and, as a matter of fact, it represents the biologically hydroxylated half of α- or β-carotene, both of which function as pro-vitamins for retinol. A shortage of vitamin A in humans may cause night blindness and some growth deficiencies. Margarines are generally vitaminized with a synthetic preparation which is either an acetic or a palmitic acid ester, which increases the fat solubility.

2. *Vitamin D or Calciferol*

The principal sources of vitamin D (calciferol) are fish-liver oils (halibut, cod), and it is present in lesser amounts in eggs, liver and dairy products. It can be totally synthesized. The two most important variants are vitamin D_2 and D_3 which regulate phosphate resorption in the animal organism. Deficiency causes illness of the bones, like rachitis in children and osteomalacy in the aged. Vitamins D_2 and D_3 are sterol derivatives, formed by irradiation of ergosterol (vitamin D_2) or of 7-dehydrocholesterol (vitamin D_3), present in the skin tissue. Margarines are vitaminized with the synthetic preparations of these vitamins.

3. *Vitamin E or α-Tocopherol*

Vitamin E (α-tocopherol) was mentioned above when discussing the natural antioxidants, the tocopherols, present in liquid vegetable oils, tomatoes and vegetables. Although its deficiency may cause some muscular trouble, there are no known cases of aberrations caused by its lack in humans, notwithstanding some suggestions of its role in supporting general health. It is produced by total synthesis industrially, mainly for fat antioxidant purposes.

4. *Vitamin K or Phytomenadion*

Vitamin K (phytomenadion) is an important factor in the clotting of blood. If vitamin K is not present in blood, then persons injured may bleed to death. Spinach and other green vegetables together with tomatoes and some vegetable oils contain vitamin K.

IV. Fat Modifying Reactions

Lipid transformation products are derived lipid compounds, which were probably either not present or present in only small amounts in the oils or fats included in the cells. The products are thus formed mainly during or after oil production as a result of external influences (reagents, conditions, processes). They are thus not necessarily the result of any maltreatment of the fats, although they are inevitably due to chemical changes. As lipid transformation products are different in structure from the parent lipids, they may give important analytical clues for monitoring quality differences. The more important reactions which cause alterations are discussed below.

A. Hydrolysis

Hydrolysis can affect all acylglycerols, sterol esters, sterol glucosides, ester waxes, tocopherol esters and esterified chlorophylls. The reaction requires an aqueous medium and usually a catalyst. The enzyme lipase acts as a biochemical catalyst, whereas mainly mineral acids or bleaching earth serve as chemical catalysts for the hydrolysis. If short-chain hydroxy fatty acids are cleaved in an acid or neutral medium, as is the case during deodorization of coconut and palm kernel oils, these acids can secondarily cyclize to lactones, some of which are important flavour carriers for butter and margarines.

B. Autoxidation and Antioxidants

Air, consisting of one fifth of its volume as oxygen, is normally freely admitted to oils and fats. Unsaturated fatty acid chains are, therefore, more or less easily oxidized.

It has been mentioned already that by its very "negative" character a double bond exerts an attractive force on the bonding electrons of the valence electrons of one of the hydrogen atoms of the methylenic groups next to it, also called an active α-methylenic group. This hydrogen atom is easily eliminated by radical formation or inducing conditions or agents, such as heat, radiation (light) and oxygen. After cleavage of the hydrogen bond, a free radical is formed on the fatty acid chain. The removal of this hydrogen atom is easier if the α-methylenic group is situated between two double bonds (1,4-pentadienoic or "skipped" methylenic structure), as is generally the case in linoleic and linolenic acids.

1. Primary Oxidation Products

The different steps of the autoxidation reaction are shown in Fig. 1.5. The initiation step (A) cannot be stopped by additives. Only the complete exclusion of radical formers can inhibit the free radical formation. The next so-called propagation step (B) consists of two parts, during the second of which a number of protective measures can be taken. In the first propagation step, oxygen addition leads to a peroxy free radical, which in the second, rate-determining step reacts with an unattacked unsaturated fatty acid. This yields a highly reactive alkyl free radical and a so-called fat hydroperoxide. The alkyl free radical reacts with an oxygen molecule in a new cycle. The first stable product of the autoxidation reaction is the fat hydroperoxide. This compound is tasteless and odourless.

$$RH \xrightarrow{\text{light, heat}} R\cdot \qquad A$$

$$\left.\begin{array}{l} R\cdot + O_2 \longrightarrow ROO\cdot \\ ROO\cdot + RH \longrightarrow ROOH + R\cdot \end{array}\right\} B$$

$$\left.\begin{array}{l} 2\,ROO\cdot \longrightarrow \\ 2\,R\cdot \longrightarrow \\ ROO\cdot + R\cdot \longrightarrow \end{array}\right\} C$$

Fig. 1.5 Autoxidation reaction: A, initiation; B, propagation; C, termination. RH, fatty acid molecule; R·, alkyl radical; ROO·, peroxy free radical.

As a rule the radical reactions do not continue until total consumption of the fat substrate. There is usually a depletion of oxygen supply: oxygen transfer is inhibited by a (polymerized) film formed in the course of the reaction at the surface of the bulk of the oil or fat. In the absence of oxygen the reaction ceases in the so-called termination step (C), which leads to different end products. The hydroperoxides, depending on whether oleic, linoleic or linolenic acids are attacked, differ as regards their stability. Oleic hydroperoxides are the relatively most stable ones.

2. Conjugation by Autoxidation

Autoxidation is initiated by radical formation at the α-methylenic carbon atoms. This situation causes mobility of all the bonding electrons in the whole group composed of three vicinal carbon (—CH—, methyne) atoms. This mobility can lead to a shift of the double bonds within this three carbon unit, the so-called "allylic" group. The initial attack on a 1,4-pentadienoic system leads to free-radical formation on the methylene group between the two double bonds. This unstable situation leads, by shifting, to the conjugation of the two double bonds, which represents a more stable state. The Z double bond which was actually shifted is transformed to an E one (see Section C below).

Isolated double bonds with an absorption maximum around 190 nm cannot be detected with a commercial UV/visible spectrophotometer. On the other hand, if, for any reason, a conjugated dienoic group is present, this can readily be detected in the UV region of the spectrum by an absorption maximum around 232 nm.

The degree of autoxidation of fats and oils containing polyunsaturated fatty acids can also be followed by the increase in the dienoic conjugation. This increase results from a rearrangement caused by radical formation and temporary stabilization in the form of hydroperoxides. If the hydroperoxide is then decomposed during the course of secondary reactions, a large part of the dienoic conjugation may still persist, thus giving some clue to the past oxidative history of a given fat sample. Conjugated trienoic systems, in which the shifted double bonds are generally in the E configuration, absorb with a triplet of absorption peaks having maxima around 262, 268 and 274 nm.

Acid-catalysed dehydration of hydroperoxides may give rise to the formation of trienoic conjugation, the dehydration extending the dienoic conjugation already present to a trienoic one. This fact can be used as, for example, a clue to previous bleaching of an oil, which has been more or less attacked by oxygen, by acid-activated bleaching earth.

3. *Secondary Oxidation Products*

Hydroperoxides decompose to secondary oxidation products as a result of heating, radiation, or the presence of heavy metal traces (Cu,Fe) and other radical initiating agents. This decomposition is usually also radical generating. Secondary lipid oxidation products can be formed either by peroxide scission alone, or by simultaneous peroxide and chain scission. The latter leads to e.g. short-chain volatiles and new radicals (chain branching) or to dimerization of the scission products (Fig. 1.6).

(a) *Off-flavour formation.* One of the most important decomposition reactions of hydroperoxides is that involving chain scission which often leads to short-chain aldehydes. The real importance of this reaction is that these aldehydes, depending on their structure and the amounts formed, can cause very typical odours, some of them being characterized as off-flavour, like (soya bean oil) reversion odours, the green, beany, painty, fishy, tallowy epithets or just plain rancidity. If present in substantially high concen-

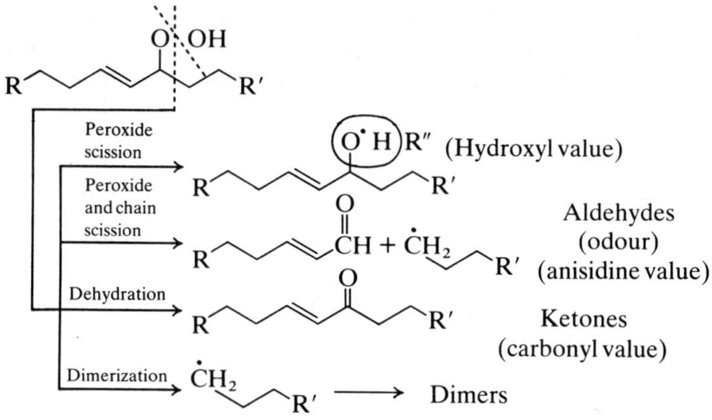

Fig. 1.6 Secondary oxidation products: R', rest of triacylglycerol molecule; R, alkyl; R"H, another triacylglycerol molecule.

trations, the aldehydes (carbonyls) formed can be determined in the bulk by specific-colour forming reagents. The individual aldehydes may be quite different, but sometimes have extremely low flavour threshold levels (1 mg ≡ 1 µg kg^{-1}), which readily explains the early off-flavour formation in fats without measurable autoxidation. (The threshold level or value is the minimum concentration of a flavour carrier, in which 50% of observers can perceive it positively.) There is a distinct difference between the sensitivity of the human nose/tongue and the sensitivity of the carbonyl group reagents towards the carrier aldehydes of different off-flavours. In spite of the confirmed organoleptic assessment of a certain off-flavour aldehyde, the bulk carbonyl reagent may fail to detect its presence. The threshold levels and detection limits can differ by a factor of up to 1000, a discrepancy which should be remembered when assessing fat quality by different means.

4. Retardation of Autoxidation

(a) *Antioxidants.* Radical quenchers, like carotenoids, act by protecting crude fats and oils (e.g. palm oil) once they have been produced only against light-induced photo-oxidation by transforming the absorbed radiation energy into heat. Storage in the dark is, therefore, a good proposition for extending the shelf-life of oils and fats. Exclusion of air by using vacuum or inert covering gases and low temperatures during storage are also excellent, but not always feasible or practical. For this reason suitable measures must be taken to reduce damage in the most critical steps of the autoxidation process.

Most vegetable oils and fats are protected by their natural antioxidants which are quite effective in slowing down the rate of oxygen absorption by reacting with the fatty acid peroxy free radical in the second part of the propagation step. These peroxy free radicals react preferentially with the phenolic hydrogen atom of the tocopherol molecule. This reaction leads to a fat hydroperoxide as in the case of the normal, uninhibited oxidation reaction. However, as the tocopherol free radical, which is produced simultaneously, has much less affinity to react with the oxygen molecules present, the reaction chain is stopped. The rate of oxidation is thus effectively reduced and the stability of the fat is increased (Fig. 1.7, A).

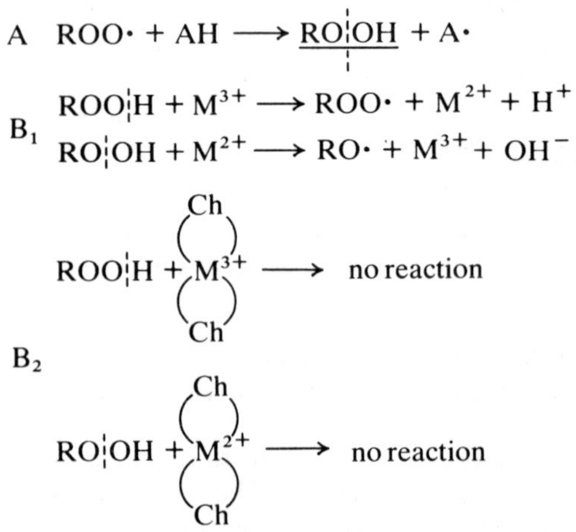

$$A \quad ROO\cdot + AH \longrightarrow \underline{ROOH} + A\cdot$$

$$B_1 \quad \begin{array}{l} ROOH + M^{3+} \longrightarrow ROO\cdot + M^{2+} + H^+ \\ ROOH + M^{2+} \longrightarrow RO\cdot + M^{3+} + OH^- \end{array}$$

$$B_2 \quad \begin{array}{l} ROOH + M^{3+}(Ch)_2 \longrightarrow \text{no reaction} \\ ROOH + M^{2+}(Ch)_2 \longrightarrow \text{no reaction} \end{array}$$

Fig. 1.7 Retardation of autoxidation. (A) Chain stopping by antioxidants ($A\cdot \ll R\cdot$; AH, antioxidant with mobile H-atom). (B_1) Scission by metal ions: M, metal ion in different valence states. (B_2) Inactivation by metal scavengers: Ch, chelators.

The formation of the same, sometimes very unstable, fat hydroperoxides, explains why, despite the presence of antioxidants, a strong and early off-flavour formation can still very often be perceived. Decomposition of hydroperoxides due to the simultaneous presence of heavy metal traces (Fig. 1.7, B) may lead to the formation of the same aldehydes, as mentioned previously.

Animal fats (except marine oils), although unprotected by natural antioxidants, are mainly composed of fatty acids which are less unsaturated than liquid vegetable oils. This characteristic renders them more stable to autoxidation. If they do become oxidized, the resulting decomposition aldehydes generally have high(er) threshold values. The addition of synthetic or natural antioxidants is, therefore, a feasible protective measure. Examples of such antioxidants are phenols and polyphenols: tertiary butylated hydroxy anisol (BHA), synthetic tocopherols, gallates and flavonoids, e.g. quercitine.

(b) *Antioxidant synergists (metal scavengers)*. Retardation of autoxidation can also be achieved by removal of the free-radical initiating impurities, generally admixed with the fats during their extraction, processing and handling. Many of these impurities are of metallic character, in either ionized or complexed form (such as the iron of myoglobin and haemoglobin) as trace contaminations in animal fats. The decomposing action of transition metals on hydroperoxides, as shown in Fig. 1.7, B_1, leads to chain-initiating free-radical formation and off-flavours.

Inorganic and organic scavenging agents, which inactivate transition and other metal ions in the form of inclusion compounds (chelates), also give some relief (Fig. 1.7, B_2). Removal of metal compounds still seems to be the best proposition for protecting final quality. Citric, ascorbic and lactic acids and (disodium) ethylene diamine tetra-acetate (EDTA) are effective metal scavenging or sequestering agents. These compounds are also the so-called "synergists", as used in combination with the true antioxidants, strengthening their primary radical-catching action.

The strength of an acid is expressed by its dissociation constant, and in this light the following series is relevant:

oxalic > phosphoric > lactic > citric > formic > ascorbic > acetic.

The strength of sequestering action can be expressed by the salt formation constants with a metallic (e.g. a trivalent ferric) cation. In this respect the following series can be established:

oxalic > citric > phosphoric > acetic > formic.

C. Isomerization

Heating oils which contain polyunsaturated fatty acids with different catalysts may cause, at temperatures above 100°C, the positional and geometrical isomerization of 1-*cis*,4-*cis*-pentadienoic double and *cis* single bonds. The process is essentially the same as the free-radical based double-bond shifting or recombination as described for the autoxidation-caused conjugations: some isolated dienoic double bonds may be transformed into conjugated 1,3-tetradienoic isomers, whereas some single bonds may undergo *cis*/*trans* (*Z*/*E*) isomerization. Catalysts which induce these transformations are nickel, copper, selenium, sulphur, sulphur dioxide and nitrogen oxides, but also bleaching earth residues which contain iron traces. More details on these processes are given in Chapter 4 (Section IV) and Chapter 5 (Section I).

Oils containing 1,4-pentadienoic unsaturation in their fatty acids, if heated with strong alkali above 200°C for a long time may undergo extensive conjugation of these dienoic acids. If heated more excessively the conjugated bonds of, for example, linolenic acid may form cyclic monomers, like cyclohexadiene derivatives.

D. Polymerization

Heating oils at higher temperatures (240–280°C) may cause shifting and Z/E isomerization of double bonds in unsaturated fatty acid molecules. Besides the use of metal catalysts, the heat-induced mobility of double bonds is also assisted by a limited access of air, both of which act as free-radical initiators. During shifting, the alkyl free radicals formed temporarily can combine with those present in adjacent fatty acid molecules, leading first to dimeric, then to trimeric, and finally to polymeric acids. By the formation of these bonds, mostly C—C bonds and some C—O—C bonds, the viscosity increases as three-dimensional cross links are formed. The resulting products are insoluble in the normal oil solvents.

The inclusion of air during heating, as in deep frying or shallow frying and baking, can lead partly to dimers, and later to trimers (polymers) of the above-mentioned three-dimensional, cross-linked structures. The dimers formed in household use are found to be completely harmless, not being resorbed by the animal organism. The oxidation of highly unsaturated oils in thin layers, as in paints, etc., in the cold leads to polymerization which is called "drying".

E. Pyrolysis

1. *Pyrolysis of Acylglycerols*

Heating fats far above 300°C for longer times leads first to the cleavage of the ester functions. The glycerol residue is dehydrated to acrolein, an unsaturated, lacrimatory C3 aldehyde. The fatty acid moiety generally first loses carbon dioxide from the carboxyl group (decarboxylation), the residual parts forming straight-chain hydrocarbons.

2. *Pyrolysis of Carotenoids (Heat Bleaching)*

Palm oil contains quite high amounts of carotenoids (400–2000 mg kg^{-1}) which are responsible for its deep orange colour. The removal of this colour can be achieved by adsorption of carotenoids on adsorbents like bleaching earth or activated carbon. Another way to remove carotenoids is by pyrolytic cleavage in the so-called "heat-bleaching" operation. Practical conditions are given in Chapter 4 (Section IV) on refining and bleaching (decolorization).

2

Vegetable-oil Separation Technology

I. Raw-material Handling Operations

A. External Transport and Storage

The great nutritional importance of oils and fats in feeding the ever increasing world population, and the diversity of geographical places where the oil- and fat-bearing raw materials are cultivated and/or produced explains the annual need for transporting and storing several millions of tons of these commodities. As the success of harvests is influenced by factors of chance, such as weather conditions and diseases, and by the agricultural methods applied, so the volumes produced per year might be quite different. Although not always possible, it is still feasible that factories should be located either in close vicinity to the raw material production areas or at least in places that are easily accessible by means of cheap mass transport. Because of the huge volumes involved it is practically impossible to forward the raw materials, which are harvested only in certain parts of the year, to the different oil-milling centres without delays. So by sheer necessity the greater part of the harvested seeds are stored for longer times in good condition. No long storage periods are needed for animal and some vegetable fat sources, like palm kernels and coconuts, which are accessible all the year round. Palm fruits, being also all-year accessible, form an exception because of the large volumes produced. The oil should be processed immediately after harvesting because of its quick fermentation-caused degradation; this is even more important if the oil is left in the fruit without sterilization. This is the reason for organizing transport and storage in the most efficient way possible. Transport and intermittent storage are not only costly and time consuming, but they also influence the quality of the seeds and thereby the oils contained within them. Economical and political unrest may cause strong fluctuations on the world fat market, with relative abundance increasing the problems of quality maintenance during long storage periods.

For the purposes of transport and storage we should distinguish between

(a) free-flowing oil seeds, like soya beans, sunflower, safflower, sesame and rape seeds, polished palm kernels and dehulled groundnuts, and (b) bulky commodities, such as undelinted cotton seeds, copra and olives.

Crude terrestrial animal fats and marine oils are produced on or close to the site of slaughtering or catching. Therefore the transport of live animals to slaughterhouses, and whaling operations will not be handled here. Palm, palm kernel and olive oils are produced, because of the extreme importance of the time of maturity on their quality, in the immediate vicinity of the oil-processing sites. Therefore the transport of these commodities is discussed when describing their production.

Growers store the seeds harvested under the possibly correct conditions of maturity and weather either on the ground, but preferably under sheds or in bins, or they send them by vans, bulk trucks, or in less-developed areas in bags by carriages, to so-called country elevators or sub-terminals situated in the centre of a producing area. From here the goods are dispatched by railroad preferably in hopper cars or on waterways in barges to the terminal elevator located at transportation terminals such as sea ports or larger market places. The terminal elevators can store huge amounts of materials (up to half a million tons) in silos and, e.g. cotton seeds and copra, in single-storey flat storage bins or multi-storey warehouses. To ensure optimal storage conditions of the seeds large-capacity cleaning, drying and conditioning installations (see later) are applied in order to control moisture, temperature and damage caused either by metabolitic (respiratory) activity in the (mainly split, broken) seeds or by mould infections, insects, rodents or foreign material such as weed seeds.

The unloading of seeds, the free-flowing ones being very similar to grains as regards their handling, from trucks and railroad cars can be carried out in covered hangars. Concrete-lined intake pits are constructed along both sides of the track(s), with mechanical conveyors at their bottom. Most of the seeds flow by gravity out of the vehicles, while the remainder are removed by power shovels. On both sides of the cars aspiration ducts hanging from the ceiling provide air suction in order to collect the dust which is evolved during the unloading operation. Bulk trucks may be discharged by a tilting platform enabling the seed to flow by gravity through the opened rear door. All these vehicles, and of course preferentially barges and ships, may be unloaded by pneumatic conveyors (suckers). Special high-capacity so-called "marine legs", which are bucket- or chain-type conveyors, are in use with grain and oil-seed elevators handling gigantic volumes of single oil-seed varieties at certain harbour terminals. For example, in the Philippines copra is unloaded from the barges by clamshell-type buckets operated by a swinging boom crane.

Flat-floored longitudinal storage buildings covered by inverse V-formed or half-cylindrical roofs and with conveyors providing top discharge and underground-tunnel reclaiming facilities may be the primary means of storing huge amounts of seeds for short periods of time (Fig. 2.1). Storage sheds with sloping roofs but without side walls just cover the seeds and the material on the flat floor can be moved in and out by power shovels.

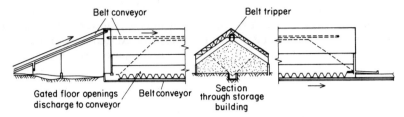

Fig. 2.1 Conveyor storage and reclaiming in a flat-floor building. [Reproduced, with permission, from Perry (1984); and courtesy of Stephenson–Adamson Division, Allis Chalmers.]

Bulk storage in warehouses on flat floors has the advantage of easy control of oil seeds, like cotton seeds, maize germs, further meals, cakes, etc. To prevent self heating or autocombustion of the seeds, recycling by horizontal and vertical transportation means is recommended. The free access of air during this operation cools the seed for further safe storage. The disadvantage of flat-floor storage is the easy access for insects and rodents to the commodities. The space utilization of this type of storage is, of course, bad, and there are fire hazards which can only partly be reduced by a compulsory sprinkler installation. As they may be situated on several levels, the vertical transport facilities, scales and cleaning devices can be placed at one side of the building. Charging these facilities by remote control mechanical conveyors is no problem, but the mechanized and remote-controlled discharge, especially of non-free-flowing commodities, requires costly equipment.

Silos are built in single or multiple side-by-side configurations. They are of cylindrical form, with a flat or hoppered bottom and are generally constructed of concrete. Cylindrical metal silos made of galvanized-steel plates or continuous metal ribbons wound up from the ground in spiral form can be quickly erected up to 15–20 m height and, e.g., 15 m in diameter. Large silos may have diameters of up to 60 m, and some 10 000-tons storage capacity. They are fitted with (many) temperature sensors and level-indicating devices. Their advantage is that they exclude fresh air and light, which might influence the bioactivity and respiration of the seeds. Silos generally protect the seeds effectively against rodents. Silos of large diameter are charged at the top through many openings to obtain a uniform spread of the different batches in layers. The reclaiming of the contents at a conical bottom is aided, if necessary, by vibrators and gyrators applied to the hopper. Flat-bottomed silos need special mechanical rest-discharge equipment. If closed properly, they constitute less of a fire hazard than spaces which are open to the air. Their disadvantage is the pressure exerted by their own weight on the softer sorts of seeds or meals, if stored higher than 30 m, and less-effective humidity control, especially in steel silos.

When storing different seeds in longitudinal flat-floor storage spaces, one must take into consideration the natural limiting angle of gliding, which determines the volume of seed which can be stapled without the danger of high lateral pressures exerted on the side walls. Rape seed and soya beans have a limiting angle below 30°, and ground nuts, sunflower, safflower and cotton seeds of above 40°. Of course the pressures exerted on silo walls are also dependent on the gliding angles, besides the apparent specific densities and storage heights.

1. *Equilibrium Moisture Content*

Seeds are biocolloids and therefore, like all other plant material, they contain moisture (water) both inside and outside their cells. Moisture in the seeds exerts a certain vapour pressure, which is generally less (p), than that of pure water at the same temperature (p_o). The proportion of these two pressures is the so-called water activity (A_w) of the seed.

$$A_w = \frac{p}{p_o} = \frac{\% \text{ relative humidity}}{100}$$

The partial vapour pressure of moist air divided by the vapour pressure of pure water at the same temperature gives its relative saturation, i.e. its relative humidity. Seeds in contact with moist air of a given relative humidity will try to equilibrate their own water activity with the humidity of the surrounding air, either by loss of moisture by evaporation or by gain of moisture by absorption. This exchange will go on until the vapour pressure measured above the seeds (in the interstitial space during storage) equals that of the surrounding air or gas. The corresponding amount of water is the "equilibrium relative humidity" (ERM) or "equilibrium moisture content" of the seed at the given conditions.

This close dependence of the moisture content (water activity) of the seeds on the relative humidity of air can be better understood by considering the different "functionality" of the moisture in seeds, which is governed by the forces through which water interacts with the seed constituents. Water molecules are asymmetrical and, therefore, the centres of their positive and negative charges do not coincide with each other. They are a typical example of (bi-)"polar" molecules. The hydrogen atoms of water are also attracted by the oxygen atoms in the hydroxyl groups of the surrounding water molecules. These weak "hydrogen bonds" cause the aggregation of water into so-called "clusters". This is the reason why water, whose molecular weight (MW) is 18, has a much higher boiling point, than, for example, nitrogen with MW = 28. Similarly, water molecules are also attracted by the polar building groups of the seed material. Such polar groups are the hydrophilic (water seeking) carboxylic, hydroxylic and amino (imino) groups of the different proteins and carbohydrates, cellulose, hemicelluloses, etc. These contrast with the ester groups of the hydrophobic (water repelling) triacylglycerol in fats.

The moisture inside the cell walls is retained in the form of liquid/solid colloid solutions or hydrogels, containing dissolved inorganic and organic material. The moisture is thus structurally bound and exhibits an excessively low vapour pressure, if any. The cell walls or membranes may bind some moisture by chemisorption via

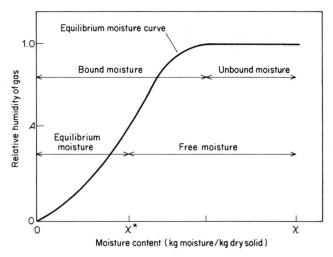

Fig. 2.2 Types of moisture. [Reproduced, with permission, from Treybal (1981).]

polar–polar bonds, whereas physical adsorption may retain moisture more loosely in the pores, capillaries and at the surface of the seed. Wet seeds may contain free moisture on the surface.

The proportion of moisture which is present inside the cells or constitutes a part of the structure is also necessary for different biochemical functions (enzyme regulated metabolism, even in dormant seeds latently present). This part cannot be removed without causing profound, irreversible changes in the seed cells (coagulation of the internal hydrogel structures). This moisture is "bound" and functional. Another part of the moisture can be removed reversibly, as these changes, e.g. shrinkage/swelling of the seeds are temporary and not fatal for any later seed functions. This moisture is "free", although not "unbound".

Based on these functionality considerations we can discriminate between the different types of moisture in the seed, manifesting itself by different behaviours (see Fig. 2.2).

(i) The part of the moisture which exerts an equilibrium vapour pressure less than that of pure water at the same temperature, is called the "bound moisture". The "equilibrium moisture" below the pure water vapour pressure is thus always "bound".

(ii) The part of the moisture which exerts in the seeds an equilibrium vapour pressure equal to that of pure water is the "unbound moisture".

(iii) The part of the moisture which is in excess of the equilibrium moisture levels and which can only be evaporated is called "free moisture".

The relation between the "wet basis" (kg moisture/kg dry seed + kg moisture) moisture content of seeds and the relative humidity of air which is in equilibrium with them has been studied and was found to be relatively insensitive to temperature differences between 0 and 50°C. Above this temperature the colloid chemical changes mentioned above are initiated and the equilibrium vapour pressures distinctly increase. Soya beans containing less oil and behaving more like grains, are somewhat more sensitive to changing external temperatures.

It has been found that the value of the critical equilibrium relative humidity of the interstitial air in stored and non-ventilated seeds at which micro-organisms (e.g. moulds) will develop rapidly is around 75%. Thus a 70% relative humidity or less is considered as safe at ambient temperatures to keep oil seeds in storage without deterioration for longer periods. The corresponding "dry basis" (kg moisture/kg dry seed) "safe" moisture content of oil seeds together with some of their main constituents is given in Table 2.1.

Table 2.1 *Composition and "safe" moisture contents of oil seeds*

Oil seed (dry matter)	Protein (%)	Carbohydrate and crude fibre (%)	Hulls (%)	Fat (%)	Safe moisture level (%)
Copra	8	25		67	6
Rape seed	26	20	14	40	7
Palm kernel	9	39		52	8
Sunflower seed	16	13	28	43	8.5
Cotton seed	19	15	48	18	10
Safflower seed	20	20	27	33	11
Ground nut (kernels)	32	20		48	11
Soya beans	40	32	8	20	12

An old rule of thumb on which the "safe" (critical) humidity for oil seeds can be calculated, is based on the practical knowledge that the "safe" humidity for starchy food grains is generally around 16%. The supposition is that the oily part of an oil seed does not absorb moisture from the surrounding air because of its hydrophobic character. The safe humidity for an oil seed containing, for example, 22% fat (soya beans) will thus be proportional to that aliquot of the 16% which is the non-oily (proteinaceous, cellulose, other carbohydrates, minerals, etc.) part, well absorbing moisture. In our selected case the humidity would be $0.78 \times 16\% = 12.5\%$. The "safe" moisture of daily practice will differ considerably in many cases: values lower by even 1 or 2% than those given above will be chosen if for any reason long storage periods (e.g. 8–10 months or longer) with large external temperature changes are envisaged.

B. Drying

1. Theoretical Background

By definition, drying means removal of a liquid from a solid by thermal means. Fundamentally two processes act simultaneously: transfer of heat to the system which contains the moisture, in order to transfer the liquid from the inside of the solid to the surface, where it is evaporated. Oil seeds contain moisture (water) in their cells or at their cell walls, as we have seen previously. The movement and addition or removal of free water causes the well known volume changes associated with swelling or shrinking of the seed. In order to remove bound water we must add to the system not only the heat required for its evaporation, but also the energy needed to disrupt polar bonds and/or the stability of the colloid system. If the temperature of evaporation is above a critical value, the process may become irreversible, e.g. because of the coagulation of proteins. This physicochemical change of the structure is critical when producing edible protein concentrates from extracted meals.

Internally the water to be evaporated must first be transported to the surface of the material by diffusion. This is a very important step in all seed-drying operations, to be achieved by indirect heat transfer to the seeds ("sweating"). No moisture should leave the surface at this stage because too rapid drying closes down the pores (capillaries) on the surface of the seeds. The moisture driven to the surface then comes into contact with (moving) dry air, the carrier medium to which ultimate transfer must be effected. To maintain this second stage of mass transfer at a constant level it has to be supported by a suitable humidity gradient between the liquid surface (Y) and the main air stream (Y_s). The air, nearly or less saturated with moisture, must constantly be removed from the system at an optimal speed. The rate of cross-circulation drying (N_c) is expressed as the amount of moisture removed from the seeds in unit time from a unit surface.

$$N_c = k_Y (Y_s - Y)$$

where k_Y may be considered to be constant for a certain stage of the process.

The dehydration or drying of oil seeds is divided into more distinct steps. (Fig. 2.3). The first step involves either increasing (A) or decreasing (A') the rate of heating depending on whether the temperature of the material is lower or higher than that of the drying medium (the air). The seed will thus be either heated up or cooled down. In the second drying stage (BC) the process has a constant rate (N_c), the surface appears uniformly wet and the temperature of the product is constant, at least if the diffusional transport can be sustained. This situation will end at the "critical" moisture content" (X_c), whereafter the surface of the seeds starts to dry out and warm up. The rate of drying will then fall, but the curve representing this part of drying is not generally linear or at least shows some breaks. This is caused by the fact that this final part of the dehydration process is a very complex one, mainly because of the inevitable internal changes in the capillary and cell structures. Finally the process should end at the equilibrium moisture level (E), characteristic of the

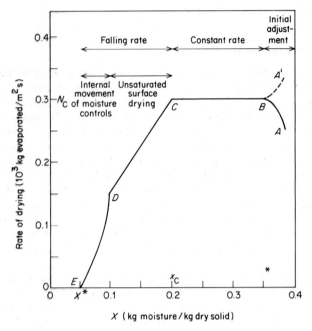

Fig. 2.3 Typical rate-of-drying curve for constant drying conditions. [Reproduced, with permission, from Treybal (1981).]

humidity and temperature of the drying medium. This should preferably not be lower than the ERH during storage.

2. Driers

Oil seeds which have a moisture content above the critical value and which are considered to be unsafe for storage or processing, must be dried. The huge volumes to be handled make the use of continuous driers imperative. The driers may work in the co-current or countercurrent fashion, but many do the work by contacting the material and the heating medium in the crosscurrent mode. The seed is heated mostly indirectly by steam and/or by directly and indirectly heated air. Microwave (radiation) heating is a novelty and is applied mainly for drying protein meals for human consumption. The forced air circulation (high hydraulic resistances caused by material to be dried and the construction of the drier) requires the application of large fans, which incur a high energy consumption. Effective dust collection is a prerequisite for the installation of fans. Another important aspect of energy saving is the recuperation of the latent heat in moisture-saturated hot used air either by means of heat exchangers/condensers or by the use of a heat pump.

The indirect rotary driers are near horizontal wrought iron or steel

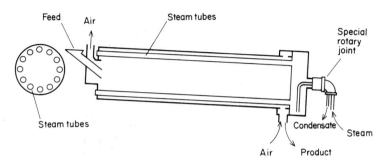

Fig. 2.4 Indirect steam tube rotary drier. [Reproduced, with permission, from Treybal (1981).]

cylinders in which either a bundle of steel heating tubes mounted in the head plates, with or without flights, rotate, or the bundle of tubes remains stationary and the cylinder equipped with flights rotates around it. Some driers employ rotary tube bundles and additional stationary double-shell steam jackets. The seeds are introduced at one end and are shovelled to the other, the air stream moving countercurrently to that of the seed transport. The draught is sustained either by a chimney above the seed entrance or by a fan (Fig. 2.4).

Drying in vertical columns by hot air in thin (40–60 mm) layers is the best way to reduce the moisture content of wet seeds. First, as the seed trickles down the columns it is heated indirectly by passing the low-pressure steam-heated tubes or radiators; the moisture is forced to the surface of the seeds (sweating) without the closing of capillaries by quick dehydration of the surface. In the actual drying section the seed comes into contact with hot air in cross flow. In the case of drying soya beans, the air is heated by using it for the direct combustion of fuel gas, for sunflower seeds indirect heating by steam is preferred. In order to produce thin seed layers the column contains thin shafts, constructed from perforated sheets and sometimes forming staggered ducts. The seeds are regularly shifted and turned by roof-shaped or other baffles in order to be dried on all sides. Only in the upper two-thirds of the column is the seed in contact with hot air (here the column is covered by a mantle), while in the lower one-third the seed is merely cooled by ambient air flowing through the perforations of the ducts. The complete and continuous filling of the column with material, the strict control of the flow rate by discharge devices, and the working temperature in the column are prerequisites for a successful operation. The amount of air passing the seeds in this system is higher than in the pipe-bundle rotary driers. This might cause some oxidative damage to broken seeds. As means of improving the heat economy of driers reuse of the cooling air and part of the cleaned heating air for the burners or a two-stage use of the hot air in the drying process are highly recommended, especially in areas with very cold atmospheric fresh air (Fig. 2.5).

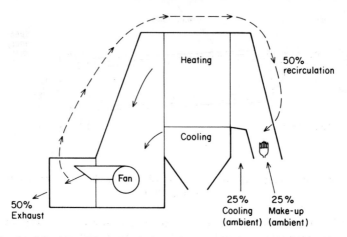

Fig. 2.5 Seed drying with recirculation of 50% of exhaust air. [Reproduced, with permission, from Dada (1983).]

Drying under vacuum is a very gentle, non-harmful way of drying oil seeds, as water evaporates at 30–50°C, depending on the residual pressure. The vacuum is produced by pumps, which may be water ring sealed or attached to a barometric condenser. The heating inside the vacuum tower is provided by low-pressure steam-heated, stacked radiators. The inlet and outlet sections are closed by automatic, e.g. pneumatically controlled, gates which make the operation semi-continuous. The seed cools down in separate compartments with ambient air which is directed to cyclons where the dust is collected. The method seems to be feasible energetically, but the capacity is limited. The investment costs are high. The introduction of microwave heating combined with vacuum drying (MIVAC) is a new concept (McDonnel Douglas) combining existing means; however this method is still at an experimental stage.

Direct-fuelled rotary driers (very large and long tunnel-like ones are called kilns) may be very efficient in terms of heat economy, if they contact the seeds directly with hot air (burnt gases) produced by burning fuel or cheap oil-milling side products such as husks, hulls and shells. The seeds are lifted and conveyed by flights fixed in cascade form to the interior surface of the shell, rotating with it. The cylinder is installed at a small angle to the horizontal. The air moves co-currently with the seeds, and the wet air is sucked above the outlet by a fan into a dust-separating cyclone. Although the fuel gas and air are cooled at the end of the drier to the so-called "adiabatic wet-bulb temperature", very careful supervision regarding the temperature, the final moisture level inside the tunnel and the exclusion of any soot formation and condensation on the seeds is still needed. The presence of carcinogenic polycyclic aromatic hydrocarbons (PAH) on sunflower seeds, on dried copra and, consequently, in the oils derived from

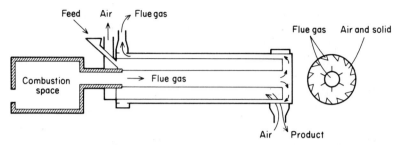

Fig. 2.6 Indirect rotary drier. [Reproduced, with permission, from Treybal (1981).]

them, may be caused by undetected partial soot condensation. Direct rotary driers are seldom applied by the oil-milling industry, although they may be in use at country elevators or sub-terminals, situated in the centre of a large production area. Double-shelled indirect rotary driers with countercurrent air flow could reduce the PAH health risks (Fig. 2.6).

All the driers described above are available from specialized manufacturers in dimensions of 1 to 3–4 m diameter and 4–40 m length.

C. Storage Defects

The condensation of humid air inside and on the walls of silos and/or containers may cause self ignition of the seeds by increased biological activity of some type. This may happen when there are large changes in external temperature during long storage periods. In addition, overseas shipment of seeds in cargo vessels for long periods or journeys may lead to spontaneous heating and quality damage, because of the sudden humidity and temperature changes encountered during the voyage.

A quick daily rise in temperature (e.g. $1°C$ day^{-1} for sunflower-seed) or temperatures above $40°C$ at the point of measurement are signals to take immediate action, as the temperature may be above $60-70°C$ at places more distant from the sensors. Such action involves the recycling of the contents, after cooling or drying, into another empty silo cell. Fig. 2.7 shows the permissible storage time of soya beans at different temperatures and moisture levels.

Incorrect handling and storage of seeds may cause, as we have seen, quality degradation by revival of cell activity (respiration) and microbial or enzyme activity. These processes influence not only the protein and carbohydrate part of the seed, but also the fats and the substances accompanying the fat, such as chlorophyll, carotenoids, gossypol, tocopherols and tocotrienols, etc. The main chemical processes influencing the oil quality are hydrolysis, autoxidation, polymerization and pyrolysis. The reduction of the chlorophyll and pheophytin levels of green, early harvested (unripe) soya beans and rape seed is advantageous, whereas the fixation of gossypol in cotton seed is detrimental to the quality of the extracted oil.

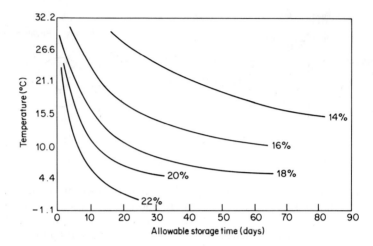

Fig. 2.7 Allowable storage time for soya beans. [Reproduced, with permission, from Spencer (1976).]

A decrease in the tocopherol and carotenoid contents accompanies the trace-metal or enzyme-catalysed autoxidation of fatty acids, primarily of the polyunsaturated, and often biologically essential, ones. This can be established by the uncharacteristically high level of dienoic conjugation in freshly extracted oils.

In order to reduce all these kinds of possible damage, the elevator or storage management must record data on the maturity of the seeds, the time between harvesting and placing in storage, and the ambient temperature at the time of placing in storage. Good housekeeping in the storage facilities demands placement of materials with different moisture levels in separate silos. Mixing of materials at low and higher-than-safe levels in order to attain equilibration at the safe level is risky and inadmissible. Aeration of higher-moisture seeds by circulation is one of the measures which can be applied, if the relative humidity of outside air makes this operation feasible. The use of dedicated microprocessors in recording data and coordinating/executing all necessary actions is self evident (see Chapter 8, Section II).

D. Internal Transport

Conventional mobile means of internal transportation in terminals, ports and oil mills are movers, like specially built tractors which can operate both on roads and rails, and various rubber-tyred payloaders, lift trucks, etc. Besides the very simple means of gravity conveying, like chutes and spouts the most typical fixed means of mass transportation are: screw conveyors, belt conveyors, flight and continuous-flow conveyors, bucket elevators, and pneumatic conveyors.

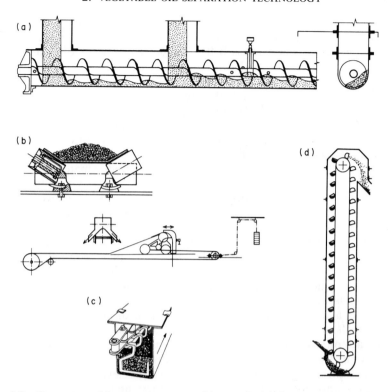

Fig. 2.8 Conveyors: (a) screw conveyor; (b) troughed belt conveyor on idlers; (c) skeleton flight; (d) bucket elevator. [(a, c, d) Reproduced, with permission, from Perry (1984).]

The screw conveyor is a full-threaded steel helix or helicoid pusher (flight) mounted on a shaft and suspended by bearings in a U-shaped sheet-steel trough [Fig. 2.8(a)]. The helix rotates at moderate, but adjustable, speeds, driven by a gear reducer coupled with an electromotor. The conveyor may be either right or left handed. Sometimes one section is right handed and another is left handed. In general, screw conveyors work horizontally, but they may be applied at a certain inclination. At 25° there is a near 50% decrease in capacity. The maximum practical length is about 40 m. Instead of full helices, helicoid ribbons or just paddles at distinct distances can be used as forwarding elements. The advantage of the latter two, to a certain extent, is their mixing and even crushing and grinding ability, but this is a disadvantage when conveying shape-sensitive material such as flakes. These types of apparatus can be charged or discharged at any place. If using hollow shafts and, for example, jacketed casings for circulating hot fluids or steam, closed screw conveyors can be converted to heating and drying devices, like the "Schnecke(n)"-type meal desolventizer/dryers of the early continuous basket-type extractors or continuous fish cookers/mashers. The conveying belts move along an end and front pulley, one of which serves as the driving pulley. The endless belt is supported by units of two or three rollers or idlers in a row at the intermittent places [Fig. 2.8(b)]. The idlers may be positioned at an angle of 20°, forming a trough and thereby increasing the capacity. If installed in an inclined position, depending on the type of seed conveyed, the maximum slope should be 30°, but preferably 18°. The material of which the belt is made should be oil resistant, at least for the transport of crushed oil seeds, oily grits and flakes. This, of course, means a higher initial cost, because the fibre-based "carcass" needs to be covered with neoprene, buna-N, vinyl or a Teflon type synthetic material instead of vulcanized natural rubber. The wear on the belt is relatively high because of the high friction. However, with regular maintenance this conveyor outlasts almost all other types. It can be used to connect sites at long distances and can be fed or discharged by special devices at a number of places, without any damage to the material being transported. Cleaners, like rotating brushes or a spring-mounted blade, are sometimes required with belt conveyors, as a clean discharge is vital to belt life. Being open to the air these produce dust and so they are not free of dust explosions. The squirrel-cage or star/delta-switch started motors are the preferred driving devices.

Flight conveyors consist of an endless chain, or twin chains, to which spaced pushers or flights are attached. The flights are either pulled from underneath or suspended clear of the bottom of the trough by sliding shoes [Fig. 2.8(c)]. The material travels with the flights, which are mostly of solid surface design. Closed-trough types may be used as elevators, or combined of vertical and horizontal sections. In this way they can replace, for example, a feeder, a bucket elevator and a horizontal conveyor, forming one single unit of moderate capacity.

In the continuous-flow chain conveyors the material moves because the internal friction of the mass is higher than the effort required to pull the flights through the mass. A typical example of this type of equipment is the Redler type with skeleton flights. If abrasion of material or deoiling of partially broken seeds by friction is a possibility, its use is preferred, as the material forwarded moves in its whole mass. The Redler-type conveyor is sturdily built, closed and can operate in different directions. All types of chain conveyors are space savers, self-regulating, non-choking and dust free.

Bucket elevators consist of a head and a foot assembly and connecting endless chain or twin chains or synthetic material (nylon, polyester, tergal) belt to which the metal buckets are attached [Fig. 2.8(d)]. The spacing between the buckets may be 25–40 cm, depending on the type and size. The buckets work in an open mode, although a casing is customary. They are the simplest and most dependable units for achieving a vertical lift of material. The "marine legs" mentioned already are heavy-duty open-bucket elevators. The discharge of the bucket is generally effected by the centrifugal force generated when moving over the head wheel at a relatively high speed. The heights covered are considerable and the power consumption is relatively low. Sticky material can be discharged badly and this fact limits their application.

Pneumatic conveyors are operated either by positive or negative air pressure and consist of a conveying duct or pipe, in which an air stream, generated by the action of a blower or a fan, is maintained. The granular material is rendered fluid by mixing it with the conveying air, the latter serving as the transport medium. In the positive-pressure system a blower forces air through the duct and the material is fed into the duct by an air lock (e.g. cells rotating on a horizontal axis in a housing with inlet and outlet connections). In the negative-pressure (vacuum) system a fan or exhaust sucks the air through a conveying duct connected to a separator and a dust collector, if required. The material leaves the system again through an air-lock gate. Combined systems are also well known. Suction-based transport is the preferred type if material is to be transported from different points to a central one, whereas positive pressure is advantageous in the opposite case. Suction-based transport moves 3–4 kg of seed per kg of air used, whereas fluidization moves 50 kg of seed per kg of air blown, so the former type of installation consumes more energy. Suction is preferentially applied for lifting materials which are also to be cooled by the conveying air. Pressure-operated transport is economical for both horizontal and vertical transportation of goods. Pneumatic conveyors can be used for almost any type of oil seed or side product, both in permanent or barge (mobile carrier) mounted installations. The power requirements (e.g. 1 kWh/ton seed) are four to six times higher than that of the mechanical conveyors described above, and pneumatic conveyors are mostly applied for unloading ships. Besides the energy problems, there is also mechanical damage to the material transported and, in connection with this, the wear of the installation in the ducts and especially in the bends of the casings. The wear can be reduced by installing bends of large radii or by using special wear-resistant material which can be replaced easily. By use of line switches the material can be routed to different destinations. Pneumatic conveyors provide for the greatest personal safety, having no moving parts in the vicinity of the operator and are also the cleanest means of transportation, being self-cleaning [Fig. 2.9(a,b)].

The disintegration of material, generally experienced in the increased amounts of splits, can also be caused by the impact of seeds, e.g. soya beans falling from a height of 30 m onto the surface of already discharged beans or hard surfaces such as the bulk carrier room or silo bottoms. Besides this there is an undesirable so-called "spout line natural separation" of splits and foreign material (broken beans are characterized as foreign material by U.S.A grading standards). The sampling might be correct and yet there might be local segregation of these components of up to 20% in certain areas of the vessel. This material is prone to quite rapid deterioration during shipping.

For gravity conveying of the seeds, chutes or spouts are used, being inclined at suitable (38–90°) angles, the minimum angle depending on the

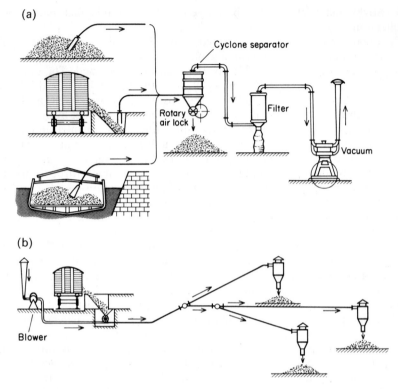

Fig. 2.9 Pneumatic conveyors: (a) vacuum type (suction); (b) pressure type (blower).

free-flowing characteristics of the goods. The minimum angle is around 45° for colza and soya beans and 60° for sunflower, safflower and cotton seeds and groundnuts. Gravity conveyors are simple for connecting short-level differences, but scatter dust.

Rotating tubes are lined on their inner wall with a helical ribbon and driven on the outside by a simple gear transport material; there is no grinding action and power consumption is low. They can be fed and discharged only at the ends.

Non-free-flowing products can be charged and discharged by vibrating, gyrating devices, by variable-pitch or fixed screws or by chain dischargers. They can be attached to hoppers, pans, tables, sloping striker plates, etc., in order to transfer seed and meal onto conveyors, elevators, scales, machines and storage means of any type.

E. Weighing

Scales (manual, automatic, batch and continuous) are the indispensable means for good stock keeping, in order to control the different steps of the oil-seed transport, storage and milling operations. The batch models are mainly railroad-track and/or truck-and-trailer scales, the tare weight influencing the accuracy. An automatic filling–emptying scale is actually a weight-batching system. The material arrives at the scale via a short conveyor belt from a storage bin or hopper and fills the weigh hopper, being supported by the scale lever system or placed on load cells, up to a preset weight. The feed is then cut off and the cell tips its contents into a feeder connected to another means of transport. (See also Section IV of this chapter.)

Continuous scales are used mainly as weight feeders and have not yet met universal official recognition, as they feed first and measure secondarily. A continuous scale may be a short-pivoted conveyor belt, which is supported together with its rollers on knife edges. The material arrives at the belt from a hopper. The load is transmitted over a lever to a scale beam which is counterbalanced by adjustable weights. The movement of the scale beam about the balance position is detected by, for example, photoelectric units, which feed back information to the feed-rate controller mechanism. The accuracy can be better than $\pm 1\%$. Continuous true-mass flow-rate meters are based on the principle of linear momentum transfer. Here the free-falling seed strikes a double-pivoted plate (sensor) in a continuous stream. The plate measures the force due to the impact (momentum) with an accuracy of 0.5–2%.

The capacity of these various devices is quite varied and should be selected according to the nature of the work, type of seed to be handled and the size of the elevator or the peak processing capacity of the seed storing/oil milling/ extracting installation. Silos may handle daily amounts increasing from several hundred up to 50 000 tons; modern oil mills are built to process up to 4000–5000 tons of seeds per day with at least 10–20 days storage capacity. In light of this, the discharging/unloading or reclaiming/loading of seeds to and from cargo vessels may require transport facilities with a capacity of 100–2000 tons h^{-1}. Expert advice on these matters should be obtained from the many companies specializing in transport installations for bulk-material handling industries, giving due consideration to the existing logistical situation for each individual project.

II. Preparatory Treatments

A general survey (principal lines) of the various preparatory and main steps of the different fat-separation processes and the different intermediate and final (main and side) products is given in Fig. 2.10.

A. Cleaning

Oil seed delivered at terminal elevators or oil mills should fulfil certain

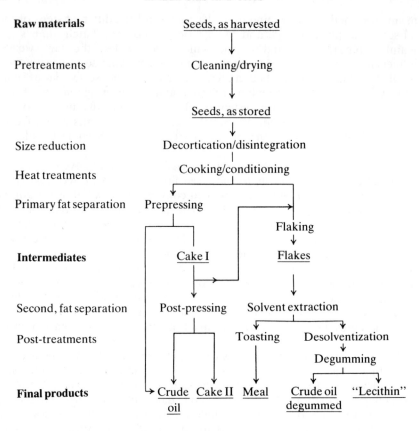

Fig. 2.10 General survey of fat-separation processes from oil seeds.

specifications according to the different trade rules of national or international agreements, under which the given commodity is handled. Harvested material might vary in its moisture level and broken-seed content, split beans and foreign material. The latter might be other oil and non-oil seeds, for example, weed seeds and waste such as hulls, linters, leaves, stems, sand, dirt, stones, strings, tramp metals and almost anything else. The level of foreign material determines not only the price and quality, but also the safe storage of the seed and, therefore, also the quality of the oil to be extracted from it. Therefore, an efficient, but of course economical, removal of these materials before transport and storage is highly recommended. The safeguarding of the processing equipment and reducing the process volume in connection with power needs makes the completion of this step prior to processing an absolute necessity.

The removal of rubbish from oil seeds is done by cleaning machines, working on mechanical, pneumatic and magnetic principles. The same machines can be used for the classification of different qualities within the same sort of seed, e.g. the separation of sunflower seeds of different sizes in more uniform fractions, thus resulting in a more effective decorication of the seeds.

1. *Screens*

The preferred mechanical devices for precleaning are rotating screens, and those for the actual cleaning the plane shaking, oscillating and vibrating screens. Screening is the separation of mixtures of different sizes into fractions; the portion staying behind being oversize, the portion passing through being undersize. Scalping is the mode of precleaning by screens where 95–99% of the seed material is considerably smaller than the screen mesh. If the screen (sieve) material is made of metal wires (e.g. 0.9-mm diameter), the distance between the adjacent wires is called mesh. The number of openings per linear inch is also called mesh. If the screen material is made of punched metal sheet (e.g. 1-mm thick) with holes, the diameter of these openings is called the aperture. Metal plates have better durability than do wire screens, but weigh more per unit surface area than wire ones.

The plane-shaking screens are brought into reciprocating longitudinal or vertical oscillation by an eccentric mechanism positioned below the plane, the eccentricity being about 10–20 mm at 400–600 rpm. The shaking boxes with metal screens are suspended on frames, e.g. with sheet-steel springs, the springs have a slope of 10–20°. The slope of the plane is about 2–5°. These screens can also be used for conveying material, in this case being called "shaking troughs". On the laterally-oscillating screens, the circular gyratory motion is generated by, for example, a motor-driven unbalanced fly wheel attached to the screen's frame, this being suspended by steel springs. Two or three screens, forming "seed-cleaning machines" (Fig. 2.12) can be positioned on top of one another, the uppermost being for large rubbish, e.g. larger than 10 mm, the middle or bottom for seeds (and seed parts), the

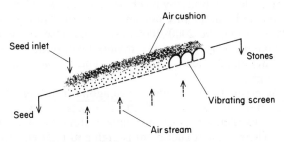

Fig. 2.11 Vibrating dry destoner. [Reproduced, courtesy of Buhler Brothers Ltd.]

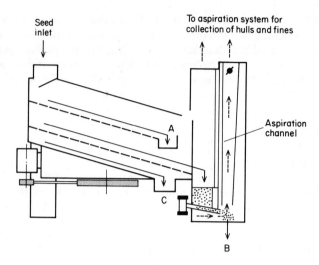

Fig. 2.12 Seed cleaner (plane shaking screen) with two sieve layers and an aspiration channel. A, Coarse offals; B, cleaned seeds; C, fine offals. [Reproduced, courtesy of Buhler Brothers Ltd.]

lowest separating out the fine trash, e.g. trash smaller than 2 mm like sand. Separation in shakers is due not only to the aperture, but also to the forces, both gravitational and sliding, acting on the particles. Uniform feeding and constant good coverage of the active surface are prerequisites for efficient operation, the spreading obtained by e.g. screws, slides, rolls, vibrating tables, etc. The shakers must be cleaned by hammers, brushes or other devices such as rubber balls below the sieve plates, to counteract clogging of the apertures.

Vibrating screens move the material in a direction normal or close to normal to the screen surface. They can be used, for example, for the separation of liquid oils from the oily drab, called "foots" which is formed during mechanical pressing. Vibrators are suspended on strong metal springs and are made to vibrate by mechanical eccentric shafts, unbalanced flywheels, a cam and tappet mechanism, or by electromagnets. The frequency of vibration may be 2000–3000 min^{-1}. Because of their sturdiness the screens can handle large volumes of material without clogging the metal wires, which in fact makes them very efficient. Their power requirements are comparatively high.

Cylindrical or prismatic hexagonal or octagonal rotating screens have a slope of 10%, and for the sake of effective separation by the screens mounted on the superficies, they cannot have a circumferential speed higher than 1.5 m s^{-1}. Their space requirement is high and their effective screening surface is only one-third to one-fifth of that of the plane ones, so they are not used very often, in spite of the durability of the screening surface and their self-cleaning character.

Centrifugal rotating screens are reels in which the horizontal screening cylinders rotate slowly, whereas the reel rotates in the inner space with 150–250 rpm, impelling the seeds to be separated mainly from coarse material against the superficies. Their space requirement is relatively high and the effective surface used is one-third of the plane ones. They are self cleaning.

Sifters are box-like machines containing a series of metal sieves on top of one another. They are kept in horizontal gyratory oscillation by eccentrics or unbalanced flywheels and some auxiliary vibration may also be caused by rubber balls bouncing against the lower surface of the screen. They may be used together with aspirators in the tail-end dehulling of extracted meals in order to separate meals and hulls.

2. *Aspirators*

Any type of screen separates material by size only. In order to separate impurities or fractions of different density, pneumatic means, like air nozzles and aspirators are applied. This sort of separation of low-density impurities or ballast material, if applied solely in a preliminary cleaning step is also called wind sifting. Aspiration channels are frequently used in combination with plane shaking screens (Fig. 2.12). If working by pressure produced by a fan, air is blown through the main stream of the seed flow leaving the screen. If working by suction, again produced by a fan or by a vacuum installation, the air is sucked through the main stream of the screened seed flow. In both cases the material must be distributed in a very uniform thin layer, falling as a curtain or film, if not done by the screen, by means of, for example, vibrating feeding tables. The air flow in the aspirator must be regulated very closely to obtain the optimum separation. This can be achieved by moving the back wall of the vertical aspiration channel or by the use of an air damper. The air box might contain a single fan, if a remote central one is not used, combined with slots, baffles, an air-expansion chamber and a discharge gate for removing the aspirated foreign material of lower density which is blown out. The heavy part might be discharged, if necessary, by finger valves or double-valve airlocks. Closed vibrator screens combined with aspirators (vacuum principle) can remove stones from seeds by keeping the uniformly distributed lighter seed material in a fluidized-bed form, the stones moving upward against the inclination of the screen, the seeds in the bed downward (Fig. 2.11).

All these appliances are, of course, connected with centrifugal dust separators, such as cyclones and/or special filters, being effective in reducing the air pollution if sized appropriately.

3. *Magnets*

Permanent or electromagnets enclosed in a rotating aluminium (not copper

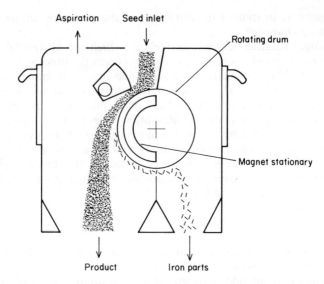

Fig. 2.13 Self-cleaning rotary magnet. [Reproduced, courtesy of Buhler Brothers Ltd.]

or brass) drum are used in feeders to distribute the material uniformly over the whole active width of the drum. The magnetic material collected is removed by a blade on the non-magnetized half and, in this way, it remains separated even in the case of power failure (Fig. 2.13). Permanent or electromagnets are used for actual seed cleaning, whereas permanent magnets are placed in front of the roll(er)s and other crushing devices in order to protect them against damage caused by tramp metal.

The connection of the above-mentioned equipment in a purposeful sequence makes it possible to clean any type of oily raw materials, the coarse-sized copra and palm kernels being treated only with magnets in order to protect the grinding equipment.

B. Size Reduction (Crushing, Grinding, Flaking)

An oil seed has linear size, surface and hardness, all this besides its internal structure. The fat is contained in the cells of the plant or specialized animal tissues, as described in Chapter 3. Histologically the plant cell is composed of the cell wall together with the cell membrane and the proto(cyto)plasm, a colloidal solution of proteins and other substances dispersed in water. The cell has its nucleus with the chromosomes, the mitochondria (which form the power house of the cell), and a complex network of strands (endoplasmic reticulum), and ribosomes, colour bodies, and, in the vacuoles or as an emulsion, the fat.

Whether emulsified or in the vacuoles, the fat is quite strongly bound within the cytoplasm. To liberate it from its surroundings it must be broken up by a more-or-less complete disruption of the cell wall. This is a mechanical operation called crushing or, more particularly, grinding. Crushing reduces the size, by which not only particles of different form, but also of different surfaces are produced. New forms and surfaces mean shorter transport ways and more exits for the oil, if pressed or extracted. Besides hardness, the size and shape requirements determine the machines which should be used in order to arrive at the goal. Water and heat, which may be partly present (partly added or formed) during the treatments and time (see Section D of this chapter) bring about not only the further rupture of the cell walls, but also counteract the forces which bind the fats inside their original surroundings.

Not only seeds or fatty tissues, but also "press cakes" and "extracted meals" are subjected to grinding operations. Press cakes are obtained by the mechanical pre-pressing of oil-seeds and are processed further for additional oil production by a second pressing or by a solvent extraction. Both cakes and extracted meals must be ground by, for example, impact breakers. As these operations are designed to grind previously crushed but later reintegrated material, it is called disintegration.

The main pieces of equipment used for size reduction in the oil-milling practice are: (i) roll(er) mills, consisting of 2–5 roll(er)s, which can be corrugated, spiked or smooth; and (ii) impact breakers, such as hammer mills, disc mills and disintegrators.

1. *Roll(er) Mills*

Crushing and grinding operations should cause the maximum rupturing of the oil-containing cells, with a minimum of fines and no separation of oil during these handlings. The most suited for primary-size reduction are roll(er) mills, because of the ease of control by adjusting the distance or "nip" between the adjacent pairs and the uniformity of the product granulation obtained.

Roller mills have diameters of 250–350 mm, widths of 800–1250 mm and are made of cast iron with a chill-hardened shell. This not only gives the necessary hardness of the cylinder surface, but also the required stability. Depending on the seed to be processed and the goal to be achieved they may have a finely/coarsely, fluted/corrugated surface or a smooth one.

(a) *Roll(er) mills with fluted/corrugated surfaces.* The flutings/corrugations are not parallel with the axis of the cylinder, but have an inclination of 10–15%. The flutes resemble an italic V, with one side shorter than the other. One side of the V is steep, the other shallow, thus forming different angles with the vertical. Height, angles and the number of flutes per centimetre differ according to the oil seed to be handled. In order to obtain a minimum of fines the corrugations on the two rolls are arranged in the so-called "sharp (steep edge) to sharp (steep edge)" configuration. This ensures that the flutes

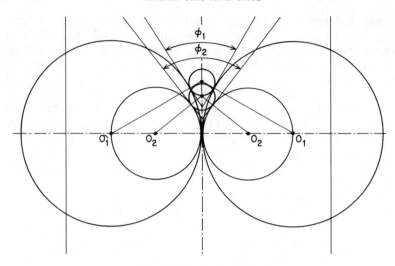

Fig. 2.14 Angle of nip.

work like scissors. Copra and decorticated palm kernels, roughly preground by special roller mills or by impact breakers (see later) can be further ground by roller mills in the configuration "sharp to dull (shallow edge)".

The "angle of nip" (Fig. 2.14) is formed by the tangents to the roll faces at the point of contact with the seed to be crushed. The angle varies according to the aim of the operation, but it seldom exceeds 30°. The bigger the diameter of the rolls, the smaller the "angle of nip" and so better their pull-in effect with increasing grinding surface. The speed difference between the rolls is 1:1.25 or 1:3. The rolls are locked at a pre-set gap by tension springs or by a hydraulic pressing system. Also the engagement/disengagement of the rolls and the feeding are automatically controlled by the product flow, which means that there is no metal-to-metal contact when running idle. The gap can vary between 3 and 30 mm.

In the oil milling it is customary to have pairs of rolls in stands or a mill body [Fig. 2.15(a)], the pairs being positioned in horizontal, vertical or transverse positions to one another. The rollers can be heated or cooled, if necessary, and are driven via multiple V belts by electromotors.

In the U.S.A. mills often use(d) five high rolls [Fig. 2.15(b)]. Here the material is crushed between the first and second roll, then transmitted by means of scraper blades between the second and third, subsequently between the third and fourth, and lastly between the fourth and fifth rolls. On these five high rolls the pressure on the seeds is exerted by the weight of rolls, except on the highest one where it is exerted by springs. Corrugated and smooth-surface rollers of different diameters may be incorporated in the same mill body.

The capacity of modern rolls may reach 600 ton day^{-1}, depending on the

2. VEGETABLE-OIL SEPARATION TECHNOLOGY

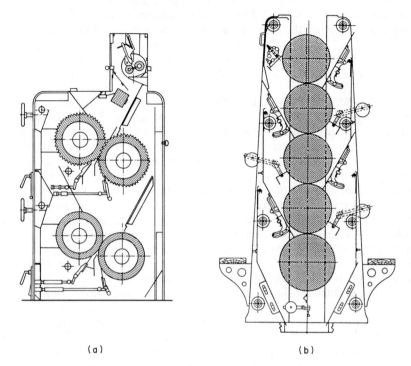

Fig. 2.15 Roll(er) mills: (a) pairs of rolls; (b) five high rolls. [Reproduced, courtesy of: (a) Miag A. G.; (b) French Oil Mill Machinery Company.]

peripheral speed, diameter, width and the size reduction needed. The only marginal disadvantage of the rolls is the wear and tear of the rollers which require frequent refluting to be done by special grinding and corrugating machines.

Special double rollers serve for cutting copra into smaller pieces. The rollers are provided with parallel circular rows of asymmetric spikes or parrot-beaked cutting elements. The parallel rows on one roller are shifted to those on the other, thus resulting in a sideways interposition of the cutting elements on the two rollers. There is a speed difference between the two rolls. The 15–25 mm pieces obtained can be ground further by conventional fluted roller mills. Similar constructions (the combination of a pair of spiked rollers above a pair of fluted rollers in the same stand) are often used for breaking cakes leaving modern screw presses.

Fluted rollers grind oil seeds into the form of grits, powders or cracked pieces, and ruptures the cells involved along the surfaces of the fracture but do not rupture their interior. The further opening of partly distributed cells is achieved by subsequent grinding using smooth surface rollers.

(b) *Roll(er) mills with smooth surfaces.* Roll(er) mills with smooth surfaces have a double use. Those of 250–350 mm diameter serve for the immediate further disruption of the cells already ground by a pair of fluted rollers. This is then followed by a "cooking" step (see Section D.1 of this chapter), which is mostly a prerequisite for the subsequent mechanical or hydraulic pressing of the seeds. In this case a pair of fluted rollers is placed in the same stand above a pair of smooth-surface ones.

Pairs of smooth-surface rollers of large diameter (6000–8000 mm) are used for "flaking" the seeds intended for solvent extraction. The material to be flaked must be previously ground by fluted rolls and/or by impact breakers, except for rape and mustard seeds. Flaking is a compressive operation and leads to the shape which has been found to be the most ideal for solvent extraction, namely platelets of 0.15–0.40 mm thickness, the thinner ones for extractors working with layers of 0.5–1 m deep, and the thicker ones for the so-called "deep bed" extractors, working with layers of 2 m or more. These porous scale-like structures are easily accessible to the solvents, shortening both the distance of diffusion of the solvent into the interior and that of the oil into the solvent. A "conditioning" step before flaking is compulsory in order to reach the necessary stability of the flakes which have to withstand the wears of transport during and after extraction.

The flaker rolls of large diameter are heavy and move with higher peripheral speeds in order to produce thin flakes and to reduce resonance in the building. They are available in lengths of 1–2 m. The rolls are closed by springs or, in modern machines, by hydraulic pressing systems. These flakers are built mostly with one pair of horizontal rolls placed in one stand.

Theoretically no difference in speed between the rolls is needed for flaking. However, although claims on better rupture of the cells by the additional shear forces produced by different speeds are not fully substantiated, nearly all flakers work with an approximately 3% differential speed. This can be achieved by one electromotor driving both rolls via multiple V-belts and pulleys or by two electromotors driving each roll separately via cogwheel transmission. Another reason for applying the differential speed is the better control of wear on the roll surfaces. An overflow system fixed sideways to the rolls instead of cheek plates or end dams may also provide for uniform wear of the entire roll, with rework of the non-rolled particles.

A special grinding device can be attached directly to the roller-mill body in order to service mainly the extreme ends of the smooth roller surface once or twice per month.

The ratio of oil directly hexane extracted from a sample of ground seed and the maximum amount of oil to be extracted after a fine grinding of the sample has been proposed as a measure of the percentage of open cells produced during the industrial grinding process. This ratio is called the "degree of treatment". The analytical procedure uses sand in combination with a mortar/pestle or a special laboratory mill to rupture all cells and the extraction is executed in Soxhlet extractors during 4 h.

The question of how many of the cells (65–85%) have to be opened by all the different crushing operations is not yet settled, because of the many other aspects such as diffusion, coagulation of proteins, inversion of the fat emulsion type and cell rupture, caused by added water and heat which all play a role in the fat-separation process.

2. Impact Breakers

(a) *Hammer mills.* The motor shaft of hammer mills is horizontal like that of the disc mills and disintegrators. The rotor carries 3–6 parallel rows of hammers which are also called beaters [Fig. 2.16(a)]. These can be T-shaped and either fixed or pivoted to the shaft or to discs fixed on the shaft. The rotor runs at high speeds in a housing containing cylindrical grinding screens or grates. The clearance between the rotor and the grinding plates determines the fineness of the product. The grinding action is effected mainly by impact and only to a certain extent by attrition.

(b) *Disc mills.* Disc mills work by attrition, one disc moving, the other one being fixed Fig. 2.16(b)]. If both move, they do so in opposite directions. Grinding takes place between the two plates. The plates have machined abrasive surfaces and are mounted on the discs. The shaft, the disc and the plate together are called a runner. The distance between the plates is adjustable. The whole is placed in a housing.

(c) *Multiple cage mills or disintegrators.* These contain at least two rows of concentric cages with bars of special alloy steel, revolving in opposite directions [Fig. 2.16(c)]. Because of the high speed attained, even relatively coarse material (e.g. hydraulic presscake) can be disintegrated to the fineness needed for flaking or end-pressing. The bolts should be easily replaceable because of the wear.

As mentioned already, these three impact breakers are primarily meant for grinding press cakes, meals, etc. and will be used for seed crushing, like that of copra and palm kernels, only in special cases. All are of simple construction. They can be erected in small spaces, because of their modest dimensions, except for the disintegrators. The big disadvantage is the large amount of fines produced and their sensitivity to tramp metal parts. Sparks may lead to dust explosions. Magnets should always be applied in the feeders of these devices as metal collectors in order to reduce such hazards.

C. Decortication/Dehulling

Morphologically seeds consist of a germ or embryo surrounded by the cotyledons and the endosperm, all covered by the seed coat or testa. This skin is sometimes surrounded by fused glumes, called husks. Fruits containing stones are surrounded by the pericarp, composed of (i) the endocarp (shell or stone), (ii) the mesocarp which in the case of palmfruit and olives is the fleshy part, and (iii) the exocarp which is the covering skin of the fleshy part.

(a)

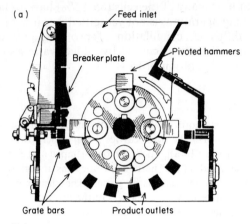

(b)

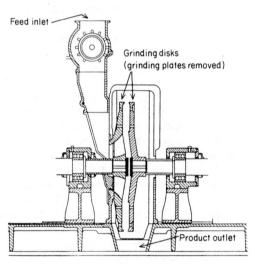

(c)

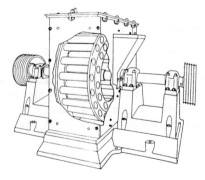

2. VEGETABLE-OIL SEPARATION TECHNOLOGY

In the case of soya beans the oil is contained in the peripheral cytoplasm of the cells in the cotyledon; the skin or testa is the hull to be removed from the two halves of the cotyledon surrounding the embryo. In the case of the non-endospermous (exalbuminous) sunflower and safflower seeds, the glumes must be removed. The hull of cotton seeds is the seed coat itself and this is called the spermoderm. This is covered outside by the linters, and it contains in its inside the germ with remnants of the endosperm. This endosperm contains the "glands" with the yellow pigment gossypol, many of the new varieties being glandless. In the case of palm and olive the main fruit-flesh oils to be extracted are in the cells of the mesocarp, whereas the stones or nuts also contain oil in the endosperm. The germ oils of maize, wheat, etc., reside in the embryos of the seeds, so in this case the endosperm and the testa must be removed. The question is, why is this removal necessary?

Oil seeds, with the exception of soya beans and cotton seeds, were primarily cultivated for their oil and only secondarily for their protein portion. Yet the need for high-quality protein sources for both animal and human food made it lucrative to increase the protein content of the extracted meals. This can be achieved by the removal of those seed parts which contain low levels of proteins and generally less than 1% oil. The removal of these cortices: shells, hulls, husks, etc., also increases the production rate and reduces the load and wear of the cooking, grinding/flaking, expelling/extracting, desolventizing/toasting/drying/cooling equipment. The crucial question is the effectivity of cortex removal and the elimination of seed parts containing oil, the "meats", from the cortex to an acceptable level. The eliminated cortices are often high-fibre structures containing cellulose and hemicellulose. Burning them in appropriately adapted furnaces of the boiler house can be useful. The husks of sunflower seeds (containing also 1–2% waxes) represent $14\,600–16\,700\,\text{kJ}\,\text{kg}^{-1}$ fuel energy, easily covering the steam requirement and a part of the electricity requirement of a fairly large oil mill. The increase in value of the meals must make up for the (weight) losses of the cortices, which in the case of sunflower seed have only fuel price. The often proposed chemical transformation of the cortices to furfural is at the moment not economically feasible.

Decortication (dehulling) operations consist of two steps: the cracking of the seeds into cortices (hulls) and kernels (cotyledons) and thereafter their separation.

1. *Dehulling of Soya Beans*

Dehulling aims at the removal of about 6–8% of the hulls, by which the protein content of the meal is increased from 40–45% to 45–49%. The

---

Fig. 2.16 Impact breakers: (a) hammer mill; (b) disc mill; (c) multiple cage mill or disintegrator. [Reproduced with permission, from Perry (1984); and courtesy of: (a) Jeffrey Mfg. Co.; (b) Sprout Waldron Co.; (c) Steadman Machinery Co.]

removal can be done before or after extraction of the oil from the seeds (front- or tail-end dehulling), a combination of the two processes also being feasible (Fetzer, 1983). In the case of front-end dehulling, optimal conditions for the release of hulls from the "meats" must first be produced by a type of conditioning, called tempering. Stored soya beans are quickly dried from 13% to 8.5–10% moisture level and kept in well-sized, temperature-controlled intermittent-tempering bins, where uniform moisture distribution is attained by storing them for 5 days or more, this being the practice in the U.S.A. and Canada. With slow drying at lower temperatures the tempering time in Europe is reduced to 1 day.

The beans are then crushed by two pairs of horizontal cracking rollers each of 250-mm diameter, placed one pair on top of the other. The beans are cracked into pieces of one-fourth to one-eighth the size of the beans. This is achieved by setting the distance of the rollers appropriately. The upper rollers may have a slit distance of 3 mm, the lower ones of 2–2.5 mm. From here the cracked pieces and the released hulls go to sifter boxes or shaking screens with aspirator channels above the upper and behind the second sieve/screen. The wired upper sieve has, for example, a 4-mm mesh and the perforated metal lower screen 1.5-mm diameter holes. The system separates fine meats, dehulled meats, uncut seeds and the aspirated hulls. If the cracking/screening/sifting/aspiration sequence is done with optimally set up machines, one grading operation is sufficient, resulting in meats containing less than 1% hulls and aspirated hulls containing less than 1% oil.

A primary wind sifting of the mixture of broken beans and husks can be executed by blowing air through the falling product stream in a cross-current way. For example, "zig-zag" type cascades inside a separator are formed by sloping baffle plates, which separate meats and conduct the fluidized husks of different size to the husk collector. After two runs with intermittent sieving on a double shaker with a 2 mm upper and 1 mm lower screen, about 85–90% meats besides 10–15% fines leave the installation separator. The husks separated with 9.6% protein content must be screened on a shaker with 2-mm sieve in order to lower the oil content of the 10% oversize husks to below 1.6% (Multiplex MZF of Alpine).

The other possibility is to remove not only the 6–8% hulls, but also the 3–5% cracked beans by aspiration applied above the upper screen and to re-separate this mixture by additional hull sifters into fine meats, picked up by the hulls, and the hulls.

As has already been mentioned, the meats are conditioned before entering the flakers in order to obtain the necessary plasticity for the flaking process and the necessary stability of the flakes for the extraction process. The hulls containing some 10% protein are either burnt in the boiler house or ground by a hammer mill and toasted in a conventional, e.g. four stacks high, live-steam-operated toaster (see Section III. B of this chapter), before being added to a (sub-standard) low protein fraction.

The tail-end dehulling of extracted soya bean flakes starts with grinding them on hammer mills which have appropriate sieve plates (2–4.3 mm openings) or liners. The oversize fraction is returned, the rest being sized by gravity tables or by dual aspiration sifters into three fractions. The fines and the hulls selected out of the high-protein (HP) fraction are added to the low-protein (LP) fraction. This is a method which is used in oil mills as an additional technology for producing high-quality animal feeds. The method produces about 25% high-protein (HP > 49%) and 75% low-protein (LP > 44%) fractions and no hulls. The flexibility of tail-end dehulling, regarding the prescribed minimum levels of protein in the fractions, is more limited than that of the front-end dehulling process. The proportion of the HP fraction can be increased by rubbish removal during preliminary seed cleaning and more severe sifting/aspiration; of course this is at the expense of the amount and protein level of the LP fraction.

A variation of this process is a combination of the two processes, described above, starting with a partial front-end dehulling. In this case the stored, cleaned, but undried and untempered beans are cracked and only the already loosened part of the hulls is removed by the process described above. The time, tempering space and heat-energy savings are evident. The rest of the hulls are removed after extraction of the meal by the additional tail-end dehulling system. As the tail-end dehulling starts with an enriched meal, so the process produces up to 40–70% HP meal, without reducing the protein level of the LP fraction below 44%. Of course these figures depend also on the initial quality of the beans.

The so-called "hot or impact (front-end) dehulling processes" use "fluidized-bed contactors" which are claimed to have a very uniform residence-time distribution for each particle and, therefore, give a more uniform end-product. The soya beans are subjected to a hot-air (150°C) treatment, drying and propelling them in a fluidized bed. The surface of the beans is heated to 75–92°C, which is enough to loosen the hulls. The hot beans are then cracked by a combination of impact splitters/dehullers or rollers and an impact dehuller, the impact of the bars or hammers removing from the two halves of the beans (the cotyledons) the hulls crimped to them. The specially adjusted impact dehuller produces less fines than does cracking on fluted rollers. The hulls are removed by aspirators. The dehulled meats can be directly ground and flaked for solvent extraction (Buhler). In another variant the hot impact-dehulled meats can be "reconditioned" by a second fluidized-bed contactor before the customary grinding and flaking operation. The interchange and reuse of air between the different units and the direct handling of wet soya beans containing, for example, 15% moisture can save 25–50% energy, as the dried beans (1–3% moisture having been removed) are not cooled but cracked and dehulled in hot condition (Escher Wyss).

2. *The Dehulling of Sunflower (and Safflower) Seeds*

This operation was originally an absolute necessity when the husk/meat proportion of the early sunflower seed varieties was 45 : 55. The new, mostly (but not always) high-in-oil and invariably more disease-resistant varieties have more favourable proportions (23–35 : 77–65). Therefore it is often a question of economics when deciding upon the type of dehulling operation. In these considerations not only the protein (and oil) content of the extracted meal, but also the residual oil and intrinsic wax content of the husks (2–3% oil, of which about 60% are waxes in the new varieties) should be included. Other aspects to be considered are the volume throughput (1% of husks removed is equivalent to a 2.5% increase in plant capacity), the wear of the different installations by silicaceous components of the husks and the wax content of the husks which are not removed. Co-extracted husks increase the wax content of the oil produced (0.05–0.3%), and the wax is difficult to remove by the winterizing operation (see Chapter 5, Section II). A special problem may arise with some small-sized low-oil varieties, in which not only is the husk content higher, but the knitting of the much softer husks to the kernels also causes difficulties with decortication.

The various machines, by which the cracking of (not necessarily sunflower seed) husks can be attained are described below.

(a) *Knife dehuller*. This machine consists of a horizontal rotating cylinder fitted with protruding longitudinally placed knives or bars. Surrounding the rotating cylinder for part of its circumference is a concave section fitted with opposing knives. The seeds are cut between the cutting edges as the cylinder rotates.

(b) *Disc dehuller*. This machine houses two vertically mounted discs with radial grooves, one rotating, one stationary. The seeds are fed to the centre of the discs, are broken by the grinding action, and then discharged centrifugally to the periphery of the disc.

(c) *Impact dehuller*. Impact dehullers generally consist of a rotor made up of paddles or bars or tubes positioned at the circumference of the reel, which fling the seeds against a corrugated impact plate, covering about half of the circumference. The gap between the impact plate and the rotor can be adjusted, as can the speed of the rotor. The machines are uniformly fed at the top by different types of feeders (e.g. fluted roll), whereas the mixture of cut and uncut seeds plus husks leaves at the bottom.

The maintenance of knife-type dehullers is more costly than that of disc dehullers, whereas the impact dehullers are cheaper and their throughput might be 2–3 times that of the former at, for example, a rated 3 tons h^{-1} capacity, thus up to 8 tons h^{-1}.

There are three main types of impact dehullers on the market: the "Russian" type, the Australian "squirrel-cage treadmill" type, and the centrifugal types.

The "Russian" type of impact dehuller is originally of U.S.S.R. design (Fig. 2.17). It has 15 steel beater bars fixed at an angle of 45° to the radius of the drum and was developed from a cotton-seed impact dehuller. It is now also produced by machine manufacturers outside its land of origin.

The "squirrel-cage treadmill" type utilizes the Australian (CSIRO) developed rotor which is composed of three parallel vertical discs on a common horizontal axis. The discs are separated at their circumferences by 16–16 metal tubes, this "squirrel-cage treadmill" serving as the impact reel. Horizontal ripples form at the middle of the corresponding impact plate an angle of 90°. At the upper and lower ends of the

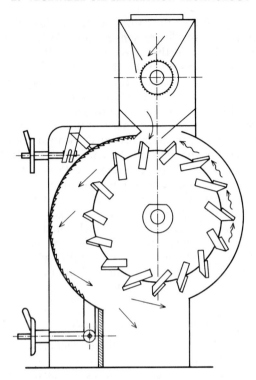

Fig. 2.17 "Russian" dehuller/decorticator. [Reproduced, courtesy of Buhler Brothers Ltd.]

plate the ripples are inclined at an angle of 120° in order not to become congested by the cracked material.

The centrifugal types of dehuller may have, for example, a central feeding cylinder which distributes the seeds into the many channels of a rotating disc, this flinging the seeds against a truncated supporting disc, on which the seeds are cracked by the impact (Buhler).

(d) *The actual process.* Before starting the actual decortication of sunflower seeds, the further reduction of the humidity of stored material from 9(9.5)% to 8% is a very important preliminary conditioning step. After partial, but at least better than 80–85%, first cracking of the seeds the separation of the different products should be done on shaker screens with aspiration channels. The husks and uncut seeds may remain on the upper screen, the husks being sucked away by a nozzle-type aspirator above and behind this screen. The uncut seeds may go on to a second impact dehuller having a reduced gap between the rotor and the impact plate. The undersize material is separated on the lower screen into two fractions, the oversize meats being aspirated for further husk removal. If less than 10% husk remains in the cut seeds, then, besides double dehulling, the second screen fractions already mentioned are brought to a second screening/(aspirating) unit, thereby further reducing the husk content of the not completely husk-free meats and the husk content of the undersize fines.

The removal of, for example, 18–22 parts of the husks by double dehulling out of 28–32 parts present in undecorticated, cleaned seed (this means 8–13% husks are left in the partially decorticated non-defatted kernels) increases the protein content of the extracted meal to 40–42% (13–14% fibre content). A less intensive, single dehulling removing 10 parts of husks may produce extracted meals with a 32% protein content. The residual oil content of the extracted meals may vary in the range 0.5–1.5%. The residual oil content in the husks can reach 1.5–2% which could be reduced to a lower level by post-screening the fines collected by the cyclone of the aspirators.

Safflower seeds may contain 30–65% of hulls, which again raises the question of whether it is economical to decorticate the low hull varieties by similar means.

3. *The Delinting and Dehulling of Cotton Seeds*

The seeds coming from the gins must be freed first from the residual 10–15% short fibres, called linters, adhering to the seed coat or skin of the seeds. This not only removes the linters, but also reduces the oil losses and the fibre ballast in the subsequent operations. Delinting is carried out with circular saws in several steps, and the last shortest remnants are removed by abrasive discs. Removal by sulphuric acid is inadmissible for seeds which are to be processed for oil and protein. The burning of lints is also an existing delinting method.

Before dehulling cotton seeds which are too dry, a "conditioning" operation must be executed by first cooling and then humidifying the seeds by live steam in order to increase the moisture levels above 8%.

The delinted seeds (2–4% residual lints, for cost reasons!) are first subjected to a general cleaning procedure by means of a shaker screen. The cleaned seeds then pass to the dehuller, generally the bar/sharp slotted knife type, or any other type of impact dehuller. A cut of about 85% on the first pass is acceptable, the uncut material being returned to the dehuller after having passed the shaker screens/aspirator type separators. In order to keep the oil content of the hulls low, rescreening of the hulls by, for example, centrifugal rotating screens is a question of good housekeeping and economics. As the kernel/hull proportion is about 55–60 : 40–45, so the production of a dehulled meat containing 10–15% hull at most and about 30–35% oil (now at less than 8%, preferably at a maximum of 6%, moisture content) is a fair goal in obtaining an extracted meal with a protein level of around 40%.

4. *Processing of Glandless Cotton Seeds*

The dehulled meat of glandless cotton seed varieties can be rolled after the cooker/

conditioner to a flake thickness of 0.35 mm and extracted directly. The pre-pressing before extraction together with cooking in moist heat were prerequisites with the glanded varieties, in order to "bind" the gossypol into the meal. When glanded meats were extracted directly using solvents, either a miscella refining or an immediate alkali refining was needed after desolventizing the miscella, in order to remove the gossypol transferred to the oil. The savings when processing glandless cotton seeds are quite impressive, as no pre-pressing is needed, the costs of refining/bleaching and the oil losses are lower, the keeping properties of the oil are claimed to be better and there is no discoloration, etc. The meal has an equivalent or better feeding quality than the glanded one and does not contain the toxic gossypol.

5. *Groundnuts and Rape Seed Dehulling*

Groundnuts may be decorticated using disc-type dehullers or special machines. The latter utilize a rotating corrugated cast-iron cylinder, which presses the seeds in husk against a surrounding metal plate with oblong holes, varying in size between 7.7 and 9.0 mm width and 50 mm length. The rotor is eccentrically positioned in the stator, the space being larger at the entrance than at the exit. The seeds are practically decorticated by the friction exerted by the husks on themselves. The 25–35% husks are separated after cracking on double shaker screens with (fine) husk aspiration above the first and after the second screen.

After the necessary conditioning, rape seeds containing around 18% hulls can be dehulled using the centrifugal-type impact dehullers to obtain about 10% hulls containing up to 18% oil in the form of fine meats. At least a part of this meat should be gained by pressing, or better, by extraction; this latter problem makes the whole operation questionable.

D. Conditioning

Conditioning is the process by which materials are brought by means of temperature and humidity control into a uniform and, for a given process, favourable condition. Some of the process steps for which the conditioning of seeds is decisive in the success of the operation in question are: front-end dehulling of soya beans; decortication of sunflower/safflower/cotton/rape seeds and ground nuts; the heat treatment of seeds to be extracted before "flaking", and the "cooking" of crushed seeds to be expelled. The "toasting" of high-protein meals after solvent-aided oil extraction and the "toasting" of the protein-containing hulls removed from soya beans can also be considered as conditioning operations; these are discussed later in Section III.B of this chapter.

The uniformity of the heat treatment, i.e. the achievement of a constant and narrow residence-time distribution of the treated seeds in any of the installations, is a design goal. However, this is not always well realized in the different constructions used.

In general, oil seeds and other fatty materials with lower ($<20–25\%$) oil contents are directly solvent extracted (soya beans), those with higher oil content, being either mechanically pre-pressed and subsequently solvent extracted or totally deoiled by mechanical/hydraulic pressing (see Section III of this chapter).

1. *Conditioning Before Solvent-aided Oil Extraction*

As already described, oil seeds of lower oil content intended for direct solvent extraction are reduced in size before the conditioning operation. To this end soya beans and cotton seed "meats" are cracked on fluted rolls. The conditioning measure for crushed soya beans is their heating up and holding at temperatures of 55–75°C (preferentially 65°C) at about an 11% discharge moisture level for 20(–30) min. For wet cottonseeds their conditioning means a heating at 65°C during 10–12 min (somewhat longer and at 85–95°C for dry ones). These measures are necessary for the subsequent "flaking" operation, giving the flakes the necessary plasticity during and stability after this operation. This conditioning is generally done in one of the simple horizontal rotary driers described below.

2. *Conditioning Before Mechanical Oil Extraction ("Cooking")*

For mechanical (pre-)pressing/expelling the different seeds or other fatty materials with a high starting oil content need to be crushed before the conditioning treatment. This is carried out by cracking the seeds on rolls and subsequently grinding them on smooth surface rollers. Smaller oil seeds, such as whole rape and mustard seeds are first heated to 40°C and are then crushed on smooth surface rollers only.

The standard conditioning procedure hereafter is the "cooking" operation in specialized "cookers" (see below). This operation has a duration of about 30–70 min at a 100–110(–125)°C discharge temperature. Varying final moisture contents are obtained, for example: a maximum 7% for decorticated, 4–6% for prepressed sunflower seed, 3.5–4.5% for groundnuts, a maximum of 3–4% for dehulled cotton seeds, a maximum 3–4% for rape seeds, 3–4% for prepressed, and 1.5% for total ("end"-)pressed copra seeds, 4% for palm kernels, and 2–3% for maize. Although infrequent, soya beans, if only pressed, are first dried in the first tray of a 4/5 high cooker at 87°C to 8% starting moisture for 20 min, ending in the fifth tray after 55 min at 115°C with 2.2% residual moisture content.

This "cooking" process serves many different purposes, which may include the following:

(i) completing the breaking down of the already crushed cells by expanding them;
(ii) decreasing the surface tension of oil droplets;
(iii) reducing the viscosity of the oil to be expelled;
(iv) partly coagulating and modifying/denaturing the proteins;
(v) adjusting the humidity in order to prevent it acting as a lubricant during pressing (see the final moisture contents given above);

(vi)   increasing the plasticity of crushed seeds;
(vii) partially sterilizing the seed (yeasts; some bacteria such as *Salmonella* need more intensive heat/moisture treatment);
(viii) destroying enzymes (lipase, lipoxygenase, urease, myrosinase, etc.);
(ix)   partly detoxifying or modifying undesirable constituents, called "antinutrients" (gossypol, haemagglutinin, saponins, thioglucosinolates, and also the trypsin inhibiting factor, a protein);
(x)   fixing some phospholipids by drying, or releasing them by adding water.

The already mentioned high final cooking temperatures and the initially high moisture contents are generally beneficial to the inactivation of the enzymes. In rape and mustard seed meals, for example, the enzyme myrosinase liberates from the parent compound isothioglucosinolates first nitriles and then isothiocyanates, the latter cyclizing to the goitrogenic isothio-oxazolidones. The maximum activity of this enzyme, and also that of lipase, the fat-splitting enzyme, are in the temperature range 37–70°C. This range must be surpassed as quickly as possible in order to minimize the damage. In the case of myrosinase the crushed rape or mustard seeds with 8–10% moisture content should be first rapidly heated in the first tray of a 4/5 high cooker to at least 70°C and then in the second tray to above 80°C. The moisture content should be kept between 6% and 10%: at levels below 6% the enzyme is not inactivated, and at levels above 10% hydrolysis of the parent glucoside is promoted (too quickly). The enzyme lipoxygenase, a prooxidative catalyst of oils, can be inactivated only at 100°C or above in a moist atmosphere. All these processes require substantial initial heat-transfer surfaces, e.g. extra steam collars in the very first stage, and different minimum moisture levels at the higher temperatures. However, longer residence times in, for example, 6–8 high cookers may give rise to dark coloured or oxidized meals and oils. Long residence times may also decrease the "availability" of certain amino acids for the animal feed (see Section IV of this chapter).

3. Conditioning/Cooking Equipment

The various types of equipment used for "conditioning/cooking" are described below.

(a) *Vertical cookers.* Vertical cookers consist of 2–10 (mostly 4–6) stacked, superimposed cylindrical steel trays (kettles), sometimes only bottom-jacketed, but sometimes also side-jacketed, which are heated with steam. At the very bottom of each tray there are centrally driven sweeper-type stirrers. Rapid heating in the first tray can be achieved by mounting an extra heating coil immersed in the seeds (Fig. 2.18). The filling level of the material in each tray can be regulated automatically (pneumatically) according to the speed of discharge from the lowest one. For the sake of the conditioning effect there are on the top of the tray (but also on the lower trays) injection nozzles for adding slightly superheated steam, or warm water, to the ground seeds. Each stage is connected to exhaust holes for venting off water vapours into a common chimney. The exhaust can be a fan-forced one with an attached

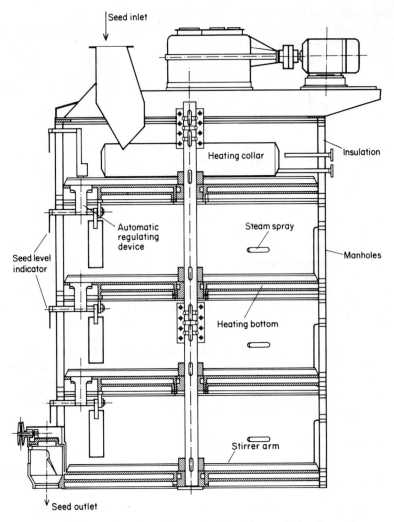

Fig. 2.18 Four-high stack cooker. [Reproduced, with permission from Homann *et al.* (1978).]

solids trap. The instrumentation, consisting of thermometers, manometers for each tray and a rotameter for water may be completely supervised by dedicated microcomputers. In the lower trays the ground seeds are allowed to dry to a favourably low final moisture. Lower final moisture contents lead to lower residual oil levels when the oil is extracted in a purely mechanical way in a two-step (pre- and post-), or one-step (final, "end"-) pressing of seeds. A 0.2–0.25 m^2 heating surface per ton of seed per day is considered to be the minimum requirement. The filling height varies from 150 to 600 mm.

2. VEGETABLE-OIL SEPARATION TECHNOLOGY

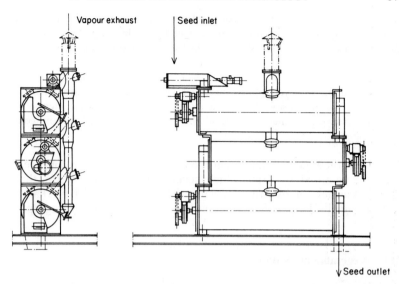

Fig. 2.19 Three-drum cooker/conditioner [Reproduced, with permission, from Homann *et al.* (1978)]

(b) *Horizontal cookers/conditioners.* These consist of a bank of 2–6 steam jacketed steel drums with special (ribbon, paddle or other type) lateral mixers/material conveyors. The cooking and drying steps are done separately, the residence time and distribution being optimized by more radial than longitudinal mixing (Fig. 2.19). Live steam and water addition provisions in the first and last drums enable the seed to be brought to the prescribed final moisture levels.

Simple indirectly heated horizontal rotary driers with pipebundles or plain closed screw conveyors with steam jackets can also be used as conditioners, if provision is made for adding sparge steam and/or water to the contents. These driers can also be used for pre-heating when dehulling soya beans or separately drying extracted meals.

(c) *Fluidized-bed contactors.* These contactors with their incorporated internal heat exchangers can be used successfully for conditioning soya beans before the customary flaking operation. Because of the excellent heat-transfer coefficient in a fluidized-bed system, the heat-transfer area required is only about 20% of that of a rotary pipebundle conditioner/drier (Florin and Bartesch, 1983).

A horizontal conditioner/drier is based on the principle of a hooded shaker trough. The whole installation is suspended on springs and is driven by an excentre device. The bottom of the trough consists of a specially constructed grid through which live steam in the first section and cooling air in the second one can be introduced in cross flow. The seeds move through the system in semi-fluidized condition, expanded on the whole surface by the gas velocities applied and by the shaking movement. Low energy consumption, little damage to the material treated, continuous operation with little maintenance and little equipment wear are among the main advantages of this type of contactor (Stork-Amsterdam).

(d) *Microwave heating.* Microwave heating in vacuum (MIVAC), already mentioned, although not yet applied on a factory scale, could be used mainly for hot front-end dehulling of soya beans. The reduction in the moisture content is only 1%, but the waste heat generated around the magnetron is used for the consequent dehulling step.

Nearly all the types of equipment mentioned in the preparatory treatments are manufactured by firms which specialize in machinery for the grain-milling business or by firms specializing in machinery for mechanical and solvent-aided oil extraction, as described in the following paragraphs.

III. Vegetable-oil Milling

A. Mechanical Oil Extraction (Expression)

The aim of all the processes discussed above is to obtain well prepared, cleaned, classified, ground, pre- and post-conditioned seed material from which the oil can be extracted. We have seen the importance of conditioning the crushed seeds before expressing and have learnt something about the changes which occur in a cooker or any other feasible conditioner. The success of mechanical oil extraction (expression) depends on both the pretreatment and the construction and sound use of the type of press applied.

Expression is a solid–liquid phase-separation method which is applied to materials that are readily pumpable. It can be executed by batch, mainly hydraulically, and by continuous, mainly mechanically, working presses.

1. *Hydraulic Pressing*

It is mainly for the sake of completeness that we discuss here the different types of hydraulic presses used in the mass production of different sorts of fats and oils by small, medium and even large-sized oil mills.

The batch-wise operation, although requiring much manual labour, was an important improvement for oil mills fitted with hydraulic presses. Besides its low-energy demands, batch-wise operation made it possible to switch relatively quickly from pressing one sort of oil seed to another.

Hydraulic presses work on the Pascal principle according to which the pressures (P, p) exerted on two different surfaces (A, a) which are hydraulically connected are directly proportional to one another

$$P : p = A : a$$

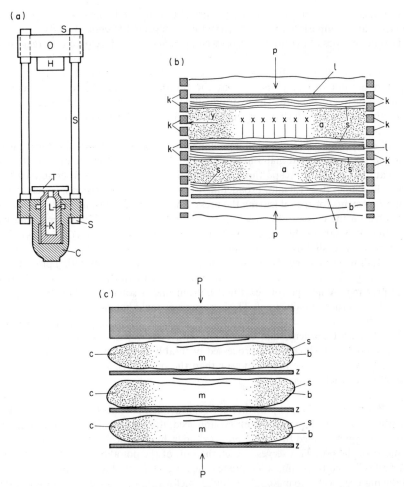

Fig. 2.20 Schematic views of: (a) hydraulic press; (b) closed press; (c) open press. (a) C, cylinder; K, ram; L, seal; T, bottom plate; O, upper end block; H, headpiece; S, steel columns. (b,c) k, Cage holes; s, press cloth; l, inserted steel plates; z, fixed steel plates; a or m, ground seed; P, pressure; c and b, direction of extension.

whereas the linear movements (S, s) of these hydraulically connected surfaces are in an inverse relation to each other

$$A : a = s : S$$

due to the equivalence of the displaced volume of fluid on both sides, and the incompressibility of fluids.

Preconditioning in cookers was at least as important as in the case of modern screw presses, but because of the batch character these generally had only one tray.

The cookers were connected directly to special portioning–filling devices in order to fill cages and compartments, which served for enclosing the cooked seeds.

Two main types of hydraulic presses were in general use: the closed type (cage and pot) and the open type (plate or Anglo-American type and box) [Fig. 2.20(a–c)]. All these presses comprised a frame of four heavy, vertical steel columns, attached at the top and bottom to heavy end blocks, the bottom one housing the hydraulically displaced piston of the ram. The hydraulic fluid was water or a suitable vegetable oil, preferably of the non-drying sort such as a well-deslimed groundnut oil.

(a) *Cage presses*. The closed presses had special, mainly cylindrical vertical cages, in which the ground seed material was charged. Compression of the material by the ram caused the oil to escape through the cage wall. The cage wall contained fine funnel-like holes (0.5–1.0 mm, broadening outward) on the superficies, which served as filter elements for the oil leaving the press. Some tens of thousands of holes had to be drilled on the cages which were up to 1.8 m height and of 270–470 mm internal diameter. In German this type of cage is called the "Seiher" (filtering) press. The cages broadened slightly towards the top, in order to alleviate the removal of the press cakes from the cage. To obtain manageable press cakes, separating press cloths and drainage steel plates were inserted in order to create thicknesses of about 25 mm after pressing. A non-perforated thick bottom and a somewhat thinner top plate formed the vertical confinement. The cage presses could handle materials of high oil and low fibre content, such as copra and palm kernels. Seeds, in which the oil cannot endure a heat treatment, like castor oil, whose ricinoleic acid dehydrates to an isomeric linoleic acid, could be processed profitably in cage presses. In general, the productivity of the cage press is lower than the open press if equal conditions are presumed for each other.

(b) *Pot presses*. The pot press has 4–5 separate compartments, each equivalent to one press cake, which are filled with ground seed, the bottom of the pot is a draining plate. The top is formed by a solid press cylinder, having the diameter of the internal diameter of the pot. This serves also as the ram of the pot above. As the cylindrical pot wall is heated by steam, so no preheating is needed and this press can be used for producing heat-sensitive materials of higher melting points such as cocoa butter. Pot presses are mainly used by factories producing cocoa powder and chocolate. Modern horizontal variants with larger capacity (more "pots" or "chambers") are also now in use, next to continuous screw presses, which also handle the cocoa beans easily.

(c) *Plate presses*. Plate presses had about 20–24 vertically suspended horizontal steel plates in the frame, the original free space between two plates being about 70–150 mm. The cooked seeds were packed into clothes made of wool, camel or horse hair. The filled clothes were placed between the corrugated steel plates and, as these clothes were more permeable to the oil than were the limited amounts of holes on the cage press, they delivered a finished product with a residual oil content of 6–8% from a final pressing in a shorter time, by about 0.5 h than the cage presses. Furthermore, the oil was less turbid, because of the better filtering action. Seed materials which tended to extrude through the cage holes were confined better by cloth presses. The residual oil content at the edges was somewhat higher, so the rim had to be trimmed, ground and repressed.

(d) *Box presses.* Box presses enclose the packed ground seeds on the sides and have filter-like corrugated drainage mats below the package. The seeds are wrapped in cloth enclosing them only at the top, the bottom and in one direction at the ends. The wrapping and filling is thus easier than that of the plate presses. Sixteen boxes take the place of the 24 in the plate presses, so the oil production at the same sizes is somewhat less.

The force developed on a press plate attached to a piston of 450-mm diameter at 350 bar is 5.5 MN (0.6×10^6 kg). Because of the considerable compressibility of the seeds in the first period of pressing, partly because of the large amounts of air occluded, fast development of force is required, which is attained by pumps of high capacity in three different pressure ranges (low, medium and high) and power (weight) accumulators at least for the low and medium ranges.

The cage presses need special filling and cake-discharging presses, and all need special (cooked seed) cake-forming machines. Their operation, in spite of many mechanized steps developed in the course of their era, is based on cheap hard labour. The capacity is around 50 ton day^{-1} and so, considering the investment costs of a newly built one, their use must be considered (except in special cases) obsolete.

2. *Continuous Screw Presses (Expellers)*

Before describing screw presses, we may recall once more the importance of the grinding/flaking and cooking/conditioning operations on the success of the pressing operation. A screw press, in most cases, consists of a screw of increasing root diameter and generally decreasing pitch, revolving in a cylindrical drainage cage (Fig. 2.21). It thus has a main wormshaft, which carries a "worm assembly". This assembly is composed of "worm sections" and of conical "distancing pieces", which space these sections apart and have no threads. Depending on the materials pressed and the construction

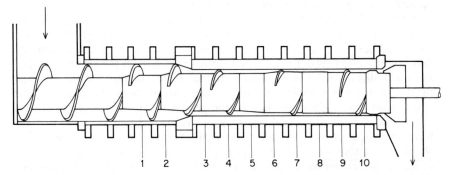

Fig. 2.21 Schematic view of a screw (pre-)press. The points at which radial pressure measurements are made are indicated [Reproduced, with permission, from Zajic *et al.* (1986).]

principles applied, the number of worms (varying between 3 and 10) and distancing pieces, the root diameter and the pitch of the worm flights are different. The wormshaft revolves in the cylindrical drainage "cage" or "barrel", which is composed of axially placed "lining bars" contained in a frame. The frame is horizontally (or vertically) split in two halves along the axial direction. The halves are held together by "clamping frames". The lining bars can be kept separate in the frame by shims, called "spacers".

Non fully axial, i.e. staggered inserted lining bars, can form a surface composed of many knives which may act as a seed breaker, and also prohibit meal rotation. Of course the cage also works as a drainer and filter, preventing the expulsion of fine seed particles ("foots") together with the pressed oil. Screw presses may also have two "knife bars", which have nibs projecting into the connecting section formed by the "distancing pieces" in order to counteract meal rotation. Instead of spacers, the lining bars may have specially chiselled grooves on one or both sides, which form with the next lining bar a screen of defined spacing. The cage is always divided into sections, where the lining bar spacings decrease in generally, three to five steps from 0.25–1 mm at the feed to 0.10–0.25 mm in the middle or close to the rear end. The bar spacing at the very end may widen again to, for example, 0.25–0.35 mm, according to the type of seed pressed. Copra may require wider bar spacings in the sections than, for example, rapeseed.

The shaft itself, if it is bored for it, can be cooled by water for "colder" pressing and heated by steam for starting up. In some constructions the cage might be cooled by oil already pressed and pre-cooled via heat exchangers.

The spacing between the worms must be adjusted according to the quality of the seeds to be processed. Soft seeds, such as ground nuts and rape seed, need wide-spaced worms at the discharge end and distancing pieces, whereas hard "seeds", like copra, decorticated palm kernels and partly decorticated sunflower seeds necessitate the use of more and close-spaced worms. Worms are generally made so that the flights do not wrap round by more than 320–345°. By this means the compressed meal may slide in any direction according to the velocity generated by the pitch of the worm.

A very important aspect of design is the cage-screw (worm) geometry, which co-determines the effectiveness of the compression resulting in a maximum de-oiling of the crushed seeds. In one cage construction the decrease in volume is achieved by a step-wise increase in the worm root (but not the flight) diameter in a constant-diameter cage, in another it is achieved by the stepwise telescopic but inverse, decrease in both the cage and the worm root and flight diameters in the sections. Any of these geometries should result in a substantial de-oiling of the meal. It is important that the layer thickness of the pressed meal should remain relatively thin and it should be newly broken and compressed in each section.

All screw presses have "feed gears" for enforcing the crushed seeds into the press, e.g. side conveyors with twin intermeshing feedworms (Rosedowns), high-speed vertical shafts (Anderson), or high-speed "quill" feedworms (FOMMCO). This feed section is a troublesome part in the construction, as the press must be able to get a "bite" on the meal, otherwise being choked by the non-conveyed meal. The "choke gear" at the rear, if present, is for adjusting the cake thickness. The shafts, with heavy thrust bearings, are driven via the gearboxes by heavy electromotors, connected to the gearboxes either directly or by V belts. The shaft speeds can be varied either by changing gear (motor pinion), or by using (DC) motors of infinitely variable speed. Additional features are automatic, hydraulic or mechanical choke adjustment, cake breakers, cake conveyors, etc. The oil draining out over the whole length of the cage(s) is collected in troughs on both sides or in just one broad trough below the press cage. The oil can be removed from the trough(s) by its slope or, for example, by a continuous screw.

The construction materials of the worm assembly, the distancing pieces, the lining bars, etc. and also those of the gears and shafts are of great importance. Here one always has to consider capital costs against running costs. The press parts should be made of hardened steel, which is less expensive, or hard-faced steel, which is more expensive. The worm shaft needs to be able to rotate freely in any material to be pressed and so it is desirable, in most cases, to use worms with low friction factors, which means, for example, Stellite-faced flights. Soft seeds, like ground nut and rape seed, require less hard bar surfaces with higher friction factors, than do hard materials, such as copra, decorticated palm pits and partly decorticated sunflower seeds. In order to avoid extreme temperature rises caused by friction, sunflower seeds must be pressed with hard-faced bars of low friction factors.

The easy general maintenance of the presses, including access for cleaning and repairs, is one of the selection criteria. For this reason an adequate runaround system needs to be designed for each installation.

(a) *The working of screw presses.* When starting the pressing operation the volume of the crushed, conditioned ("cooked") meal entering the feeding section of the press is strongly decreased. This is partly due to the displacement of air occluded in the material, which is an absolute necessity. Under the relatively low pressure exerted, the meal particles thus become closer together and the total external surface area decreases. The interstitial space is diminished and the oil starts to flow out. This partly feeding, partly deaerating and pre-compressing/de-oiling task is thus done either in the first section of a horizontal press or in a vertical cage section situated on top of it, giving the whole assembly an L form.

The crushed seeds are transported in the cage by the fact that they have a greater friction relative to the cage bars, inserted in a slightly staggered manner, than to the worm surface. During this transport the screw produces a plug of compressed meal along the discharge end of the cage. The meal particles are now pressed themselves, leading to renewed oil outflow. The cake, as the compressed meal is called, is removed at the end. The new cake is produced continuously without a "choke gear" at the inner end of this plug, because of its frictional resistance. In newer constructions the "choke

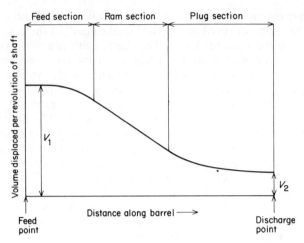

Fig. 2.22 Compression curve: compression ratio = V_1/V_2. [Reproduced, with permission, from Ward (1976).]

gear" may be absent, although if present its setting can help determine the cake thickness and porosity. The pressure built up in the cage during the flow of the seed and the heat due to friction, causes a narrowing and, ultimately, closing of the different oil-delivering structures, like capillaries. For this reason screw pressing should be done at the lowest possible pressure, in order to obtain maximum oil removal with optimum residual porosity of the cake. This sealing-off effect is the reason why seeds cannot be de-oiled completely, by mechanical presses, which in certain cases makes solvent extraction after pre-pressing an economically attractive step.

The reduction in the volume of seed displaced per revolution of the shaft results in its compression, the pressure being built up gradually or by steps from worm section to worm section. This volume reduction stops at a certain level close to the final worm section(s). A compression curve of the press can be constructed, which is a plot of the volume displacement of the seed to a base of length along the cage. A term often used is the "compression ratio", this being the volume displaced per revolution at the feed end (V_1) divided by the volume of cake displaced at the discharge end (V_2) (Fig. 2.22). Theoretically the compression ratio can be 4:1, but because of rotation and gliding of the seed mass around the worms it is usually about 10:1.

Axial pressures are between 200 and 500 bar, depending on whether "partial" or "full" de-oiling is required. These high pressures on single worms cannot be sustained for long, as the result would be that the worms would act like breaks. The section can be overflown by too-high volumes of seed material, which are not being conveyed to the next section. The material rotates along the axis and by this the rear flank of the worm flight is

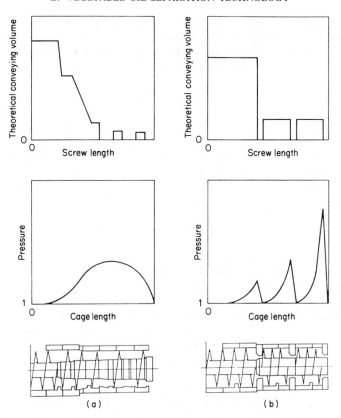

Fig. 2.23 Conventional (a) vs. new cage and screw (b) designs with the corresponding volumes of seed theoretically conveyed and radial compression development curves along the cage length. [Reproduced, with permission, from Homann et al. (1978).]

subjected to higher wear and tear than its pressure flank. The pressure build up and the theoretically conveyed volume per revolution of a conventionally equipped press is given in Fig. 2.23(a). Besides the application of the correct worm geometry possible solutions to this problem are applying the correct combination of friction factors to the bars and worms, and placing additional and cheap distancing pieces into the second half of the shaft.

In some new presses manufactured by the Krupp and French Oil Mill Machinery Company throttle rings or orifices are placed in the cage. Furthermore, instead of many different diameters only a few different-diameter worm sections are used, of which the last two have constant conveying volumes. This latter amendment has two consequences. (i) According to studies made by Krupp (Homann et al., 1978), although the throttles offer resistance to the seeds, the pressure generated by the worms is completely released through the throttle because the following worm conveys considerably more material than the worm in front of the throttle is capable of pressing through it [see Fig. 2.23(b)]. (ii) The shear developed between the worm and the throttle ring also produces a strong grinding/milling effect.

As already mentioned, the finished cake can be removed or expelled without the use of a choke gear if the meal is conditioned to the right moisture and temperature level and the load of the press is chosen optimally. In such a case the load (for a given size of the motor) is determined by the correct worm assembly for a given seed, in balance with the feed flowing at a certain speed. Only limited importance is ascribed to the choke gear by Simon-Rosedowns (Ward, 1976) and this mainly when starting up and stopping the press. With small oil seeds such as rape seeds, the choke gear can help to counteract the formation of foots. In general, if the oil contains a large amount of foots there is trouble with the feed volume.

(b) *De-oiling techniques.* As already mentioned, screw presses can be used for "partial" de-oiling, pre-pressing and for "full" de-oiling, final- or "end-pressing. Partial pressing is a low-pressure operation, leading from, say a 30–50% initial oil content to a residual oil level of 15–25% in the cake ("cake I"). Although full de-oiling solely by extraction has been proposed (e.g. the Direx process proposed by Bernardini, 1976) for raw materials with relatively high initial oil levels, similar processes seem to require 50% more (mainly heat) energy, if compared with the energy costs of the "classical" process of "warm" mechanical pre-pressing combined with solvent extraction (Jurisztowsky, 1983). If investment and depreciation are also considered, the cost differences are less divergent. Another possibility is full mechanical de-oiling by means of expellers. The pre-pressing/extraction combination is only economical above a certain production volume (100–150 ton day^{-1}) because of the high investment costs of a solvent-extraction installation. The combination of "cold" prepressing with solvent extraction is discussed below.

Full mechanical de-oiling, being a safe operation with no explosion hazards, is interesting for oil-seeds with a high initial oil content. Besides the local transport and power problems, the higher residual oil content in the cake must also be taken into account. De-oiling can be done in one or two runs. End-pressing in one run requires longer pretreatment in cookers at high temperatures in order to reduce the final moisture level of the meal to quite low levels (see Section II.D.2 of this chapter). This step is followed by a high-pressure expelling operation, ending with 3–8% residual oil levels in the cake ("cake II"), After cooking in a pre-press the meal is pressed in two separate runs and the cake obtained is disintegrated by roller-mill grinding, reheated in a cooker if necessary, and once more pressed in an end-press.

Although screw presses were originally designed specially for specific purposes, the so-called full ("end"-)presses with capacities of 20–120 ton day^{-1} can be used, if required, in a dual-purpose way. Their capacity as a pre-press is then generally at least doubled, but of course at the cost of a higher base load. Not only is the speed of the shaft increased from, for example, 20–35 rpm (older types 9–12) to 40–70 rpm (older types 25–35), but the shaft diameter at the feed end is also somewhat smaller than that of a genuine end-press, whereas the shaft taper is made somewhat more gradual.

(c) *Description of some widely used types of expellers.* In this section, instead of recalling most of the well-known firms and all of their machines (Tindale and Hill-Haas, 1976), some selected presses which have certain special design features are described. The general tendency for building mammoth-size factories specialized in one or two oil seeds has led to the manufacture of single presses with pre-pressing capacities of up to 460 ton day^{-1}. The capital costs of such presses are high, but the running costs may be low. Presses of 50–200 ton day^{-1} pre-pressing capacities are the ones which are generally installed where not just one or two but many more types of oil seed are to be processed.

The L-form connection of two cages (Duo-, now Duplex-expeller) was the speciality of Anderson, who in 1906 also produced the first single-cage screw press, called an expeller, for flax seed extraction. Even now this firm produces a range of single- and double-cage expellers, of different sizes and configurations for different oil seeds and with provisions for e.g., the heating or cooling of the shafts and oil.

Krupp was the first company in Europe to produce single-cage Anderson expellers under licence. Probably based on this tradition they now produce a screw press with two cages positioned in the L form, to be used as a dual purpose machine. The vertical cage has two worm sections and the horizontal one has five. No wedge bars are applied and the machine, being very robustly built, has a cage produced of cast steel.

Damman-Croes also produce expellers of the L-form configuration and with single cages.

The analysis of empirical volume-decrease and pressure-increase data induced Krupp to construct a new horizontal pre-press with a special cage. The cage consisting of two hinged halves can be opened horizontally at the bottom. Of the cage wall 180° in the feeding section is formed by a solid half cylinder, the opposite half being formed by normal lining bars. The other nine sections in the cage also consist of lining bars with spacers. The bars are not placed fully parallel with the worm shaft, but at an angle of 30° to it. This not only improves the feeding, but the edges of the inclined and staggered bars also act like knives on the uncrushed cold seeds. Therefore the seeds are intensively sheared, giving a high grade of cell rupture. All the worms have high conveying capacities per revolution. They work against the resistance of several deep neckings in the free cross-section, the throttles. The pressure on the meal can be adjusted continuously during operation by moving the whole cage in an axial direction either mechanically or hydraulically. This movement changes the width of the slits (1–7 mm) between the throttles and the conical rings on the shaft. The pressure-relieving action of the throttles works as described previously. The material is thus first loosely stacked and then pressed by the following worm.

The uncrushed seeds entering the machine are at a temperature of 20°C. The oil leaving the press may have a temperature of 70–80°C, and the cake, a very thin strip, of 95–105°C. In spite of some moisture evaporation the weight of oil removed causes the meal's final humidity to be increased. In order to prevent any clogging problems during extraction due to oil-weight-loss caused humidity increases, the starting humidity of seeds is limited by Krupp to <9%. This press is the basic machine for their VPEX process. The pre-pressed (German: Vor-Pressung, VP) cake produced with 18–22% residual oil content from any type of cleaned seed (even partially dehulled sunflower seed) without previous crushing and conditioning (thus "cold") is, after breaking and rolling of the cake, submitted to immediate further de-oiling by hexane extraction (EX).

The French Oil Mill Machinery Company (FOMMCO) produces the biggest of the expellers already mentioned, as well as a range of other one-barrel machines. For one of their bigger machines they report the pre-pressing of whole flax and sunflower seeds at ambient temperature without previous crushing, and with a preconditioning treatment of only 40°C. The machine has a patented worm with bevelled notches on the faces, like a shingled roof, for mastication of the seed. A fixed restrictive orifice (throttle), already mentioned, is located at the mid-point of the press and a variable one at the discharge end; this is called a vented-cone mechanism. A speciality is the two-speed main shaft, with a higher speed axis for the feeding section and a much slower one for the pressing section. Some of the other features offered by this manufacturer are: water-cooled/steam-heated inserts in the cages, which carry the lining bars, are easily removed by two cap screws on each insert for changing the bars; further water-cooled shafts; patented shaft-speed change; motor cooling by built-in high-capacity blower; bottom-hinged main cages; variable-speed non-clogging feeder; combinations or three or four section cages; and no wedge bars.

Speichim, besides manufacturing some FOMMCO presses and extractors under licence for the European Economic Community (EEC) and the French-speaking areas of the world, also make single-cage screw presses of their own design for oil seeds and special presses used for palm-fruit-oil extraction. The palm oil presses have perforated steel cages instead of barrels with bars. Because of the absence of knives or breaker bars the worms can have continuous flights. The shafts have perforations at the discharge end through which live steam can be injected into the cage to improve the mechanical oil extraction.

Olier and Mechanique Moderne produce single-cage presses for oil seeds of high initial fat content (groundnut, sunflower seed, rape seed etc.). As most of the oil leaves the press close to the feed end, these presses have, before the part of the barrel with bars, a solid cylindrical cage containing over 10 000 small holes. This design is similar to those of the classical "Seiher" filtering-type hydraulic cage presses, except that the funnel-like holes bored into this cage have a 1.5 mm starting inside diameter widening at the outside surface to 3 mm.

Stork-Amsterdam constructs, amongst other types, special vertical twin-screw presses with pretzel-shaped perforated cages for the palm oil industry, and a special press called Rotopress 400 (Fig. 2.24) for pre-pressing high oil bearing seeds, like copra, palm kernel, peanuts, and, as claimed, also rape and sunflower seeds. The principle is quite different to usual screw presses, in that a fluted roller, surrounded for about three-quarters of its circumference by press bars, expels oil at the bottom, and gives a continuous sheet of cake with a residual oil content of 20–25%, e.g. for

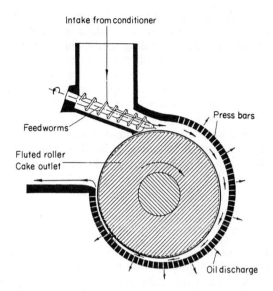

Fig. 2.24 Rotopress 400. [Reproduced, courtesy of Stork-Amsterdam.]

cotton seed. The gap between the rolls and the wall can be adjusted according to the quality of the material. Low power consumption, better quality oil and less downtime for maintenance together with less need for spare parts are the advantages claimed for this machine.

The last, but not least, of the four biggest in the field (besides Anderson, FOMMCO and Krupp), Simon-Rosedowns produces, among other machines, single-cage presses containing up to three divided sections. Among the features of this press, all of which are offered in wide ranges, are easily removable shaft; mechanical lifting and lowering of the main cages; oil and shaft cooling; special feeding devices; a cake cutter with reject and normal cake discharge; and an oil/foots conveyor, driven from the main motor.

3. *Compactors*

Compactors, which are a new type of device, are in fact extruders, which press conditioned, flaked soya bean or cotton seed meats of 0.4 mm thickness by means of an extruder spiral through a die plate. Anderson calls them expanders, and FOMMCO calls them enhancers, although similar equipment is produced e.g. in Brasil also by Tecnal. The flakes are compressed by spirals and are then masticated by inverted pins which replace the knife bars of the expellers. Through some of the hollow pins the mass is first wetted by water and/or sparge steam to a moisture of about 18% at 60°C. The ± 20-mm diameter sausage-like strands of compressed meal leaving the apertures are cooled from 85°C to 60°C. This compacted (pelletized) meal is quite dense although its porosity is increased. The pellets are dried to 10–

14% residual humidity and extracted by hexane. One of the claims is the retention of less (20–23% versus 30%) miscella in the flakes, which means notable heat energy savings. According to the claims of the manufacturer, the capacity of a given extraction plant can be increased by 20 to 50%, if these extrudates are charged to it instead of flakes. The electrical power needs of these machines, rotating at approximately 300 rpm, is about 8–10 kWh per ton of flake for enhancers densifying 200 ton day^{-1}. The live steam requirement is about 8 kg per ton of flake.

Presses which produce, after the humidification and mixing of crushed meals, the well-known small cylinders (pellets) of animal feed can also be used for producing a more compact and regular feed for extractors, although of course without the improved porosity created by the compactors.

4. Products and their Further Handling

The handling of the crude oil is described in Section IV of this chapter. The foots removed by any straining operation (see later) are returned (after post de-oiling) to the cookers of the screw presses or, after being mixed with pre-pressed cake, to the extractor. The pre-pressed cakes are first disintegrated by (impact or other types of) breakers to smaller pieces of 5–8 mm. They are then further ground in fluted roller mills. If directed to solvent-aided extraction, the fluted-roll ground material can be reconditioned in "cookers" or horizontal conditioner drums before the usual flaking on smooth surface rollers. Alternatively, it is not uncommon to cool down the broken pre-pressed cakes to 60–65°C and, after fluted-roll grinding, pass them directly to the solvent extractor.

B. Solvent-aided Oil Extraction (Leaching)

The lowest levels of residual oil after pressing are 3–8%; exhaustive removal of the oil present in the seeds by mechanical means alone is impossible. The main reasons for this are that (i) the increasing pressure and the frictional-heat release close the capillaries, and (ii) the physicochemical adhesion of oil to the cell walls under pressure opposes bodily movement. The residual oil can, therefore, only be removed by a different approach, this being solvent-aided extraction.

1. Mass-transfer Operations

All the methods for separating oils from their surrounding matrix described so far are based on the combined application of heat and mechanical energy. The temperature differences for heat transfer and the pressure differences for bodily movement served as the "driving forces" of the processes applied. Aiming at a near total de-oiling of the flakes, we need a process based on a third and different driving force. This is delivered by the process of diffusion, i.e. causing the molecular migration of matter through boundaries, e.g. cell-membrane-like resistances by a sort of "chemical pressure". The driving force for this diffusion is called the chemical potential and is based on the mass energy of the matter involved. The creation of, for example, a concentration difference between the oil-containing inert matrix and a suitable auxiliary medium, the solvent, can and will induce a molecular transport of oil and solvent from sites with higher chemical potentials to those with lower chemical potentials, most generally expressed as mass-concentration differences or gradients. We will not discuss here the chemical potential as the driving force for chemical reactions, which is also a form of energy transfer.

2. Extraction

In general extraction is the separation of a liquid or solid substance from a mixture of other solids and liquids by means of a selective solvent, in which the preferred substance is more soluble than all the others.

Any substance, introduced into a mixture of two preferably immiscible liquids, in both of which it is directly soluble, distributes itself between the two. The distribution, practically a mass transfer process, will end when the substance exerts the same chemical potential in both phases, i.e. it is at equilibrium. The proportion of the concentrations in the two phases at equilibrium and at a given temperature is constant.

$$y = m \cdot x$$

where y and x are the equilibrium mole fractions (concentrations) of the substance in the two phases, and m is the distribution of partition coefficient. An "equilibrium (distribution) plot" is described ideally by a straight (practically a curved) line through the origin with slope m.

3. Leaching

The operation in which we are interested could more appropriately be called "leaching" (old nomenclature: lixivation). In the present case this means the removal of just one substance, the liquid oil and some less oil/solvent-soluble components like phosphoacylglycerols, from the interior of a solid matrix, the flakes. This is carried out by dissolving and removing the components in one selective solvent which, as we shall see, is nearly always hexane. Before entering into the details of this process (Schoenemann and Voeste, 1952), we will discuss some assumptions which influence the calculations of leaching or solid/liquid extraction.

(i) The oil and the accompanying substances to be extracted from the flakes are emulsified in the protoplasm of the cell but are not molecularly dissolved in it. The oil is assumed to be retained in the cell-wall material only by the porous and capillary structures formed during the grinding and flaking the seeds. With this prerequisite the flakes are considered to be an "inert" carrier material or matrix for the oil present in their already partly ruptured cells. The technical calculations are more simple, if the composition of all the phases and components of the system are related to a unit weight of the "inert" carrier.

(ii) The oil in the flakes has unlimited solubility/miscibility with the most commonly used hexane type solvents. After the flakes have been brought into contact with the solvent and left for an adequate time for leaching at a given temperature, a varying part of the solution is drained out by gravity settlement from a contacting device. Both the dissolution and the drainage are time-dependent steps of the operation. This "free liquor" removed in equilibrium has the same composition as the remainder of the solution, this is in contrast with substances which are distributed between non or restrictedly miscible phases. The "free liquor" is thus the first so called "pseudo-phase" of the mass-transfer operation. Considering its mobility it is also called "overflow".

(iii) After the drainage step a varying amount of the miscella^a is thus retained between the interstices of the flakes. This oil-containing solvent is the so-called "bound liquor" or holdup, which forms together with the oil still residing in the flakes the second "pseudo-phase" called the "underflow". The amount of holdup depends on many factors, such as the preparation of the flakes, the temperature, the times of contact and drainage and, last but not least, the oil concentration of the miscella removed. The oil content determines the surface tension of the solution and by this its ability to wet the "inert carrier's" surface. By adding portions of fresh "overflow" (solvent or miscella of different concentration to the flakes) this operation leads to the near-complete depletion of the oil content of the "underflow" ("bound liquor") and the residual oil in the flakes.

(a) *Practical calculations of leaching.* The compartments (kettles, cells, baskets, belt segments, etc.), in which the flakes and solvent are mixed, settled and finally separated constitute a stage. If contacting of the phases and the time of contact were chosen optimally, in order to establish true equilibrium, this stage would represent an ideal or theoretical one. If, for example, all the oil could have been transferred to the solvent in one step, and the miscella totally removed, we could speak of an equilibrium stage of 100% efficiency. Practically all contacting devices have less than 100% efficiency and therefore more stages are needed (multiple stage) and/or special ways of contacting/separating the solvent and the flakes in order to arrive at an economically acceptable exhaustion of it (Treybal, 1981). More leaching stages are required due to (i) the short contact times prohibiting complete dissolution of the oil, (ii) the imperfect phase separations, and (iii) the varying amount of "bound liquor" or holdup. In addition, the amount of fine flake particles, removed with the miscella, if not reiterated in the contacting device, reduces the efficiency of the operation.

---

^{a} An often used technical term in the seed extraction business goes back to old Italian tradition. The solvent which contains the dissolved oil is called in Italian "miscella". It should not be confused with "micelle", the colloidal molecular aggregates which are described in Section IV of this chapter.

The primary aims are either the calculation of the ideal (theoretical) number of stages required to reduce the initial oil content of the flakes to a given residual level with a certain amount of solvent, or the calculation of the extent of leaching in an actual apparatus with a known number of stages. As the counter- or cross-current contact of the two material flows is the most economical and most used method of operation, the calculations described below are relevant to these cases.

First, the usual "mass-balance equations" for the solvent and the oil are written. These equations predetermine the amounts of solvent and flakes introduced into the leaching system and the initial and required residual oil contents and the initial and required final miscella concentrations. For example, the leaching operation starts with (soya bean) flakes with a 20% oil content (25 parts oil in 100 parts oil-free flakes) and, oil-free hexane. The prescribed residual-oil content is 0.5% (0.5 parts oil on 100 parts) in the extracted meal and the final miscella concentration is set at 30% oil ($25 - 0.5 = 24.5/30 \times 100 = 81.7$ parts of miscella); all values are given as weight per weight (w/w). The final and initial levels of oil, "Δ oil" ($25 - 24.5 = 0.5$ parts), and solvent, "Δ solvent" ($81.7 - 25 = 56.7$ parts), are constants for the system. The varying parts are the actual amounts and oil concentration of "static" holdup as a function of the amount and oil concentration of the miscella separated in a previous or following stage of the process.

The actual "balance equation" is given by

$$Y_n = X_{n+1} = \frac{\text{"Δ oil"} + \text{oil weight in nth stage "underflow"}}{\text{"Δ solvent"} + \text{solvent weight in nth stage "underflow"}}$$

The balance equation determines the points of the so-called "operating line" in a rectangular y/x diagram where y represents the oil concentration of the "overflow" and x that of the "underflow". Only the points of the "operating line" actually determined represent real phase compositions in the contacting device, the rest of the compositions being in the space in between the real ones. On the same diagram also the "equilibrium (distribution) line" is depicted. This line is straight and has a slope of 1, because of the ideal equal oil concentrations of the two pseudo-phases which are in contact. The difference between the two lines discloses how far the two (moving) pseudo phases are from equilibrium at the different stages, the concentration differences being the "driving force" of the mass transfer.

It has already been mentioned that the holdup retained by the flakes is not constant, but increases with increasing oil concentration in the miscella. This causes the "operating line" of leaching to have a slightly curved form. The actual form (curvature) of the "operating line" has to be determined by laboratory experimentation.

(b) *Graphical calculation.* The results of above case are depicted in Fig. 2.25. The number of the theoretical or ideal stages is graphically calculated as described by Armstrong and Kammermeyer (1942). Starting on the "equilibrium line" with the 30% final "free liquor" (2) this could be enriched to 50% "bound liquor", if contacted with the entering oil-rich "dry flakes" (1). This contact was actually only executed in the laboratory. The final "free liquor" is formed during the contact of an approximately 15% equilibrium "free liquor" (4) with "wet flakes" to which the 30%

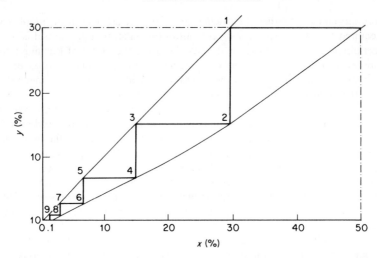

Fig. 2.25 Graphical calculation of the theoretical leaching stages. [Reproduced, with permission, from Schoenemann and Voeste (1952).]

"bound liquor" adheres (3). The vertical line (2) to (3) represents the first contact or stage. An approximately 6.5% equilibrium "free liquor" (6) contacted with the "wet flakes" containing the 15% "bound liquor" (5) delivers the second stage of the leaching, given by the vertical line (4) to (5). Thus starting this graphical calculation at level (1) to (2) leads, in this case, to five steps of decreasing height, which give the number of equilibrium stages needed to arrive at the required exhaustion of the flakes. The amounts of holdup and its oil concentration were found to be (depending on time of drainage) 0.58–0.45 kg of oil-free hexane miscella ("bound liquor"), 0.66–0.53 kg of "bound liquor" of 20% oil content, and 0.7–0.6 kg of "bound liquor" of 30% oil content, all for 1 kg of "inert" carrier, the formally oil-free flakes.

(c) *Collection of laboratory data.* The experiments are carried out in a large cylindrical glass column of not too narrow diameter, in which fresh ("dry") and partially exhausted ("wet") flakes are deposited at heights occurring in the daily practice. The two types of flake are contacted in different stages of their exhaustion with the same amounts of the solvent or miscella, having mostly lower oil concentrations than the actual holdup. The amounts to be used can be calculated, but should be close to the real practical values. At the bottom of the column is a discharge pipe with a stopcock and above this a perforated plate, on which the flakes can be deposited. After a substantial time has been allowed for contact the solvents (miscellas) are removed. The whole routine is continued until the preset final oil concentration in the flakes (say <1.0%) has been achieved.

The amount and the oil concentration of the "free liquor" or "overflow" is measured directly at every stage. The amount of "bound liquor" or holdup is the difference between the amount of solvent or miscella entering the system and the "free liquor" drained from it. The oil content of the "wet flakes" is that of the residual oil in the flakes and that in the "bound liquor" combined. The "wet-flakes oil content" is calculated by the difference between the original amount of oil in the flakes and the oil already removed by the solvent. This gives the oil content of the "underflow".

The amount of residual (undissolved) oil in the flakes can be calculated by subtracting the oil content of the "bound liquor" from the oil content of the "wet flakes". It has been found experimentally that the amount of residual oil depends only on the time of contact and is independent of the oil concentration in the hexane used as solvent (Coats and Karnofsky, 1950). This finding is also valid for acetone, but not for other solvents (Othmer and Jaatinen, 1959). This independence means that oil should be dissolved at nearly the same rate in an actual extractor as in the laboratory experiment, a fact substantiated by the daily practice (Karnofsky, 1986).

In addition, the rate of percolation through a given height of flakes under the liquid head (pressure) of the solvent ($l\,dm^{-2}\,h^{-1}$) and the rate of drainage of solvent and miscellas of different oil concentration must be established. In the case of soya bean flakes a practical percolation rate is $360\,l\,dm^{-2}\,h^{-1}$ or around $1\,cm\,s^{-1}$, although lower rates have also been mentioned in the literature.

The number of calculated ideal or equilibrium stages must be adjusted by stage-efficiency factors. This can be done by measuring/calculating the so-called height equivalent to a transfer unit (HTU), which expresses the length of apparatus required to accomplish a separation of standard difficulty. In the case of leaching this length depends strongly on the relative rate of flow of the quasi-phases.

Equilibrium-stage calculations cannot give an answer to the influence of the solvent (miscella) distribution in space and time through and below the flake beds in an actual extractor with horizontally moving cells and/or belts. In addition to the already measured static holdup relationships some dynamic aspects must also be considered. In most of the actual devices it is impossible to separate distinctly the moving miscella belonging to different stages, as back-mixing is inevitable. Calculations (Kehse, 1974) substantiated the practice that the dragging away of miscella can be compensated for by the positioning of the (partial) miscella spraying liquid manifolds in distances of not less than twice the width of the cells being separated by the stage divider (walls) or, for example, stopping an endless belt at fixed intervals. The extra number of cells (stages) actually applied discounts both the back mixing and the non-100% stage efficiency. A greater number of cells also minimizes the mechanical damage occurring to flakes when they are charged to and discarded from the extractor by free fall.

4. *Special Aspects of Oil Extraction*

In practice the leaching/extraction process is actually performed in two main phases. In the initial phase the oil which has oozed out to or close to the surface of the flakes (about 30–50% of the total) is nearly mechanically "washed away" by a relatively concentrated miscella. About 80–90% of the oil is dissolved in a short time. The final phase lasts somewhat longer as this involves the removal of residual oil and some other lipid constituents from the nearly exhausted flakes by addition of fresh hexane. The latter are phospho- and glucolipids, very sparingly soluble in hexane, but more easily in the analytical solvent diethyl ether. Their low solubility in pure hexane explains the low numerical value of the diffusion coefficients calculated for this phase of the extraction.

The importance of the correct pretreatment of seeds before extraction with respect to the production of flakes with a high percentage of ruptured cells, with high plasticity, rigidity and internal porosity has been duly stressed above. The primary increased "surface (diameter) to unit volume" proportions of the flakes with decreased axial distances to be traversed within the solids are expressed as thickness. The secondary internal surface structures can be characterized by the porosity, which is expressed by the volume ratio of the pores and capillaries present to that of the total volume of the flake. Besides the individual structure and porosity of the flakes the importance of there being a high bulk density of them should be stressed; this depends on their size distribution and the specific density of the seeds pretreated. The uniformity of the bed with a narrow particle-size distribution and less than 5% loose and too-fine particles provides a substantial free volume between the flakes, resulting in a low flow thus in high percolation and draining rates.

It is evident that the size (thickness) of flakes is mainly responsible for short-mass transfer distances. Among many others, Coats and Wingard (1950) have studied the influence of flake thickness on the contact time versus residual oil content. Using the most commonly applied solvent, hexane, it was found that, although different oil seeds differed markedly when extracted, the decreasing flake thickness (limits 0.15–0.55 mm) caused an increase in the rate of extraction. Therefore, because of the decreased surface-to-volume proportions, grits are less easy to extract than are flakes. Cotton seed is the commodity which is most easily extracted (0.35 mm thickness) and flax seed is the most difficult. The correlation found states:

$$T = K D^n$$

where T is the time (in minutes) required to reduce the residual oil content to 1%, D is the flake thickness or grit diameter (in units of 0.25 mm), n has a value between 2 and 8 which is oil-seed dependable, and K takes a value between 1.4 and 3600 representing the ease of extraction.

Another important point is the inevitable increase in the moisture content of the meal during leaching. Meals with high initial oil and moisture levels may be converted, after the removal of oil by the solvent, to a sticky mass which clogs compartments and screens and causes other operational troubles. This problem can be encountered in systems where seeds are not pretreated and are pressed to a 20% residual oil and a 10% final moisture content. This cake, if extracted without an additional drying step before extraction, may deliver a meal with more than 12.5% moisture. This results in an increased residual-oil content in the meals or a reduced rate of extraction in non-deep-bed extractors, because the too-wet particles cling together.

The amount of hexane type solvent used is also an important parameter which influences the solvent losses. The amount used is generally between 0.65 and 0.95 kg per kg of seed entering the countercurrent system, this being determined by the adherence of solvent to the exhausted meal (holdup) before solvent removal and concentration of the miscella withdrawn for distillation. In general, 30–35% bound liquor or holdup of hexane in the extracted meals, originally containing 18–25% oil, is considered to be normal. In removing a so-called full miscella containing 30% oil from the extractors, the amount of hexane to be removed from the meal is 0.45–0.58 kg per kg meal. If a more concentrated miscella, e.g. 40%, is produced, the hexane ratio will be 0.45–0.52 kg per kg meal, which means some saving in the distillation costs.

The minimum residual oil content in the meal is not lower than 0.1–0.2%. Depending on the final (low) oil concentration of the average 30–35% hexane holdup in the extracted flakes, the residual oil content in the desolventized, dried meal will amount to 0.3–1.0%. Mechanical pressing of the meal or gravitational pre-desolventization by longer drainage times/distances are two possible ways of reducing the high costs involved in thermal desolventization of the meal. On the other hand, it has already been stated that compacted (expanded, enhanced) soya bean "sausages" have an increased porosity, by which the capacity of a solvent extraction plant is claimed to be increased by 30–50%. The hexane holdup might also be lower than normal, say 20–22%, because of the better and quicker draining characteristics of the porous structures, when leaving the extractor.

5. Solvents

Good solubility of the oil in a solvent of low viscosity is a prerequisite for a successful operation. As the wetting of water-containing flakes by hydrophobic solvents is more difficult, so a high water content of the flakes is generally not advantageous. As the diffusion is enhanced and the viscosity of the oil/solvent solution is lowered by raising the temperature, so extraction is generally carried out at a temperature just below the boiling point of the solvent. High boiling solvents are less advantageous, because at higher temperatures more-polar lipid components like the waxes of sunflower seed husks or rice brans are also dissolved and these must then be removed by winterization. A low boiling point solvent reduces the costs of solvent removal from both the meal and the miscella. In any case, no solvent with a boiling point above 100°C should be used. A narrow range of boiling points is advantageous because of the losses encountered during recovery of the solvent.

The ideal solvent should have the following characteristics:

(i) a narrow and not too high boiling point or range, and should remain liquid even at very low temperatures;
(ii) be neutral to the oil when dissolving it easily and selectively;

(iii) be stable and inert when in contact with metal surfaces;
(iv) a low specific heat, a low heat of evaporation, and a low viscosity and density;
(v) insoluble in water;
(vi) non-toxic;
(vii) inflammable and non-explosive; and
(viii) available at low prices in adequate quantities.

It is obvious that there is no solvent which fulfils all these criteria. For practical purposes only three solvents are feasible for use, only one of which has any real importance.

(a) *Light petroleum hydrocarbons, "naphthas"*. The miscibility or solubility of nearly non-polar tri(acyl)glycerols in non-polar petroleum hydrocarbons is unlimited. This is due to the strong solute–solvent interactions, expressed popularly as "like dissolves like". This total miscibility is invalid for the more-polar lipid components, for which the rules of solubility are applicable. The aliphatic hydrocarbons pentane, hexane and heptane and cyclohexane all come from the same feedstock as the gasoline fuel for vehicles. Therefore, their production is only feasible if they are sold at a premium to fuel gas. The hexane type solvents, containing about 45–90% n-hexane with the rest being, for example, 2- and 3-methyl pentane, 2,3-dimethyl butane, methyl-cyclopentane and cyclohexane, are the most widely used, having a boiling range of 63–69°C and fulfilling most of the above criteria, except inflammability and non-explosivity. Pure n-hexane extracts oil slower than the commercial mixtures. The commercial hexane mixture dissolves, e.g. gossypol, the colouring pigment of cotton seed which is toxic to non-ruminants, and polar lipids such as phospholipids and glucolipids, less completely. It is thus feasible that there is some difference in the polar lipid content of oils obtained using a more apolar lean miscella or pure solvent, against a more polar and more concentrated one, or in other words those obtained from more-or-less exhausted meals. This fact has consequences on the quality of the crude oils, which contain different types of phospholipids and hydratable and non-hydratable phosphatides, to be removed before the refining operation. These are present in the miscella in the form of "micelles", the molecular aggregates already mentioned, surrounded by the less-polar tri(acyl)glycerols (see Section IV.C of this chapter).

(b) *Trichloroethylene*. Trichloroethylene has been used as a solvent mainly because of its inflammability and non-explosive character, which makes its use in delocalized, small-scale extraction factories quite popular. Less-strict safety precautions are needed for this solvent and this, besides its easy availability through its large-scale use in the dry-cleaning business, reduces the costs of investment. However, because of the lethal consequences ("bloody nose disease", a haemorrhagic aplastic anaemia) of feeding the extracted meal to cattle (only ruminants have been really affected), the use of this solvent has been abandoned. The precise toxic mechanism involving meal, solvent and cattle has not yet been clarified.

(c) *Ethanol.* Ethanol was used as a solvent during the early 1930s in Dairen (the pre-First and Second World War Manchuria), as a consequence of Japanese shortages of light hydrocarbon fractions. The solubility of triglycerides in hot 99% ethanol is high, because of the miscibility of the two in all proportions. However, no, or at least less than 3%, water should remain in flakes extracted with ethanol, because the azeotropic mixture of 95% ethanol can only be used for the purpose if work is carried out under pressure, which was the case in Dairen. On cooling the extract, it separates into two fractions, the upper one containing ethanol with some oil, and the lower one containing 95% oil and 5% ethanol. The upper phase was reused and the lower distilled. The quality of the meal was characterized as excellent, free of the flatus inducing sugars and components causing bitterness and beany flavours. The disadvantages were the high levels of phospholipids and glucosides present in the crude oil, which had to be removed before refining, and the costs of distillation, the heat of evaporation of alcohols being three times that of hydrocarbons. If cooling was applied without distillation some 25% of the energy could be saved, but the upper layer becomes so saturated in reuse that it ceases to extract oil from the seeds.

(d) *Other solvents.* Some special solvents have proved to be useful in special processes, like acetone in removing gossypol from cotton seed meal in the Vaccarino process. Isopropanol, isopropanol and ethanol azeotropes, mixtures of alcohols with hexane and acetone are among the many solvent variants proposed and used for the removal of oil and non-oil constituents (Johnson and Lusas, 1983).

(e) *Supercritical gases.* The use of liquid or supercritical gases, like liquefied carbon dioxide (critical pressure 72.9 bar, critical temperature 31.3°C, boiling point −78.5°C at 1 bar) as non-flammable solvents for oil seed extraction is in an experimental pilot-plant stage (Mangold, 1983). The high pressures (200–350 bar) required to achieve reasonable solubility of the oil in liquid CO_2 need very costly installations. Because of this it is doubtful whether the method will ever be competitive, despite its many advantages, with any other normal oil-seed extraction solvent.

6. *Extraction Systems*

In the different practical systems, the freely flowing overflow (the miscella) depletes the "bound liquor" of the fixed and/or the moving underflow (the flakes). The systems applied can be characterized by the direction of the material flows (co-, cross-, hybrid- or counter-current), by the number of stages (single, multiple or differential), by the cycles applied (batch, multibatch, intermittent or continuous), and by the method of contact (sprayed or immersion percolation), all of which enable suitable, but variable, concentration differences (mass flow) along the path of the two flows, to be achieved.

The five systems which are most often used for seed or animal-tissue leaching operations are described briefly below.

(a) *Single-stage batch contact.* In the single-stage batch contact system the fresh flakes and the fresh solvent are contacted in a kettle ("pot" or batch contactor).

Whether the two phases are mixed or not, after arriving at equilibrium, the solvent containing the oil is removed and the solvent is distilled off. This type of operation is sometimes used for extracting cotton seed, but it is not considered to be very effective as large amounts of fresh solvent are required.

(b) *(Cocurrent) multiple-stage single-batch contact.* In the (cocurrent) multiple-stage single-batch contact system the first volume of solvent is drawn off after equilibrium with the flakes has been attained and the second fresh portion of solvent is added. This process is repeated once or twice more. Every portion of drawn-off solution is continuously distilled in order to reuse the solvent immediately and to separate the oil further. This method is employed in the laboratory Soxhlet extractors and in some industrial batch contractors (e.g. the Merz type) for defatting bones, cracklings, garbage and used bleaching adsorbents. The final oil solution is very dilute and, therefore, the costs of distillation make such repeated "displacements" of the oil prohibitive.

(c) *Countercurrent multiple-stage single-batch contact.* A more efficient variant is the countercurrent multiple-stage single-batch contact system. First, a nearly saturated miscella is added to the fresh meal in, for example, a rotating extractor. After withdrawal the saturated miscella is distilled. The second contact is done using a leaner miscella and the last one using fresh solvent. The exhausted meal is desolventized *in situ* by heating with live steam at 100°C. After removal of the contents the extractor is refilled with fresh meal and contacted with the second miscella which was withdrawn and stored from the previous treatment.

(d) *Countercurrent (multiple-stage) multi-batch contact.* This system uses at least four interconnected batch contactors (the fifth is emptied and refilled) with the fresh flakes placed in the fourth one where they are contacted with linear solvent coming from the third. The third contactor obtains its solvent from the second contactor, and the second obtains its solvent from the first where almost exhausted flakes are brought into contact with fresh solvent. The fresh solvent is able to remove the remaining oil from the nearly exhausted flakes, whereas even a "saturated" miscella may mechanically wash away some oil which has oozed out to the surface of the fresh flakes. This system enriches the solvent before distillation by up to 60% because of its quadruple reuse.

(e) *Countercurrent multiple-stage true continuous contact.* In all previous systems the ground seeds or flakes were stationary and the solvent and miscella (the overflow) were in motion, thereby providing the necessary gradient for sustaining mass transfer. If both the flakes move in one direction and the solvent in another; the operation is called the countercurrent multiple-stage true-continuous-contact process and is proven to be the most economic. The process is mostly used in hybridized modifications, such as local cross- or cocurrent flows, because of the problems of the instability of the flakes under the mechanical wear of transport and the geometry of the equipment. In practical operations the percolation principle of the multiple-stage batch contact systems described above is used, or the flakes are moved through the system by totally immersing them in the solvent.

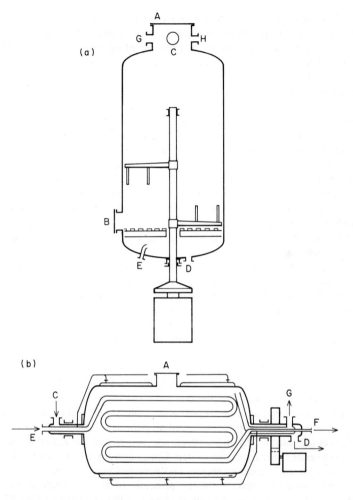

Fig. 2.26 Batch extractors: (a) vertically stirred; (b) horizontally rotating. A, charging; B, discharging; C, solvent in; D, miscella out; E, steam in; F, condensate out; G, vent gas outlet during meal desolventizing; H, deaeration manifold. [Reproduced, with permission, from Gemeinschaftsarbeiten der DGF (1975).]

7. *Extraction (Leaching) Equipment*

(a) *Batch extractors.* Batch extractors are generally cylindrical kettles (pots) of 2–10 m³ volume, which can be installed vertically or horizontally (Fig. 2.26), used as single units or in batteries of 8–12 units. They are provided with built-in filter bottoms or separations, with a charging door at the top and an unloading one at a lower point, or with just one door for both purposes. The material is transported by conveyors and the extractor is filled by gravity. The solvent is pumped in or through the vessel. In general, these extractors have no internal mixing devices if they are vertical, but may have rotary or oscillatory mixers if placed horizontally. If used as single units, the extractors work on the principle of percolation either by "displacement" or "enrichment". They may be heated by means of jackets and the residues of solvent from the spent meal are blown out by live steam. After finishing the desolventization all the vents can be closed down, except for the live-steam one. A slight overpressure of the live steam build up can be used for the discharge of the meal when the door is opened. Screw or other types of conveyors deliver the meal after grinding via the driers/coolers to the storage places. If combined in batteries (at least four units working in series), the extractors work on the percolative "enrichment" principle, i.e. by batch countercurrent/crosscurrent/multiple-stage contact. Because of the large volumes of material to be handled manually, these extractors are now considered to be more-or-less obsolete, except for the small-scale (medium, up to 500 ton day^{-1}) production of special oily materials. Completely mechanized and thus updated older installations, which exclude most of the manual labour, are suitable for use in developing countries, as the kettles can be easily made by local workshops/factories.

The miscella (up to 60% oil concentration) obtained is partly filtered in the extractor (sieve plates) and partly in closed-filter presses or leaf filters by means of clothes and filter aids. With the exception of the Merz type, which distils the miscella continuously in the bottom part of the extraction kettle, like a Soxhlet apparatus, these extractors have separate miscella reservoirs and distillation units. Solvent recovery and distillation is dealt with separately.

(b) *Continuous extractors*. The present-day volumes of oil and fat production demand the use of fully mechanized and largely automated medium- to very-large-scale continuous extraction installations with capacities of up to 4000 ton seed day^{-1}. Continuous extractors transport huge volumes of meal and solvent, bring them into intimate contact for substantial periods of time in order to reach equilibration (30 min to 2 h) and separate the two streams of material without too much carryover of either from step to step. Continuous extractors can be classified as one of two types: immersion extractors and percolation extractors.

Immersion extractors operate on the principle of true countercurrent material flow. A screw as a conveying element as shown in the extractor of Fig. 2.27 is not always necessary. The most successful predecessor of this type of extractor was the U-form tube extractor of Hildebrandt, first used in Hamburg, Germany, by Harburger Oelwerken Brinkmann and Mergell in the 1930s. In the U-form tube extractor material was forwarded in three interconnected tubes (down, horizontal, up) by perforated helical (screw) conveyors. It could be used for smaller volumes of even, fine or coarse seed and other material, for high oil content materials, like copra, peanuts, cotton seed and safflower seed, which tend to disintegrate during extraction. However, at least 0.5–5% of fines were formed by the mechanical wear of the screws and these had to be removed by filtration in special filters. The relatively large inner volume of the tubes and the short residence time of the solvent, as compared to that of the seeds, did not produce substantial concentration gradients and thus higher miscella concentrations could not be attained.

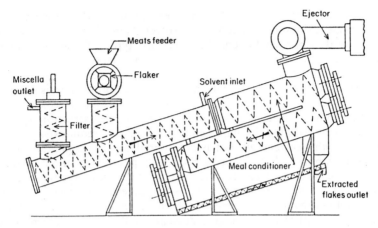

Fig. 2.27 Ford-type immersion extractor. [Reproduced, with permission, from Bailey (1948).]

The newer countercurrent types of immersion extractor use stirred continuous mixer/settlers at the meal entrance section and different discharge options (Speichim, Olier). They are claimed to be less sensitive to liquid/solid ratios than are the percolation types. They are proposed to be used as a finishing step, if interconnected with a percolation extractor, e.g. of the horizontal basket type. Between these two extractors internal screw presses or a smooth-faced roller mill may be applied in order to squeeze out any adhering miscella, thus creating new concentration gradients (Direx, Bernardini). Others apply a partial or full loop continuous "*en masse*" (flight) conveyor in a casing, where the flakes are dragged along above two sections of wedge wire screens (Crown, EMI, Bernardini) or apply a semi-submerged series of slowly rotating cages with peripheral paddles on a common horizontal axis, placed in a horizontal cylindrical hull (Satellite, Bernardini). There is a countercurrent immersion extractor consisting of five interconnected U tubes, in which the (not necessarily flaked) material is conveyed by (short) screws and the adiabatic evaporation of the solvent under vacuum produces new gradients (differential pressure extractor of Stork). The claimed advantage of this type of extractor is the removal of air from the interstices and pores of the material, thereby providing better access of the solvent to the oily matrix.

Percolation extractors all operate on the same principle: the solvent and/or the miscella is pumped and distributed (sprayed) through moving beds of flakes of varying thickness, flooding and percolating them. A substantial number of pumps serve the different bed sections with solvents/miscella of different concentration. The pumping capacity must be variable in order to deliver the necessary volumes for flooding the meals of different porosity: sunflower seed flakes percolate at double the rate of, say, soya bean flakes. The liquids leave the beds via sprinklers, perforated plates, metal screens, or grids of (wedged) bars. In most cases the operation is countercurrent (in the trendsetting original vertical Bollman extractor in the downstream leg cocurrent), continuous, as regards the solvent, and the flakes are continuously moved in batches (except the single-belt types). Being compact, they save space, as a bed of flakes is necessarily more dense than the suspended particles in an immersion extractor. They are able to produce highly concentrated miscellas, as the solvent/meal weight proportion is easily kept well below 1. This means greater savings in the distillation costs, which is one of the more heat-energy consuming parts of the extraction operation. Desolventizing/toasting/drying/cooling of the meal requires at least 4–5 times as much heat energy, but is considered to be a separate operation which is common to, but independent of, all types of continuous extractors.

Percolation extractors can be classified into one of four types according to the method of the internal transport of the flakes: (i) the vertically and/or horizontally moving-basket type; (ii) the horizontal rotary type, with slow movement of either the flake confinements (cells) or the solvent/miscella transport system; (iii) the horizontal double-belt type, with moving cells; and (iv) horizontal uncompartmented single-belt type. Each of these four types is discussed below.

The first industrially important prototype of *the moving basket type* of extractor was the Bollman designed vertical-basket extractor of the "bucket elevator" principle (Fig. 2.28) which was installed at the premises of the Hanseatische Muehlenbau A. G. (Hansa Muehle) in Hamburg, Germany. The baskets (30–40, or up to 60 in non-Bollmann designs) have perforated bottoms or sprinklers on the bottom and are suspended from trunnions on a twin string of endless chains. The chain is moved by two large pinions. The height of flakes in the baskets is not more than 0.5 m, this height determining the number of buckets or the speed of chain movement. The fresh flakes are charged via a continuously filled screw conveyor, which acts as a plug-seal, into a, consequently, "gas-tight" hopper. The hopper deposits meal into the first basket on the descending leg. This basket and all the others move down by continuous cocurrent percolation of the so-called "half-miscella". This half-miscella, on leaving this leg of the extractor at the bottom, is highly loaded with oil (20–40%) and is now called "full-miscella". Because of its repeated "self-filtration" through the unagitated flakes into the baskets below, the full-miscella is relatively clean and requires only filtration on simple "bag" filters before distillation. The baskets in the ascending leg obtain fresh solvent from near the top and the residual oil is eluted from the flakes countercurrently. The partly loaded solvent is collected at the bottom as "half-miscella' and is pumped on top of the first basket which is loaded with fresh flakes. Just before reaching the top, the exhausted flakes are drained, the basket is inverted, and the contents are removed by a screw conveyor. The rate is about one full revolution in one hour, but this can be altered according to need, i.e. the required residual oil content of the meal. The percolation rates are not adjustable from stage to stage. Another disadvantage is the possibility of channelling of the solvent in the bed which cannot be stopped should it happen. The possibility of chain breakage,

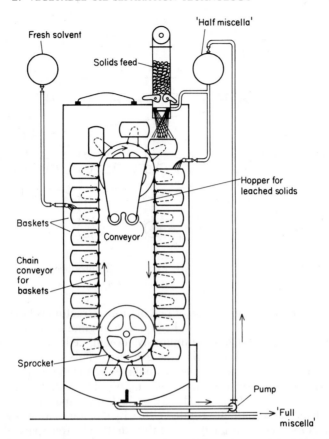

Fig. 2.28 Vertical-basket (Bollmann) type extractor. [Reproduced, with permission, from Treybal (1981).]

together with the already mentioned disadvantages has made this type of extractor virtually obsolete.

The newer horizontal-basket type extractors (Fig. 2.29) are convenient in their layout in that they do not need high buildings. These basket-type extractors manipulate the recirculation and advancement of the miscellas of different concentration by the adjustment or positioning of the liquid manifolds and the miscella-collecting troughs with stage dividers, which are also a characteristic of the rotary systems. The baskets in the upper row are all contacted with "nearly full" miscella collected under the second lower row of baskets. Some of the horizontal designs use hinged perforated cells, with one longitudinal side serving as a wall for two neighbouring cells, instead of buckets, from which the flakes in one cell can be overturned at the end of the top row, falling from there into an empty cell in the lower row (Bernardini, HLS, Gianazza, HLS-Tom).

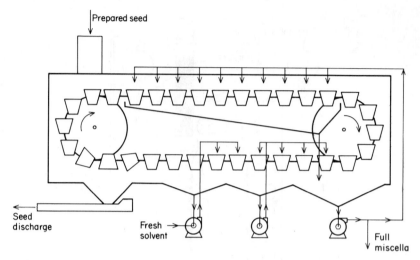

Fig. 2.29 Horizontal-basket type extractor. [Reproduced, with permission, from Treybal (1981).]

The horizontal rotary type of percolation extractors is housed in a vertical cylindrical shell of convenient height. The original Blaw-Knox (now Dravo) Rotocel (Fig. 2.30) has continuously moving cells with hinged perforated bottom doors for discharging the exhausted flakes at the end of the cycle. The cells on this rotor are filled with flakes at a feed point by a continuously full screw conveyor, forming a "gas tight" plug seal. The layer is 0.5–2.5 m thick. The fresh solvent is fed only to the last compartment. The 18 or more cells fixed on a chain are rotated either by a chain drive at the rim or, in the smaller extractors, rotated by a central shaft. The free space above the flakes is flooded with solvent which is sprayed on top of it. The cells move slowly (1–10 mm s^{-1}) horizontally around the divided stages collecting the miscellas of different concentration, which are pumped and sprayed over the flakes in countercurrent fashion. The collecting stages have twice the width of the percolation cells, in order to compensate for back-mixing of the miscella and to give ample opportunity for draining. The cells are interconnected horizontally by holes in order to induce a flow of miscella opposite to the movement of the flakes. Before being discharged above the last section, the exhausted flakes have ample opportunity to drain a part of the adhering solvent above a drainage section (Simon-Rosedowns, EMI, Krupp, Bernardini, Revoilex).

Some manufacturers use rotating bottomless cells on fixed metal screens with one opening for the final flake discharge. The miscella movement and the collecting stages are fixed (Extraktionstechnik, FOMMCO). In their Carrousel extractor Extraktionstechnik use concentrically fixed special wedge bars of trapezoidal cross-section with a top slit width of 0.8 mm as the bottom screen. The partition walls of the cells widen conically downwards to prevent the flakes from sticking on the walls.

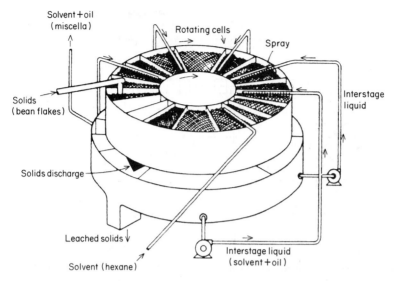

Fig. 2.30 Horizontal rotary (Blaw-Knox Rotocel) type extractor. [Reproduced, with permission]

A third variation of the horizontal rotary type percolation extractor has stationary cells with perforated hinged bottom doors and the manifolds and the miscella-collecting stages are suspended on a shaft and rotate under and above the rest (FOMMCO, Speichim).

In the latest design of this type of extractor some manufacturers stack two rows of cells, one on top of the other, forming a "deep-bed". This positioning increases the volume and contact time, making it possible to use thicker and more stable flakes and to obtain highly concentrated full-miscella (FOMMCO, Extraktionstechnik).

The horizontal double-belt types of percolation extractor with moving cells are the longitudinal variants of some of the rotary types. A continuous series of cells (bucket frames) move in two tiers above two rows of perforated moving endless metal screen belts, which serve as false bottoms to the cells (Fig. 2.31). The cells are emptied at the end of the top row by overturning them, whereby the flakes fall into an empty cell at the start of the second row of cells and belts. In the upper row the relatively concentrated miscella is not transferred to the next stage, but recirculated within the same section in order to obtain equilibrium conditions. The lower row handles more exhausted material and works on the percolation principle shifting the flows countercurrently. Separate solvent and miscella spraying and collecting devices are fixed under the two separate rows of belts (Lurgi). Lurgi now produces, instead of moving screen belts the same cells (bucket frames) moving above two stationary (upper and lower) screen plates, as false bottoms.

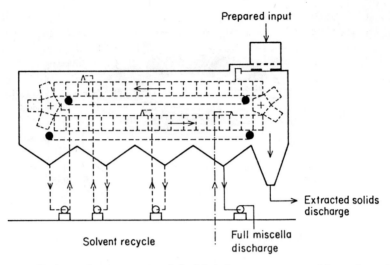

Fig. 2.31 Horizontal compartmented double-belt type extractor with moving cells. [Reproduced, with permission, from Hamm (1983).]

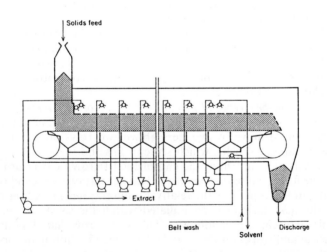

Fig. 2.32 Horizontal uncompartmented single-belt type extractor. [Reproduced, with permission, from Hamm (1983).]

The horizontal uncompartmented single-belt type of percolation extractor comprises a single horizontally moving perforated endless belt with no compartments which carries the flakes on its surface (Fig. 2.32). Rakes penetrate the upper 150 mm of the up to 2-m high flake bed to form temporary ridges at intervals (quasi cells) and to break up the upper layers of the feed. The belt may move discontinuously in order to provide discrete percolation and drainage periods (e.g. 6 min between two sprays), before the new spray of miscella of higher concentration is applied. Under the belt are troughs separated by ridges which serve as barriers for the backflow of the miscella. Each trough has its own pump, sprinkling the flakes on the belt continuously in a short circuit with the miscella transported and drained by the flakes and collected below this section of the belt. The outgoing flakes receive fresh solvent in front of the last trough; the last trough serves as recipient for the drained solvent, pumped (as in other systems) back to the previous section and/or over the incoming flakes. The flakes serve as a natural filter for clarifying the full-miscella, which leads to less sediment and thus less clogging (De Smet).

8. *Miscella Filtration*

Because of the quality criteria for crude oils and lecithin, but also to ensure the least possible fluid transport defects (clogging in pumps, pipes, etc.) and heat transfer resistances (crust forming on the packings, jacketed wall and pipe bundles of distillation units), the miscella must be freed of solid meal particles before proceeding to distillation or refining. We have already seen that the miscella of the original vertically built Bollmann-type extractors did not need heavy post-filtration, because of the "self-filtration" effect on the full miscella of the flakes in the cocurrent descending leg. In this system the closed streamline filter pots with filter cloth "bags" covering the onflow pipes were sufficient to remove the fines. If more filtration is required special closed filter-presses, tanks with vertical leaf filters and a conical bottom with a butterfly valve for discharging solids into a vacuum tank situated beneath it (Lochem), closed continuous rotating-screen filters, or Kelly-type filters without a filter aid can be used, or the more recently proposed sludge ejecting continuous centrifuges or even decanters.

In present-day practice mostly horizontal extractors are used which outlay a short bypass of the almost full-miscella in the cocurrent way and can lead to the required low fines content, an effect achieved previously only with the Bollmann-type extractors. In one execution the miscella collected in the third-to-last stages is added to the fresh flakes and the filtrate of the second-to-last stages which is discharged as the full miscella. This trick effectively solves the problem of heavy filtration, the miscella containing only mg kg^{-1} levels of fines. On-line bag filters, filters with replaceable paper filter cartridges, or closed continuous rotating-screen filters can all cope with these traces.

9. Solvent Recovery

(a) *Miscella distillation (evaporation)*. As has been mentioned above, besides the desolventizing/toasting/drying (DTD) process, distillation is the most energy consuming part of the total extraction process. With meals or seeds and a miscella, both containing 20% oil, the volume of distillate is equivalent to the volume of the raw material. Miscellas with a 35–60% oil content are, of course, more economical. Because of the large volumes of hexane to be evaporated, except in small factories, some form of steam-energy saving is an essential requirement. The heat of vaporization of hexane is about 80 kcal kg^{-1} (335 kJ kg^{-1}), its specific heat being 0.5 cal g^{-1} °C^{-1}.

In general the evaporation is executed in two or three stages, mostly in long-tube type evaporators with a vapour head. In the first pre-evaporation stage, the warming up of the filtered miscella to its boiling point and the evaporation of the main portion of solvent can be done by heating the solution indirectly in heat exchangers. Instead of fresh steam, this can be achieved by using the hexane vapour arising from the desolventizer/toaster, added on the shell side of the chest of a vertical-tube type evaporator, if the pre-evaporation is carried out under vacuum. The vapour from the shell goes via the so-called excess-vapour surface condenser to the solvent/water separator tank. The vapour distilled off goes via the primary surface condenser to the same hexane/water separation tank. The fresh hexane entering the extractor can also be heated by a part of the desolventizer/toaster vapours. Before entering the second or mid-evaporation stage, the pre-concentrated miscella of 65–85% oil content is heated to its boiling point by a flash-steam heat exchanger or by indirect heat exchange with stripped oil leaving the system for storage (Fig. 2.33). In the second stage the miscella is concentrated to 90–98% oil content.

Instead of these single vertical-tube evaporators, dual-effect vacuum pre-evaporators are offered by some manufacturers. The residual pressure (vacuum) at the various stages, depending on the cooling-water temperature (perhaps 27°C), may be about 350–500 mbar, resulting in a hexane solution with a boiling point of 45°C or lower. In the past, pot stills with condensers were used for the whole evaporation operation, stripping being done in a separate vessel.

Finally, the oil goes via a heat exchanger to the third stage, where the solvent residues are stripped out by live steam under vacuum (around 250 mbar residual pressure), as only (vacuum)/steam distillation sufficiently reduces the boiling point of the mixture (danger of overheating) and the residual hexane level of the oil close to or even below 500 mg kg^{-1}. The flash point of oil containing 500 mg kg^{-1} hexane is 160°C, that of 1000 mg kg^{-1} hexane 120°C, and that of hexane -26°C; the ignition temperature is 247°C. This stripper column might be provided with packings (e.g. Raschig rings) or bubble cap trays, "discs and doughnuts", and the recipient at the bottom with indirect heating pipes and sparge steam inlets. A drier column under

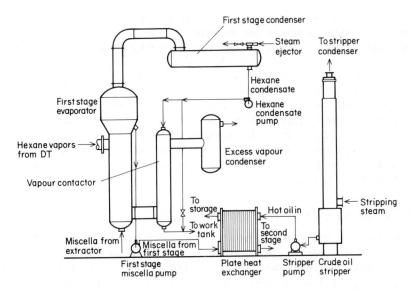

Fig. 2.33 Heat savings with a vapour contactor and a plate heat exchanger. [Reproduced, with permission, from Dada (1983).]

40 mbar residual pressure can be attached to the bottom of the stripper column in order to deliver a "dry" or moisture-free and subsequently cooled oil for storage. Two-stage oil strippers with air-cooled condensers are in use in countries where a shortage of cooling water might exist.

The vacuum is generally produced by one- or two-stage steam ejectors and/or by barometric condensers with mechanical displacement or rotary water-ring sealed pumps. The non-condensed vapours from the second mid-desolventizer stage might also be used, for example, the direct heating of the fresh solvent to be added to the oil-rich flakes. It is also possible to achieve contact with the vapour before the so-called excess vapour condenser of the system (Fig. 2.33). The steam from the vacuum ejectors is sometimes discharged to the deodorizer part of the meal desolventizer, to the waste-water boiler or to the base of the stripper column, instead of using fresh sparge steam.

If no condensation of steam occurs in the oil the crude oil contains all its phospholipids in the form of large micelles, i.e. as a colloid chemical solution called a "sol". The phospholipids are removed in a later step in either the oil mill or in the refinery (see section IV. C. 1 of this chapter on "Pre-desliming operations and lecithin production").

(b) *Desolventizing of the meal.* The extracted meal leaves the extractors with 30–35% of adhering solvent which must be removed; at the present time this is done mostly by evaporation. Any feasible pre-desolventization device would imply energy savings. Of course the horizontally built percolation extractors provide some degree of drainage before discharging the meal from the system, whereas the immersion type extractors generally use horizontal closed-screw-like pressing and conveying devices to remove part of this lean miscella mechanically. More drainage means more extracted oil and a lower residual oil content in the meal, which, however, is not always the final goal. Proposals of using solvent tight-screw presses, flaking rolls, Rotopress 400, or centrifugal decanters for pre-desolventization are all described in the literature. A practical method which has already been applied is the extension of the drainage surface, e.g. by an extra closed desolventizing conveyor outside the extractor. Another alternative is the extraction of the porous extrudates produced by a compactor which have better drainage properties.

10. *Conditioning of the Meal after Solvent Extraction (Toasting)*

In general soya beans, and sometimes even cotton seeds, are not prepressed mechanically. Therefore, apart from the drying, head-end dehulling and flaking operations soya beans do not undergo the extensive heat treatment in "cookers" or the heat developed by friction in the expellers. This lack of an extensive heat treatment is compensated for after solvent extraction by toasting.

Toasting is a time, heat and moisture-content dependent conditioning post-treatment of extracted meals, which may or may not contain the extraction solvent. Toasting before the meal drying/cooling step provides the possibility of influencing the protein dispersibility (PD) and nitrogen solubility (NS) of the proteins in water. The PD index (PDI) and NS index (NSI) are both quality measures of the meals to be used as animal or human protein sources.

The toasting of soya-bean meals is specifically necessary for the deactivation of the so-called trypsin inhibiting factor, a protein present in the crude meal. This factor lowers the direct utilization of the proteins present in this meal. A glucoprotein called haemagglutinin, a goitrogenic protein, and soyasapogenins which have a complicated glucoside structure and haemolytic (red blood cell hydrolysis) properties are also inactivated by toasting. Toasted meals are described in more detail in Section IV of this chapter.

The extent of the heat treatment of soya-bean meals can be monitored by measuring either the increasing cresol red-dye binding capacity of the treated meal (Olomucki and Bornstein, 1960) or for example, the decrease in urease enzyme activity after toasting [by the method of pH change as described by Caskey and Knapp Frances (1944) and modified by Bird *et al.*, (1947)]. A residence time of the meal for, e.g., 15 min under live steaming conditions at 17.5% moisture content of the meal reduces the urease activity to 0.1Δ pH unit, this being an acceptable extent of heat treatment (Fig. 2.34).

A separate toasting of soya-bean husks, containing about 10% protein, which have been removed by front-end dehulling is an absolute necessity, if

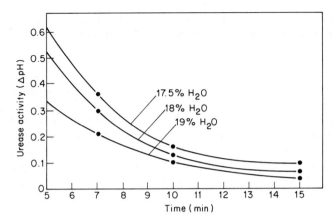

Fig. 2.34 Urease activity vs. residence time for different moisture contents. [Reproduced, with permission, from Lebrun et al. (1985).]

they are to be used for animal feed and not for burning, which would be a waste.

11. Meal Desolventizing/Toasting/Drying/Cooling Equipment

The extracted (soya bean) meal entering the toaster, with a 11/0.8 = 13.8% moisture level in the case of soya beans, has the temperature of the solvent (about 60°C) and contains about 30% of its weight as solvent (42 kg of solvent per 100 kg of meal + water). Because of the similar conditions applied, toasting is generally connected with the removal of the residual solvent. Originally, when batch "pot"-extractor batteries were used, desolventizing and thus "toasting" was carried out in the extractor pot itself, heating the contents through the jacketed walls by steam and also by introducing live steam through the meal until it became free of solvent at a final temperature of 120°C. Additional drying of the wet meal in order to achieve "safe" storage moisture levels (see below) was carried out separately.

(a) *Vertically stacked horizontal heating cylinders.* The introduction in the 1920s of the continuous types of extractors gave rise to the use of cylindrical jacketed and externally heated horizontal heaters with internal screw or ribbon conveyors, called "Schnecken" heaters. These conveyors were packed in banks of six or more, the lowest (the "deodorizer") having a higher internal diameter and an internal pipe bundle and flights and a sparge steam inlet at the bottom. The solvent-rich, wet meal entered the top conveyor and, as it had already been dried and thus warmed up on its way down, it could endure the countercurrent passage of live steam without condensation, the steam driving out the solvent into the higher compartments. Similar stacked heaters (without the live steam inlet) are still considered to be quite economical if used for pre-desolventizing the meal, reducing live-steam requirements of the post-desolventizing/toasting operation.

(b) *Vertical column heaters with stacked trays.* Instead of the horizontal desolventizer/toaster/driers (DTD) the use of steam-jacketed vertically stacked trays, working on the design principles of the previously mentioned "cookers", have found wider introduction, based mainly on U.S.A. practices. The first DT was developed by the French Oil Machinery Company and installed by Central Soya in 1950. Modern DTs have double bottoms for indirect heating, level regulation and compressed-air regulated meal valves. The principal difference between a DT and a cooker is that the

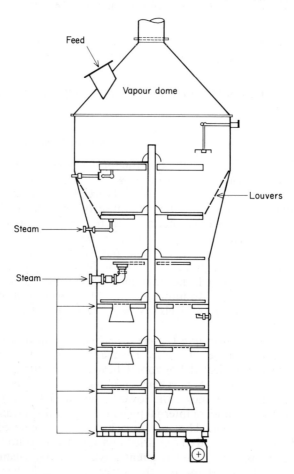

Fig. 2.35 Improved model desolventizer/toaster. [Reproduced, with permission, from Milligan (1976).]

former must be absolutely vapour tight. The headspace above the first tray may be conically expanded and is sometimes covered by a dome in order to reduce vapour velocities and entrainment of the fines. The evaporation is enhanced by the injection of live steam, by which about 98% of the solvent present in the meal is evaporated in this area. Live steam can also be added at the second and/or third trays and at the lowest, mostly the sixth, one. The upper trays are the desolventizers and the lower ones form the cooking zone. In order to obtain a uniform residual moisture distribution, thorough mixing is achieved by using sweeper-type stirrers. Some stirrers also act as live-steam distributors, having a hollow body with perforations for the steam inlet (Fig. 2.35). The non-condensed motive steam of the vacuum ejectors of the solvent distillation system can be used as a live-steam source, this being also a means of better heat economy. Toasters intended for producing light-coloured cotton seed meals, are constructed with much larger indirect heating surfaces, as the live-steam use must be reduced to a minimum to fulfil the colour norms.

(c) *Meal drying and cooling.* During toasting the meal has a temperature of 105–110°C and becomes wet due to the condensation of live steam (17.5–22%). The level of live steam can be reduced first to 13–15%, in order for the meal to be dried further to the final moisture content: 11–13% for soyabean, 8–10% for sunflower-seed and 6–8% for rape-seed meals. The drying is done by separate rotary pipe bundle driers and horizontal or vertical rotary coolers with cold-water cooling or with cold-air aspiration, lowering the temperature of the meal to 10°C above ambient temperature. Of course the pneumatic systems require adequate air filters. Combined vertical, cascade tray rotary air driers/coolers are also available (De Smet).

(d) *The desolventizer/toaster/drier/cooler system.* Steam savings can be achieved by providing more equal residence times for the meal in the different trays, by more indirectly heated drying surfaces, higher filling levels in the trays and the deletion of intermittent transport between desolventizer/toaster and the dryer and cooker sections. This is realized by a combined desolventizer/toaster/dryer/cooler (DTDC) system. Such a system is provided with perforated trays, which makes possible a countercurrent sparge steam addition in the lowest desolventizing/toasting tray, which is the fourth in the row. The fifth tray serves for drying the meal by compressed air at 50°C, and the sixth one for cooling the meal by air at ambient temperature (Fig 2.36). Special rotary valves extracting the meal along the whole radius of the tray are driven by hydraulic motors, their speed being controlled by a float in the upper part of the corresponding compartment (Schumacher design).

(e) *Flash desolventizers.* Equipment applying the so-called flash-desolventization of meals uses hexane vapour superheated up to 160°C in order to eliminate the residual hexane adhering to the meal: 1.5 kg of superheated hexane vaporizes 1 kg of liquid hexane. The flakes entering the continuous-

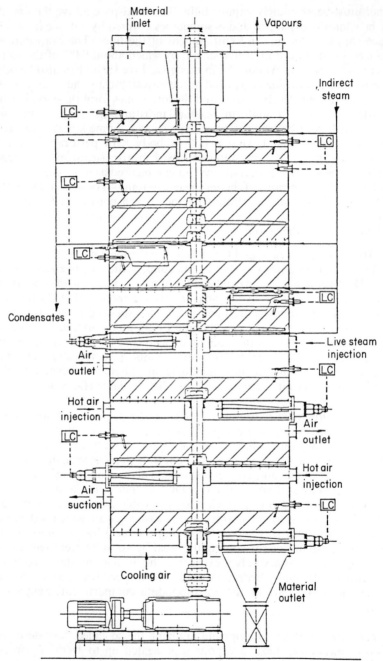

Fig. 2.36 Desolventizer/toaster/drier–cooler, Schumacher type. [Reproduced, with permission, from Lebrun *et al*. (1985).]

loop contacting tube (a.o. EMI, Extraktionstechnik) or the rotating contacting cage (Dravo) of the installations are fluidized in the saturated-vapour stream and have the temperature of the solvent (about 60°C). After a very short contact time with the hot hexane vapours, the flakes arrive at a maximum of 80–90°C. The solids are separated by cyclones and the solvent vapours are compressed by a ventillator. The flakes with 1–1.5% residual solvent content are then further treated in continuous live-steam stripper/deodorizers. Low jacket temperatures and superheated (non-condensing) live-steam treatment deliver light-coloured meals with high PDI and NSI. For this reason flash desolventizing is mainly used when producing edible proteins intended for human foodstuffs. Animal feeds with no urease activity are produced by treating the flash-desolventized meal in separate cookers, where wet live steam addition and high jacket temperatures produce meals of low urease activity but also of low(er) PDI and NSI. Drying/cooling finishes the operation. The hot hexane vapour desolventizating of flakes was occasionally used in the earlier pot-extraction period.

(f) *Fluidized-bed driers/coolers.* Fluidized bed (desolventized/toasted) meal drier/coolers consist of six consecutive compartments, in three or four of which hot air and in two or three of which cold fresh air is blown through a perforated bed, the whole system being inclined at about 6° to the horizontal (Fig. 2.37). The meal is introduced into the system by rotary air locks and

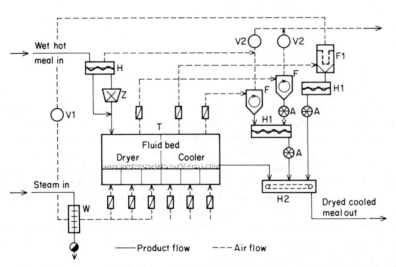

Fig. 2.37 Soya bean meal drier/cooler. T, Fluidized bed drier/cooler; V1, recirculating air fan; V2, exhaust air fan; W, air heater; H, sieve screw conveyor; Z, crusher; F, cyclone; F1, exhaust air filter; H1, collection screw conveyor; H2, finished-product conveyor; A, rotary valve. [Reproduced, with permission, from Florin and Bartesch (1983).]

discharged at the end by a discharge valve. These are the only moving parts of the system. The residual heat of the meal is also used for the cooling/drying process. Obviously, efficient dust collection is crucial to these systems (Buhler, Escher-Wyss).

The desolventizing/toasting operation, considering its high energy needs in steam and electricity and the possible solvent losses, is one of the most costly steps in the whole oil-processing sequence. These energy needs and the above-mentioned quality aspects can be balanced only by very careful design of the relevant heat- and mass-transfer units, with as much as possible steam and solvent recovery in different ways. As already mentioned, the steam/hexane vapours leaving the desolventizer/toasters can be re-used for, for example, the indirect heating/vacuum-mediated pre-evaporation of the hexane/oil mixture (full miscella) obtained during solvent extraction.

12. *Waste Recovery*

(a) *Vent-gas treatment.* Solvent losses of more than 0.1% from the incoming non-dehulled oil seeds are considered to be uneconomic today, although a loss of double this value is reported with very oil-rich seeds and under warm climatic conditions. The solvent loss on the meal can even be <0.1% in well-constructed desolventizer/toaster plants. In order to achieve low solvent losses, the solvent vapours from the extractor, miscella or solvent tanks and from the excess vapour condenser go via a water-cooled vent condenser to a cool mineral-oil rinsed absorber/scrubber in a counter-current fashion (Fig. 2.38). The starting temperature of the mineral oil rinsing is 15°C which is achieved by water or ammonia cooling. The column is filled with Raschig or Pall rings. The recommended capacity of such a system is 140 $m^3 h^{-1}$ of vent vapour per 1000 ton day^{-1} of plant capacity. The non-condensable gases are vented through a flame arrester to the atmosphere. The mineral oil saturated with hexane is stripped from the hulls at 110°C (<100°C under vacuum) with sparge steam and the hexane eliminated is recondensed in the vent condenser. A vent blower, connecting this condenser with the absorber/scrubber or steam ejectors, produces a slightly negative pressure in all the process equipment in order to direct all the air leaking in the system to the vent condenser already described. This air is cooled with the coldest water available; thus the vent condenser is placed first in the row of other water(air)-cooled devices. As 1 m^3 of air can carry 150 g of hexane vapour, the negative pressure is generally only a few millimetres of water column. Air condensers installed for first-stage shell-side evaporators and for hexane and waste water strippers vapours increase the economy and safety of the process.

A peculiar problem encountered with rape seeds and soya beans during solvent extraction is the formation of hydrogen sulphide (H_2S) gas from the sulphur containing seed components, such as the isothiocyanates, and/or the sulphur containing amino acids of seed proteins in general. Hydrogen

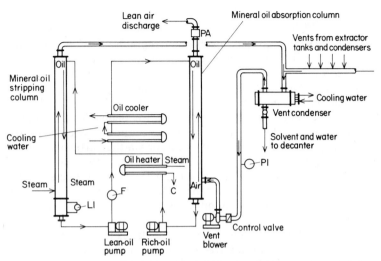

Fig. 2.38 Vent-gas recovery system. [Reproduced, with permission, from Milligan (1976).]

sulphide gas is liberated mainly during the contact of the meals with live steam in the desolventizer/toaster. The gas is easily oxidized via sulphur dioxide (SO_2) to sulphurous and sulphuric acids, which cause the corrosion of the vapour condensers and other metallic parts of the system. The result is the precipitation of ferrous sulphide in the form of a black powder in the mineral oil of the absorber/scrubber, reaching the pumps of the mineral-oil circulation system. The remedy is, for example, to wash the mineral oil with a 10% aqueous borax solution of pH > 9 before it is stripped from the hexane.

(b) *Water treatment*. The frequent mixing of live steam with solvent(s) necessitates the condensation, separation and re-use of both components as far as possible. Direct condensation has been described previously. Separation of the condensed mixture of hexane and waste water is executed in "florentine" or other types of closed gravity settlers, directing the lighter (upper) solvent layer back to the pure-solvent tank. The separated lower water layer is stripped by live steam for 5–10 min in order to remove any traces of hexane. The waste-water phase is led to separation swamps or containment basins, leaving a substantial time for decanting spilled light phases or oily miscella to settle within the factory. This process demands capacities of about 1.5 times the largest volume of hexane or miscella applied within the solvent extraction plant. The settled water having prescribed total fatty matter and chemical and/or biological oxygen demands (COD/BOD) can be led to the public sewer.

13. Solvent Losses

The non-condensable vent-gas losses may represent 15–25% of the total losses, being estimated at around 0.1% for extracted meal. Lower amounts are attributed to the oil (7%) and to the waste water (1%). If this is true then up to 20–40% of the losses are mechanical losses, such as those through seals, packing glands, gaskets, etc. These losses could be reduced by better (of course much more costly) equipment and/or better design. The critical places at which solvent losses should be closely monitored are: the headspace air above the fresh meal entering the extractor via the particular conveying/feeding device, the meal leaving the plant, the air from the vent, the finished oil from the oil stripper, the water leaving the waste-water stripper, the ambient air in general, and any hexane leakage (gas or liquid). The human nose is a quite sensitive hexane monitoring device, being able to detect levels of 500 ppm.

In a completely filled feeding conveyor the flakes form a plug seal which works as a good gasket for the extractor. However, this still does not exclude the problem of air entering the system. This air acts as a ballast which must be transported through the whole process and must be freed from hexane vapours before exiting to the atmosphere.

14. Safety

The ample dimensions of the main desolventizer, distillation condensers, of vent and ducting lines are a prerequisite for working without heat accumulation or too high pressures in the system. Admission of air to inflammable solvents above and under the critical explosion limits [in the case of hexane 1.2% for the lower and 7.2% for the upper limits by volume (3.4% and 18% respectively by weight)] should be prevented. Of course no direct flame or sparks, no generation of static electricity (e.g. from too fast pumping of solvent or miscella), and no heating of the components to their ignition temperatures are permitted. This is relevant also to the places and commodities (seeds, dried meals), where dust formation might cause explosions; these explosions being at least as severe as those with inflammable solvents. The lower and upper critical explosion limits for dust are 50 and 3000 g per l^3 of air. It should also be emphasized that dust is a health hazard.

Hexane may be stubbornly retained via chemisorption or dissolution by the residual phospholipids in the meal: at the beginning of storage extracted DTDC meals may retain up to 0.2% (2000 mg kg^{-1}) hexane. The measured hexane levels vary because of some sampling (solvent flashing) difficulties and the reproducibility problems of present-day analytical techniques. If after 1–2 days of storage 3/4 of this amount of hexane is released, the dangers of explosion, because of the low flash point of the solvent in the meal, become imminent. Better desolventizer designs should take care of this "solvent loss", amounting to 15–50% of the total losses. Real residual levels of 500–1000 mg kg^{-1} hexane are more safe, whereas with the more costly latest designs of plant a residual hexane level of 0.03% (300 mg kg^{-1}) can be achieved. Some additional measures, like headspace atmosphere dilution during the discharge of stored meals from silo cells, are described in Section IV of this chapter.

The start-up and shut-down periods of the process are serious hazard occasions, wherefore, for example, purging with hot hexane vapours instead of inert gases or live steam (U.S.A. practice) seems to be preferable.

Hazard areas and codes of prevention practice are well publicized and are generally based on the IEC 79 recommendations. The U.S.A. conditions are given in the *Regulations for Solvent Extraction Plants* (National Fire Protection Association, booklet no. 36, revised many times).

IV. Main and Side Products of Oil-milling Operations

A. Crude Oils

Three types of crude oil are regularly obtained from milling and extraction: (i) undegummed mechanically pre- and/or end-pressed oils; (ii) undegummed solvent-extracted oils; and (iii) mixtures of the above two oils.

The main differences between the three types of oil are in their colour, their phosphoacylglycerol (glycerophospholipid) content and the types represented, their free acidity (FFA) and their moisture, impurities and unsaponifiable-matter (MIU) levels. Pre-pressed oils generally contain less phospholipids and have a lighter colour than do the extracted ones. If their free acidity is low, pre-pressed oils might be sold as virgin oils, provided that the odour and taste correspond to the locally preferred taste pattern. Although it is not customary, the hydroperoxide content and the UV extinctions ($E_{1\,cm}^{1\%}$ values at 232 and 268 nm) of pressed oils can be used as characteristic measures of their pre-oxidation history. The oxidation level of (pre-)pressed oils might be higher than that of extracted oils, due to the direct air contact at the relatively high temperatures applied during the processing of the former. The colour-change problems may be caused by excessive temperatures, e.g. during cooking/conditioning and hexane stripping, or by unnecessarily long heating times during delivery into or out of the storage facilities and tank yards.

Before storage the mechanically pressed oils must be freed of the coarse drab by vibrator screens and/or from foots by sedimentation in screening tanks. The screening tank settles the foots in a quiet zone of the tank and removes them by means of buckets or paddles attached to a chain conveyor, whereas the oil is pumped out for further clarification (Fig. 2.39). Filtration by plate filter presses or other conventional filters demands much manual labour, the modern automated metal-gauze based leaf filters being the more economical. Vertical sludge-ejecting (nozzle) centrifuges or horizontally mounted desludging centrifuges, called decanters, might also be used. The latter, having characteristic cylindrical/conical bowls and an internal scroll

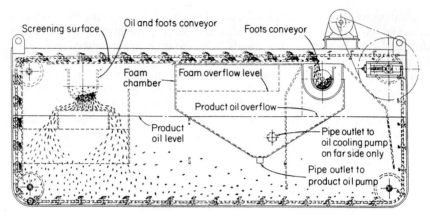

Fig. 2.39 Oil-screening tank. [Reproduced, with permission, from Latondress (1983).]

for removing the sludges, are more often used for the purification of animal- and plant-tissue (see Chapter 3). However, centrifuges cannot remove the very finest sludges, a 0.5–1% residue being quite normal. The oils are dried, if necessary, in vacuum kettles at 80–90°C and 80 mbar residual pressure produced by, for example, water-ring pumps.

The drab and sludges separated from pressed oils can be mixed with the pre-pressed and disintegrated cake intended for extraction. Mixing can be done continuously and directly from the vibrator screen or screening tank, directing the drab to the cake. Another option is to collect the drab and sludges from the decanter and add them to the cake in special animal-feed mixers, which have efficient mixers in the form of horizontal beaters, double helixes or some other type.

B. Meals and Cakes

As we have already mentioned, extracted meals are used extensively as valuable animal feeds. In particular, soya bean processing is aimed primarily at the protein and not at the oil content of this seed, which is graded and sold as grain under the provisions of the United States Grain Standards Act. After physical and/or chemical enrichment to different protein levels, meals also form the basis of various human food products, for example as texturized vegetable proteins. In both cases the simultaneous desolventizing and toasting of the meal determines the final quality of the product. The details of these operations have already been discussed in Sections III.B.9.(b) and III.B.10.

The heating of the extracted meals in the live-steam created moist atmosphere for a longer time, involved in both operations, might cause changes in the quality of the proteins. Such changes are:

2. VEGETABLE-OIL SEPARATION TECHNOLOGY

(i) the breaking of existing "natural" bonds between the helical polypeptide chains, the so-called "tertiary structure" of proteins. This may expose originally buried "primary structure" elements (the amino acid sequence in the chains) for better access by gastric juices. These changes and the "unwinding" of folded helical threads which form the "secondary structure", are also known as denaturation. Denaturation is not detrimental to the nutritional value of the proteins.

(ii) Maillard type reactions, the condensation of the free ε-amino groups of certain amino acids, particularly of lysine, with the carbonyl groups of reducing sugars cause the non-enzymatic browning of the meals.

(iii) the actual destruction of "essential" and other amino acids.

1. *Protein Dispersibility and Nitrogen Solubility Indexes as Primary Quality Measures*

The protein dispersibility and nitrogen solubility in water (both expressed in percentages) are the two factors, which (analysed by standard methods) determine the practical applicability of meals to human food applications. Protein dispersibility index (PDI) and the nitrogen solubility index (NSI) values of 10–25% are considered to be low, of 40–60% to be medium, and of 75–90 to be high.

The animal-feed quality of the meal depends primarily on the intended levels ($>44\%$ or $>49\%$) of total proteins ($N\% \times 6.25$ or 5.46) and oil content (it is not indifferent, whether the latter is determined by hexane or by diethylether, as solvent). In addition, the amount of ballast material ("crude fibre") must be closely scrutinized. This ballast consists mainly of cellulose, other polysaccharides, hemicelluloses and lignin residues which do not reside solely in the hulls or husks. Monogastric animals can endure only a limited amount of fibre, generally $<10\%$, in contrast with ruminants which can effectively decompose a great part of fibre.

2. *Antinutrients*

The customary low PDI and NSI levels of 20–40% are a compromise to be accepted because of the necessity of heat treatment in most cases. The destruction of "antinutrients" (adverse substances from the point of view of feeding) depends, to different extents, on the duration, temperature and the moisture levels during the heat treatment of the meals. Fortunately, toxic or retarding proteins, such as the trypsin hydrolysis inhibiting factor, haemagglutinine (a goitrogenic factor) and other toxic substances like soyasapogenins (in the case of soya beans) or gossypol (in glanded cotton seeds), are generally more quickly inactivated in moist heat than proteins are totally denaturated/decomposed/insolubilized. This is also the case with the inactivation of lipase, lipoxygenase and urease enzymes, the destruction of the latter being a measure of the excess of heat treatments in soya bean meals.

The special attention given to the careful deactivation of the trypsin inhibiting factor during the toasting of soya-bean meals needs some explanation. This protein, which is present in the crude meal, paralyses the hydrolysis of the amino acids lysine and arginine by the proteolytic enzyme trypsin in the gastric tract. Lysine belongs to the group of essential amino acids which must be introduced into the feed because animals are unable to synthesize them. Soya-bean meal proteins contain substantial amounts of lysine, but are deficient in methionine, another essential amino acid.

The protein levels (scores) in feed (food) proteins as analysed chemically are not equivalent to their nutritive value which depends, among other things, on the type, concentration and sequence of the component amino acids, being a characteristic property of the type of feed proteins. The synthesis of tissue proteins serving for the growth and health of the animal is governed by the availability (presence and a given minimum concentration) of certain essential amino acid(s). Inadequate availability of one or more of these "limiting" amino acid(s) means that the other (essential) amino acids present cannot be used for tissue protein synthesis; they are only metabolized for energy.

The active trypsin inhibiting factor renders lysine unavailable for resorption and so this factor becomes (besides methionine) limiting for the other amino acids. This defect is also experienced with mixed animal feeds, containing non-toasted soya-bean meals. Mixing meals which are deficient in different essential amino acids, but which complement each other by containing the missing ones, the resultant nutritive value can be much improved: for example, by mixing toasted soya-bean meal with sunflower seed meal (rich in lysine, but deficient in methionine).

3. *Aflatoxins*

Primarily ground-nuts, but also other seeds, may be infected by the mould *Aspergillus flavus* which produces the carcinogenic protein aflatoxin. To remove the mould the heat-treated meals should be post-treated, e.g. by solvents such as methanol or ethanol, or by ammonia. When using ammonia gas the meals are kept for at least 20 min under 1–3 bar in jacketed batch autoclaves of 16-m length. The temperature is 90–95°C, the moisture content of the meal 15–17% and the mass (5–5 ton) is mixed by anchor-type mixers (60–110 rpm). After pressure is released the ammonia residues are removed by vacuum. In this way 95% of the aflatoxin is destroyed, but both nitrogen solubility and the available essential amino acid content is reduced.

All the above-mentioned processes are heat, moisture and time dependent, and many compromises are necessary in order to obtain a safe balance between "nutritional value" and "antinutrients".

4. *Storage of Cakes and Meals*

The ground, sifted meals are generally stored in silos. The admissible storage heights depend on the apparent specific density and the cohesion of the meals, which determine the pressure at the bottom and at the walls of the silos. Soya-bean meals of 12% humidity and sunflower-seed and rape-seed meals of 9% humidity can be stored at heights of up to 50 m, whereas ground-nut meals of 7% humidity are better stored at maximum heights of 20 m.

A problem encountered during the transport of meals before and after storage is dust formation. Dust filters working with slight aspiration should be mounted above the points, where the free-falling meal (ex chutes) enters the means of transport at angles of 80° to the horizontal. Free storage of meals in piles, or better packed in bags on the flat floors of the warehouses, is possible, but in any case there should be a thorough control of temperature, moisture content and hexane release (< 500 mg kg^{-1} being considered safe). The corrective measures for the temperature and moisture content are circulation, if necessary after drying and/or cooling. Residual and released hexane vapours can be removed in a closed horizontal conveyor by diluting the headspace above the moving meal with a countercurrent stream of fresh air. The dust and hexane aspirated by the air is removed via vertical dust filters, adding the filtered solids back to the meal in the conveyor and transferring the hexane vapours to the atmosphere.

Full-(end-)pressed oil-seed cakes (cake II), after having been ground by disintegrators and sifted to a meal, contain about 3–8% residual oil. They can be used (except aflatoxin) safely for animal feed, because the heat treatment applied to them before and during pressing changes only slightly the availability of the different amino aids, whereas the PDI and NSI are not primary measures of their quality. Because there is no dust formation there is little or no trouble with storing cakes and pelletized meals free in piles or packaged in sacks.

C. Lecithin Production

1. *Pre-degumming of Oils*

Hot, filtered, extracted oils, as obtained after a careful stripping of the last traces of hexane, contain, besides free fatty acids, partial acylglycerols, the normal accompanying lipid components and some contamination, and the phosphoacylglycerols, commercially called "lecithin".

The main source of lecithin is soya-bean oil, although other oils, like crude cotton-seed, rape and sunflower-seed oils, also contain more-or-less similar amounts of lecithin. Soya bean lecithin is most commonly used because, apart from its high concentration in soya beans, it has a better taste than that from other sources. This latter fact is obvious because the phosphoacylglycerols of this oil have suffered less heat damage than those of other oils, having been isolated from oil seeds which were not only extracted, but also mechanically pre-pressed.

Lecithin has a very broad field of commercial use: as a food emulsifier, antispattering (margarine) agent, wetting and dispersing agent, viscosity modifying agent (chocolate), (pan) releasing/phase parting agent and as a dietary supplement. It also has other, non-food applications.

Lecithin should be separated from the crude oil as early as possible for a number of reasons (apart from its many uses): (i) oils with reduced lecithin content form less sediment during transport and storage (self-hydration by moist air in the headspace); (ii) oils freed of lecithin cause less "refining losses" during alkali refining. If present, their co-emulsifying effect increases neutral oil inclusion in the "soap stock"; and (iii) the surface water pollution caused by acid split soap stocks is decreased, because the stubborn emulsions formed by partly decomposed lecithins, accumulated steroids and sugars in the interphase layers of this process are evaded.

(a) *Water-only process*. The expressions "degumming" or "desliming" are traditional misnomers for the removal of the phosphoacylglycerols or "lecithins" from the oil. Strictly speaking, gums and slimes are polyglucuronic acids which are generally not present in oils and fats.

Lecithins are separated from oils by their affinity to water, this rendering them oil "insoluble". The process is called hydra(ta)tion. The success and completeness of the hydration process can be explained by the differences in the surface-tension reducing ability of the component phosphoacylglycerols and by this their specific inclination to stabilize given types of emulsion (see Chapter 4). Therefore, phosphatidyl choline (PC) stabilizes oil-in-water emulsions, whereas phosphatidyl ethanolamine (PE) and phosphatidyl inositol (PI) favour water-in-oil emulsions. The type of emulsion formed also depends on the oil/water ratio and the temperature of the system. The phosphoacylglycerol molecules form a colloid solution with the oil, forming large molecular aggregates called "micelles" Micelles are globular structures in which the hydrophobic acyl chains point outward and the hydrophilic phosphoric acid esters inward. The added polar water molecules have an affinity via their polar groups to penetrate inward and form a monolayer on the micelles. An inversion then immediately takes place, the hydrophylic parts with the water molecules turning outward. The hydrophobic regions are now facing inward and they include some oil. Because of the general water/oil incompatibility and the preference of PC for stabilizing oil-in-water type emulsions, the phases separate. The flocks (a so-called "lyotropic mesophase"; see Chapter 4) have a (liquid) crystalline structure which is lamellar at room temperature and cubic or hexagonal at higher temperatures. Native or partially hydrolysed phosphoacylglycerols, like PA and part of the PE, if already reacted with the calcium and magnesium ions present either in the cells or in carelessly used hard water, are excluded from this hydration process. Together with close neighbour native phosphoacylglycerols the partially hydrolysed ones are rendered non-hydratable and are not or incompletely removed by this "desliming" or "degumming" operation.

In the actual hydration process an amount of hot (preferably demineralized) water equivalent to the lecithin content (1–3%) is added to the hot oil. If necessary, the oil is reheated to around 80°C. This addition is done in a steam-jacketed batch kettle with a conical bottom, where the two phases are mixed effectively by gate-type mechanical stirrers. Instead of water, wet steam can be introduced, which on condensing in the oil heats and mixes it. The dispersed water starts to become opaque and flocks of white "crystals" separate after 10–15 min of batch mixing. After 20–30 min of treatment the flocks are either left to settle out, their specific density being close to that of water, or centrifuged. In a variation of this method the flocking and separation occur after 1 min, if the hot water is added to the hot oil in a knife mixer, a static mixer or similar. Finally, the slurry is kept in dwell mixers for 10–15 min after which the hydrated oil is centrifuged.

Care should be taken to avoid the development of high shear stresses during pumping of the slurry in order to avoid the formation of stubborn emulsions. Either vertical sludge-ejecting (nozzle) centrifuges or horizontal decanters are used. The sludge obtained contains about 40–50% water, and the average pre-degummed crude oil contains about 0.5% (0.2–0.8%) residual phosphoacylglycerols.

The separated water phase is dehydrated under vacuum in batch or continuous working, preferably in horizontal film evaporators (Fig. 2.40), after which the lecithin is cooled and packed. The curious viscosity increase and culmination at about 7% moisture content causes difficulties in vertical film evaporators, the film breaking easily between 5 and 15% water content of the mass.

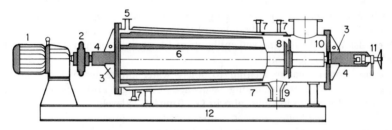

Fig. 2.40 Horizontal rotating thin-layer evaporator for lecithin. 1, Power drive; 2, clutch; 3, sealing; 4, socket; 5, inlet; 6, rotor; 7, heating medium; 8, separator; 9, outlet; 10, vapour space; 11, setting gear; 12, frame. [Reproduced, with permission, from Pardun (1983).]

The final product contains about 60% lecithin (acetone insoluble) besides some residual water, the rest being crude soya-bean oil which contains all the components originally present and enriched amounts of some decomposed sugars derived from sterol glucosides.

Soya-bean oil is actually pre-degummed by this "water only" process, the original phosphorus content of 800–1000 mg kg^{-1} being reduced to 150–250 mg kg^{-1}, in rape-seed oil the phosphorus content is reduced from 500 mg kg^{-1} to 250 mg kg^{-1}. The pre-degummed oil is dried under vacuum and cooled before storage.

(b) *Special degumming processes.* The presence of the varying amounts of non-hydratable phosphatides is the cause of the incompleteness of the "water only" pre-degumming process. The yield of lecithins can be improved by all means by which either the hydrolitic and/or oxidative damage of phospholipids before or after extraction can be counteracted (time of bean harvest, storage, drying, processing, and, mainly, the moisture/heat/time relationships during DTDC) or the hydratability reducing effect of calcium or magnesium salts or complexes can be counteracted. These actions do not necessarily improve the quality of the separated product, but might be useful from the point of view of the refiner for the reasons given above.

The addition of acids [organic acids such as acetic acid anhydride (the Staley process) or citric acid (the "superdegumming" reagent), or lactic acid, etc.; inorganic acids, mostly phosphoric acid of edible quality] at 60–70°C, besides the necessary amounts (about 1.5%) of water, causes the decomposition of metal ion salts/complexes and the formation of adducts with polar groups or even reacts with them (acetylated PE of the Staley process), all of which reduce the residual phosphoacylglycerols content. It was established 50 years ago by Russian workers that 25°C is the "critical temperature" of lecithin/water agglomeration and is thus an optimum. Therefore, if acid addition is made at this optimum temperature, and the mixture is then heated up to an intermediate temperature, in order to reduce the oil viscosity during centrifugation of the separated lecithin, low residual levels of phosphorus ($7–30$ mg kg^{-1} levels are cited) can be achieved. This temperature profile coupled with citric acid as reagent is called the "superdegumming process" (Ringers et al., 1977). The Staley process gives 10 mg kg^{1-} phosphorus level without the need of the special temperature sequence. Acetylated PE promotes the formation of oil-in-water type emulsions, reducing the susceptibility of PE against calcium and magnesium salts present in hard(er) water.

The addition of native lecithin to solely water pre-degummed oils and the execution of the "superdegumming" sequence on it might also reduce the lecithin content of these oils to low phosphorus (< 30 mg kg^{-1}) levels.

In the case of the Staley and the "superdegumming" processes, the oils obtained are easily subjected to alkali refining, without the known difficulties of soapstock removal, high losses and heavy surface-pollution problems (lower total fatty matter and COD). In certain cases the oils can even be refined physically, if the residual phosphorus is reduced to at least below 3 mg kg^{-1} before deodorization. The unremoved oxidized glycerides, which are removed in an alkali deacidification step with the soapstock, must be destroyed or transformed and partly eliminated by bleaching. The flavour stability of solely physically refined oils depends on, besides the starting crude-oil quality (oxidative damage), the removal of phophoacylglycerols and oxidized compounds. The final oil quality (flavour acceptability) is determined solely by the local taste patterns which vary strongly from place to place.

Under the name of Alcon a process has been introduced as a means of arriving at very low phosphorus levels ($10–17$ mg kg^{-1}) after a simple water-only degumming of a conventionally extracted soya-bean oil (Kock, 1983). The method involves taking special measures before the extraction of the flakes: these are extra moistening/heating (conditioning) plus extra drying/cooling of the normally prepared flakes. This is, of course, an additional operation which does not relieve the process of the final TCDC operation which must still be done on the extracted meals. The yield of hydratable lecithins and the proportion of PC in extracted meals is increased. An explanation for this finding is supposed to be the inactivation of phospholipase (?) and lipoxigenase enzymes before extraction, the treatment being executed by moist heating of the flakes for a certain time. Phospholipase action should be the reason for the increased levels of non-hydratable phosphatides (NHP) in normally extracted flakes, the lower yields of phosphatides being caused by their autoxidation (lipoxigenase) and the binding of the resulting peroxides by the meal proteins. There are problems with higher residual oil levels in the meal, which can be solved by deep-bed extraction, and the initial higher investment costs, due to the additional conditioning and drying/cooling machinery, needed, which must be taken into consideration. The advantages of this method are the easy hydratability of the oil and because of the removal of most of the phospholipids, the direct applicability of the oil to physical refining. Like the Staley or "superdegumming" processes there are less problems with surface water pollution in the case of subsequent alkali refining.

2. Commercial Lecithin Preparations

The composition of a commercial soya lecithin is approximately:

(i) phosphatidyl choline (PC, the real lecithin) $c.$ 20%;
(ii) phosphatidyl ethanolamine (PE, cephaline) $c.$ 20%;
(iii) phosphatidyl inositol (PI) $c.$ 10%;
(iv) phosphatidic acids (PA) $c.$ 10%;
(v) phosphatidyl serine (PS) $c.$ 2%;
(vi) carbohydrates and steroids $c.$ 3%; and
(vii) oil, including additives, if any, $c.$ 35%; and low levels of moisture.

Commercial lecithins, whether from soya beans or other sources, may contain fluidizing agents, like fatty acids and calcium chloride (fluidized lecithin), different food additives, mostly fatty acid esters or derivatives (compounded lecithins), may be de-oiled (acetone extracted), fractionated, partially hydrolysed or bleached (van Nieuwenhuyzen, 1976; Szuhaj, 1983).

The solubility of PC in 90% ethanol makes it possible to produce a PC enriched extract, in which the PC:PE ratio is >5:1. The removal of PE reduces the calcium and magnesium salt sensitivity of the concentrate, which makes possible the use of the PC enriched product in saltless "health" margarines as an antispattering agent. PE, but also PA, are very sensitive to the hardness of the water with which a margarine is produced. Sodium chloride ions protect PE against the destabilizing action exerted by calcium and magnesium salts. In the absence of sodium chloride, as in most "health" margarines, the calcium and magnesium ions of hard water react with PE and PA when heated, causing them to flocculate. This flocculation does not only mean that PE and PA are inactivated as emulsifiers, but also that a drab sediment is formed when a "health" margarine is used for frying.

Another method for increasing the oil-in-water type emulsion stabilizing action of lecithins utilizes their selective and partial hydrolysis by the enzyme phospholipase A. This enzyme removes the β-hydroxyl substituting acyl (fatty acid) chain and forms a "lyso-lecithin".

The colour, an important quality index, may be regulated by very careful addition of hydrogen peroxide solution, which also acts as a bactericidic agent during the storage of the separated aqueous sludge, before its dehydration. Unduly high dosages of hydrogen peroxide cause the oxidation of the colour-giving compounds and the unsaturated fatty acids, which leads to the undesirable development of an off-flavour (soapyness).

3

Water- and Heat-promoted Fat Separation from Animal and Plant "Fatty Tissues"

There are many operational similarities in separating fats (oils) from high moisture level animal or plant "fatty tissue". Because of the aqueous environment of fatty tissue, all the processes discussed in this Chapter are based on the water-promoted separation of fat from its natural enclosure, the cells. Apart from modern olive-oil processing, all the processes traditionally use heat. The process which is applied specifically in the animal-fat industry is called "rendering". A general survey (principal lines) of the fat-separation processes discussed here is given in Fig. 3.1.

I. Theoretical Background

A. Selective Wetting of Surfaces

The common feature of all these fat-separation processes is the theory of "selective wetting of surfaces" proposed by Rehbinder (1933). This theory explains the displacement of fat from its high-moisture surroundings solely by water, wherein the higher temperatures decrease the viscosity of the fat and promote the denaturation/coagulation of the proteinaceous surroundings (see below). According to this theory, if two immiscible liquids compete for wetting of the same surface, the one having the lower surface tension will preferentially wet it. The polarities of the components of the system fit the well-known polarity consideration popularly expressed as "polar seeks polar". Therefore it is obvious that above a certain optimum temperature the polar water will selectively wet the polar protein/carbohydrate structures thereby disrupting the weak bonds between them and the apolar fat molecules. Water thus displaces the fat, which coalesces to a continuous fat phase, leaving moist, swollen protein/carbohydrate residues.

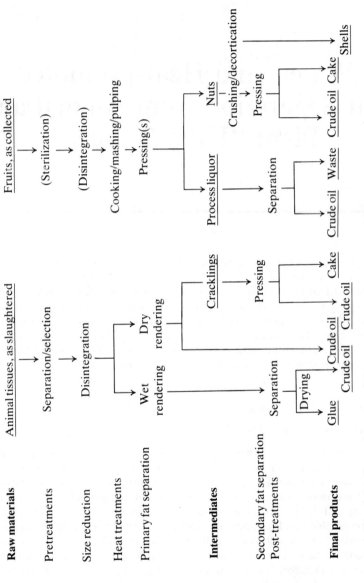

Fig. 3.1 General survey of fat-separation sequences from animal and plant fatty tissues.

3. WATER- AND HEAT-PROMOTED FAT SEPARATION

1. *Animal Fatty Tissues*

Animal fats reside in the white adipose (fatty) tissue in polygon- or round-shaped cells, the nuclei of which are strongly pushed away to one side. The fat is located in the cell in one single globule (unilocular cell). The cell has no wall, but only a surrounding membrane, consisting of glucolipids and phospholipids. The membrane may be covered by some mucous polysaccharide material. The connective tissue between the animal-fat cells is composed of inelastic collagen and elastic elastin, both of which are insoluble fibrous proteins. A large part of the cells consists of water which contains soluble organic components.

If a preliminary (generally mechanical) size reduction disrupts a part of the cells and the tissues are then heated to, say, above $60°C$ (with or without the addition of water), the proteinaceous material is denatured and coagulates, i.e. an irreversible change occurs in the chemical and/or fine structure of the cell proteins. At these temperatures the fat is in the liquid state, has a lower viscosity, and expands together with the water, which causes a quasi-spontaneous deformation/rupture of the non-cell-wall protected animal cells. The separation of fat can be explained by the theory of selective wetting of surfaces as described above.

A new process, Elcrack (Krupp), claims to distort the original (polarity) conditions by developing a polarizing electrical field around the previously disrupted membrane structures of steam-heated, ground animal fatty tissue (bones, oily fish). The field is strong enough to liberate the weakly bound fat for subsequent mechanical pressing.

It is worth noting here that ultrasound devices, which were designed with the aim of invoking microphysical changes at the cell level, have found no practical application in fat-separation processes.

2. *"Fatty Tissues" in Plant Material*

The fat in plant cells generally resides in the peripheral parts of the cytoplasm in small globules. Besides a cell membrane plant cells also have a cellulose and hemicellulose based second wall. Because of their high water content, the fruit fats of palm and olive mesocarp leave their surroundings during pressing or centrifugation more easily if contacted/heated with (more) water during mashing/pulping. When removing fats from olives heating is out of the question as it causes hydrolitic damage (high free fatty acid levels) to the oil, thereby decreasing the oil quality.

The situation is more complex when considering primarily water- and heat-induced fat-separation processes applied to crushed seeds of high oil, but low water content. The theory of selective wetting of surfaces can explain the displacement of fat by water, being the basis of the oil-separation methods invented by Skipin and by Iljin. These methods were described more extensively by Goldowskij and were used from the 1920s in sunflower seed, cotton seed and linseed processing in the U.S.S.R. for some decades (Heublum and Japhe, 1936).

B. Practical Application for Seed-oil Processing, a Sideline

1. *The Skipin Oil-separation Method by Water Displacement*

The Skipin oil-separation method involves adding 20–25% water to crushed (decorticated) seed meal at the first (top) level of a five-high stacked cooker. This top tray has a double bottom and is heated to 60°C under heavy stirring by special stirrer arms. The water is taken up by the meal, about 70% of the original fat being displaced and let off through the upper perforated (screen) part above the second solid bottom part. The remaining crumbs, containing about 30% water, are dried in four stages and then pressed mechanically.

2. *The Iljin Oil-separation Method by Selective Wetting*

The Iljin oil-separation method is carried out in a five-high stacked cooker, in which wet live steam is condensed on the colder surface of a crushed meal, previously dried to a 3% moisture content. The meal sucks up the water and is thus wetted by it, raising its moisture content in the top stage to 8–8.5%. The other stages in the process serve to provide uniform conditioning. The meal is pressed mechanically at about 25 bar (instead of the customary 200–400 bar), the low pressure displacing about 80% of the original oil content of the meal.

Both the Skipin and Iljin methods were used when there was a dearth of mechanical presses and solvent extractors in the U.S.S.R. for seeds processing.

II. Edible-fat Processing (Rendering) from Terrestrial Animals

A. Raw Materials for Lards, Tallows and Bone Fats

The supporting and protective tissues of freshly slaughtered swine, cattle and sheep in good health, contain in their cells quite high amounts of fat; these are the main sources of edible animal fat. Some of these fatty tissues, the killing fats (those around the kidney, heart, caul and intestines), are collected immediately after the cleaning, skinning, etc., of the animals. These tissues can be sent for rendering quite quickly, and yield a higher quality fat. The fats removed after dismembering of the chilled animal carcass are the skeletal fats (leaf, back fat) and the trimmings of ham and shoulder, called cutting fats. Chilling to below 2°C is necessary for the separation of the trimmings from the carcasses and, because of the longer time needed for this process, a certain amount of damage occurs to the fatty tissues. Bones containing fat in the marrow are collected separately for rendering. The killing, leaf and back fats have a high fat content, say 60–95%, whereas trimmings and especially the bone fats may contain much lower levels, around or below 20%.

The selection of raw materials in slaughterhouses is the veterinary inspection, whereas classification is the sorting of fatty raw material into fats of similar anatomical origin. Selection and classification make it possible to produce more uniform quality, as killing fats are harder, less unsaturated (lower iodine value), than cutting fats (leaf, back fat), which are more unsaturated (higher iodine value). The fat hardness also depends on the feeding pattern of the animals.

B. Transport

Living animals are transported by rail or road. The animals are weighed on railroad-track or truck-and-trailer type (automatic) scales or smaller scales for single animals or parts.

The internal transport of fatty tissues from the slaughter houses to the fat-producing department can be executed by hooks, fixed onto a continuously moving chain, the big chunks of bacon and fatty tissue being hung on the hooks, by small carriages or by any of the other transport methods we have already described when discussing oil-milling technology.

C. Pretreatment

Cleaning of the fatty animal tissues is carried out in open stainless-steel, concrete or wooden troughs by cold water. Washing not only removes dirt, blood and foreign material, but also precools the tissues to the water temperature. All the washings are accompanied by fat losses (1%) to the washing water, which must be recovered by fat traps.

Considering the enormous size of the fatty tissues separated, for example that from whales or even the back fat or killing fat of bovine animals, coarse mechanical size reduction is necessary because it cannot be handled unless it is cut into smaller pieces. Heat transfer is also much better to smaller pieces, reducing the time needed by the raw material to attain high(er) temperatures. Coarse cutting is done by machines with parallel- and vertically-mounted knives on a horizontal axis, slicing the material into pieces of suitable size for the cooling and/or rendering operations.

After the coarse cutting, the tissues are cooled (if not immediately rendered) by immersing them in water at 0–5°C for 6–8 h; the water is cooled by circulating it around the expansion elements of a cooling aggregate. Cooling/dripping out onto trays (racks) in separate cooling rooms or cooling towers by cold air (0–5°C) is very advantageous, because not only is the temperature kept low, but the wet material loses moisture, which after washing normally amounts to 7–24% (the latter figure is relevant to fats around the intestines).

Fine size reduction of fat is done immediately before rendering by means of wolves which are large industrial meat grinders. (It should be remembered that fat rendering was originally a small-scale household operation.) The cutting knives are fixed at the end of a horizontally mounted continuous screw (transport helix) and work against one or two perforated orifices, the diameter of the perforation determining the size and thus the primary yield of fat rendering (a 5-mm diameter leads to a better yield than a 20-mm diameter). Both the rendering process (heat transfer, more uniform filling of space) and the secondary fat yield on the subsequent mechanical pressing of the cracklings is promoted by fine cutting.

D. Rendering Processes

The rendering of terrestial-animal fats by heat or heat and added water (live steam) is based on the combined action of water and heat. The swollen proteins coagulate to cracklings (30% or more water) and the proteinaceous fibrous tissue partly decomposes to gelatin (glue) in the process liquor. Similar considerations are also relevant to marine-fat separations. The cooking of oily fish delivers a more dense process liquor, from which the fat is removed by mechanical pressing. The systems can be classified as batch or continuous and the principal methods as dry- and wet-rendering processes, either of which can be carried out in open or in closed vessels.

1. Small-scale Batch Operations

Open-kettle dry rendering is done in stirred, steam-jacketed kettles, preferably using high fat content tissues. This operation generally consists of two steps: in the first step the water present in the tissues is removed, and in the second the cracklings formed are dried to a 10% residual moisture content. The maximum allowable temperature is 115°C. The fat is separated from the cracklings in a strainer box, and is then filtered through cloth and sent for storage or distribution without any further treatment. The cracklings can be pressed and/or directly consumed.

Dry drip rendering obtained its name because in this process the fatty tissue is placed in an apparatus similar to an old-fashioned drip coffee pot. The apparatus is thus a two-stage stacked kettle, the stages being separated by a perforated bottom. Each stage is provided with separate stirring arms, steam-heated jackets and both stages work under vacuum (being practically an autoclave). The finely cut material is put into the first (top) stage where it is heated under vacuum (e.g. by a water-ring sealed pump) to a temperature not higher than 115°C. The fat drips into the lower compartment, the cracklings theoretically remaining behind, which in practice is not strictly true. When the lower so-called refining section is heated, some of the added active carbon and dry sodium hydrogen carbonate reduce the colour and free acidity to very low levels. Release of the vacuum is followed by cooling and filtration on suitable filters, to remove solids. The cracklings can be pressed and/or consumed.

Open-kettle wet rendering uses some 20% added water, and the fat is obtained at a maximum temperature of 95°C. The fat must be settled (separated) from the lower aqueous layer, containing the wet cracklings. Adding powdered salt on top of the fat layer eases the separation of the phases. The cracklings must be dried, preferably under vacuum, from 80% to 10% moisture content in order to be pressed for the residual fat.

2. Large-scale Batch Operations

Dry rendering in vacuum kettles (autoclaves) is a definite improvement upon the open-kettle methods (reduced air contact at high temperatures, less autoxidation). The horizontal kettle is provided with mechanical agitation (paddles), ensuring good contact with the steam-heated jacketed wall, and preventing the finely cut raw material from burning on it. The material is first cooked under steam pressure, which develops from the moisture present in the tissues. (Variants of the method involve first removing more than 30% water either during heating with the charging door open or under constant vacuum.) In the second (or third) stage a vacuum is applied and the temperature raised to not higher than 115°C in order to remove the rest of water from the fat and the cracklings formed in the first step. Samples are taken to establish the residual water content of the cracklings either by hand proof pressing or by dielectric or any other suitable instrumental moisture-determination method. Modern variants apply more intensive mixing, which considerably reduces the cooking time to, e.g. 3–4 h.

The fat obtained is discharged into strainer boxes and, after removing the cracklings, is filtered and, if necessary, bleached. Without bleaching the flavour and free acidity of the fat resembles that of open-kettle dry-rendered lard. The cracklings are generally intended for further pressing and not for (direct) consumption.

Wet rendering by live steam is mainly carried out in vertical cylindrical autoclaves (although horizontal models are also known), called digesters, which are provided with charging and emptying doors at the top and the bottom, vent lines, a safety valve, live steam inlets, and draw-off cocks on the side at different levels. The raw material can be charged either uncut or coarsely cut and steaming is started during the charging. When filled and deaerated, the vent line is closed and the material is cooked at maximum of 140°C under pressure for 2–6 h. After the pressure is very carefully released (there is a danger of emulsion formation), the large amount of condensed and cell water is separated by settling. The fat is removed and freed from the residual water by centrifuging or settling in tanks where it is kept at 15°C above its melting point. The fat is neither bleached, nor filtered. This lard (about 75% of the lard produced in the U.S.A.) is called "prime steam lard", provided that it was produced from raw material which did not contain any bones, detached skin, organs, scrap fat or pressings.

As the burning of fat in contact with water is impossible, "prime steam lard" is much milder in flavour than open-kettle (or vacuum, see later) dry-rendered lard, but because of the longer contact with water under pressure (3 bar), its free acidity may be somewhat increased (0.06% h^{-1}). The proteinaceous solids at the bottom (tankage) are separated from the process liquor (glue or stick water) and dried under vacuum before pressing. Tankage may contain about 10–12% fat.

Although this process was mainly intended for high-fat land-animal tissues, it has proved to be very versatile and, therefore, high-fat whale tissues and even bones can be effectively processed by it.

The Titan wet-rendering process is done in a stainless-steel autoclave (the "expulsor") into which the ground (minced) fatty tissues are charged in batches. On the top of the dome are appliances such as a pressure gauge, safety valve and a vent line for deaeration. The dome is closed with bolts, live steam is admitted and, after release of

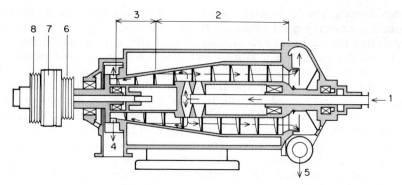

Fig. 3.2 Cylindrical/conical decanter. 1, Inlet; 2, settling zone; 3, drying zone; 4, solids outlet; 5, clarified fluid outlet; 6, drum drive; 7, cyclo-drive; 8, scroll drive for solids. [Reproduced, with permission, from Düpjohann and Hemfort Jr. (1975).]

the air, the vent line is closed and the contents are kept under 3 bar pressure for 1.5 h. The condensing steam and the moisture in the tissues cause the cells to break up; the proteins coagulate and the fat is released. The fat is skimmed off through a pipe, and then strained and purified by desludging in (three-phase) centrifuges.

3. *Continuous Large-scale Operations*

Besides these batch processes different continuous processes have been developed. Only the wet processes are of interest for edible-fat processing purposes. The finely cut (minced) materials are cooked continuously with live steam, the mechanical action of the steam serving to mash and pulp the fatty tissues. The method might be a two-stage process, the temperature of the tissues first being raised (e.g. in the mincer itself) by live steam to 50–60°C, at which temperature some proteins coagulate. This primary mash may be cooked in a closed, stirred buffer kettle of 500–1000 l volume with live steam, its temperature being raised to 90°C, and simultaneously deaerated.

Instead of using this kettle, the heating and further mashing can be executed in a continuous masher provided with rotating discs and live-steam nozzles. The heated material is then pumped into a generally horizontal decanting centrifuge (4000 rpm), where solids are separated from the fatty process liquor and discharged by means of an inside scroll rotating at a lower speed than the decanting cylinder. In the second stage the process liquor, still containing some solids, can be further disintegrated by a (second) high-speed rotating (mashing) disc and reheated to 90°C. The liquor obtained is separated in nozzle (solids ejecting) centrifuges into liquid fat, aqueous glue and residual to 45°C. The residence time is about 1 hour or, in the fully continuous solution, less than 0.5 h. The well known major

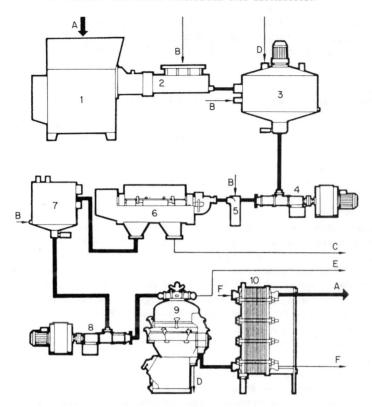

Fig. 3.3 A continuous wet fat-rendering line, the Centriflow: 1, mincer; 2, preheater; 3, cooker; 4,8, pumps; 5, intermediate heater; 6, decanting centrifuge (Dekanter); 7, holding tank; 9, nozzle or solids discharging centrifugal separator; 10, plate heat exchanger; A, crude fatty tissues and rendered fat; B, steam; C, solids (cracklings); D, residual solids from the separator; E, aqueous glue; F, cooling water.
[Reproduced, with permission, from Reuter (1982).]

firms, Alfa Laval, Westfalia and Sharpless, not only produce the normal vertically mounted centrifuges but also the horizontally rotating continuously decanting/desludging centrifuges (Fig. 3.2), and offer these other types of equipment mentioned as complete lines with all the necessary control devices (Fig. 3.3).

4. *Finishing Operations*

Rendered animal fats (mainly lards) generally should be packed after straining and settling (centrifuging, if available) with no further refining treatments. If, for any acceptable reason, flavour, colour and free acidity are not up to standard, but the fat was obtained from raw materials well up to standard, the necessary refining can easily be executed, if it complies with local legislation (see Chapter 4). The fats to be packed should preferably be cooled and treated as described in Chapter 6, Section III.A.

E. The Rendering and Fractionation of Edible Tallows

The rendering of prime-quality edible beef and mutton tallows (premier jus), and also the preparation of oleostock is carried out in equipment similar to that used for open-kettle wet rendering, the main difference being the lower maximum rendering temperature, say 50°C. Tallows rendered solely from killing fats are salted at the end of the process for better separation of the fat/water layers. The fat is siphoned off by a side tap and the residue from the kettle is transferred to autoclaves for further steam rendering.

The fat obtained without further separation is called "premier jus". This fat (or any low-temperature rendered tallow type tissue) is kept at a temperature about 15°C above its melting point in order to remove any traces of water. The fat can then be packed, or filled into shallow "pans" (several litres in volume) on racks or into vans (300–400 l in volume), all kept in rooms at 30–35°C for 1–3 days. The solid triacylglycerols (the oleostearine) crystallize in large crystals the interstices of which are filled with liquid fat (oleomargarine) which is the historical raw material for margarine production (see later).

On mechanical filling tables, small portions (a few kilograms each) of the crystallized fat (e.g. the contents of a pan) were put into press cloths placed in cavities and totally wrapped up to produce "tablets". These tablets were placed on the stages of an open-plate filter press (about two rows of four tablets, i.e., eight in total). Two-stage (low and high pressure) pressing was used to separate the liquid oleomargarine from the oloestearine in the cloths, which were then emptied and reused. Modern continuous or batch fractional-crystallization systems (see Chapter 5, Section 2) produce tallow fractions with a considerably reduced need of human labour.

F. The Production of Bone Fats

After cleaning and washing the bones of freshly slaughtered animals (cooled to 0°C if not immediatly processed, maximum storage not exceeding 24 h in chilled conditions) are broken by hammer mills to pieces of 20 mm size in order to ease the fat separation from the cells. (Tubular bones, which are used in the production of combs, button, etc., are not broken into small pieces.) A large metallic basket filled with bones is placed in an autoclave, the dome of which is totally removable. The lower space in the autoclave is filled with water, the basket residing on a separating grid. The dome, fitted with all the necessary appliances, is clamped to the body and, after deaeration, the vent line is closed. Direct steam or indirect heating is started. The pressure is increased to up to 4–5 bar; if tubular bones are treated the pressure used is not higher than 0.8 bar. From time to time the fat collected in the bottom is discharged into a special fat/glue separator. The glue solution (stick water) formed is evaporated after due settling in vacuum evaporators. The fat is treated with powdered salt, and settled and packed as edible bone fat. Direct steam heating in wet-rendering digesters (normal autoclaves) is also practised, instead of the above-described special Truested-type process.

G. Fat Separation from Cracklings

The amount of cracklings depends on the type and purity of the original tissue rendered, the process used and the final temperatures applied. Dry rendering delivers about 4–20% cracklings and wet rendering about 30–35%. The water content also varies: that of wet-rendered tissue being (before drying) up to 80%. Vacuum drying in steam jacketed autoclaves should give cracklings with 8–10% moisture content, this being an optimal level for mechanical (hydraulic or screw) pressing. The optimum temperatures are in the range 85–90°C. Hydraulic pressing delivers 15–30% residual fat levels, screw pressing 6–10%. The rest of the material consists of about 5% moisture together with proteins or partly decomposed proteins. The press cake from the hydraulic pressing can be further defatted after disintegration by solvent extraction, as was described in detail above for oil seeds. Pressed fat obtained from dry renderings can be consumed without refining only if it was processed immediately after the rendering step.

Only cracklings of the open-kettle and drip "dry" rendering methods, if pressed immediately after rendering, are suitable for consumption. Those obtained by other methods are generally not. Dry-rendered fats, their cracklings and the fats pressed from the cracklings, are generally preferred by consumers in geographical areas where the production of animal fats is carried out on smaller scales. Consumers in some highly industrialized areas prefer the white-coloured, low or no-flavour "prime-steam lards" (and tallows), produced by the wet-rendering (steam) process.

III. Edible-Fat Processing from Marine Animals

Sea animals, like whales, sea elephants and seals, contain quite large amounts of fat (oil) in their adipose tissues (up to 70%), whereas the so-called oily fishes (anchovy, herring, menhaden, pilchard, capelin, sardine, cod) may contain about 10–20% oil in the whole body.

The number of whales which can be caught is now regulated by treaties, as most species are close to extinction; in this light whaling should be totally forbidden, not only around the Antarctic. Dolphins and certain seals are also protected animals. Notwithstanding the very important aspects of fauna conservation, the problem is of course also a question of maintaining a balance between the consumption of fish in the biological food chain by sea animals and the amount of fish caught by the competing commercial fishers.

Whaling is carried out by small specialized vessels. The animals are harpooned with detonating charges, a terrible, but effective, method of killing. The whales are (were) skinned, dismembered on the deck of the mother (factory) ship, the fatty tissues (called blubber) separated and the meat processed either for human food (frozen, including dolphins, for some regional delicacies) or for animal (pet food of the Western World) or poultry feed.

Fish are caught by trawlers, short-distance vessels delivering the catch to factories on shore, and long-distance ones being factories for preparing either frozen/prepared (even canned) lean fish for human food or deoiled fish meal for animal feed.

A. Whale-oil Rendering

The blubber is first cut by special cutters, fixed on the deck, into huge slices. The material is then sent down a chute to below deck for rendering into autoclaves similar to those of the wet fat-rendering process. Semi-continuous systems are also useful: the precooking of the material in batches is done by indirect steam in closed vertical kettles with jacketed walls. The material is transferred by sluices into the second horizontal cylindrical autoclave by helical forwarding devices inside a slowly rotating inner, perforated cylinder. In the autoclave the material is cooked and mashed under 3.5 bar pressure developed by live steam. This stage not only renders the fat, but also decomposes the proteins to a glue solution. The mash is transferred through a mechanical sluice into a separating vessel, releasing the pressure to the atmosphere and discharging the separated fat and glue (e.g. the Schlotterhose/Hartmann or the Sommermeyer/Hartmann systems). The fat is settled, centrifuged and stored in tanks in the ship (not in the emptied fuel tanks!), the glue and solid proteinaceous parts being dried and used as high-protein (60–70%), high-fat (15–20%) animal feed.

B. Fish-oil Production

Oily fishes are generally cooked in pressure-resistant, or at least closed, horizontal tubular vessels, in which the fish material charged at one end is continuously moved by a screw conveyor. The system is generally heated by live steam (wet rendering). The mashed fish pulp is brought into vessels, where it is cooked at 90°C for 0.5 h, before being dewatered/deoiled by screw presses (see below) and/or by horizontal decanter centrifuges. The process liquor containing the oily emulsions is reheated to 90°C and the oil is separated by (nozzle) centrifuges, the remaining solids being continuously or periodically discharged. The separated aqueous fraction (stick water) can be evaporated for glue in double-effect evaporators or can be added to the crushed, wet, solid press cake and dried (not necessarily under vacuum) to a moisture level below 10%. After reheating the separated oil can be clarified further in dewatering centrifuges.

Rotary vacuum cooker/driers with steam-jacketed heating (dry rendering) are also customary in the menhaden industry. A disadvantage of these cooker/driers is the formation of scum which adheres to the heating surfaces.

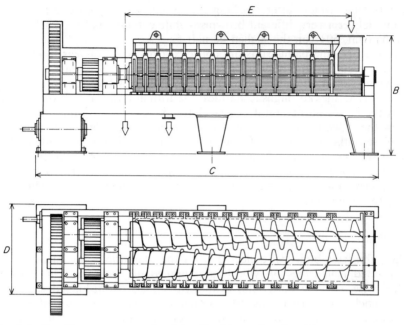

Fig. 3.4 Double screw press for fish oils. [Reproduced, with permission, from Barlow and Stansby (1982).]

The scum must be removed by internal rotary scrapers. The dried meal is deoiled by screw pressing. The "barrel" of these presses is very similar to that used for palm-oil pressing: it consists of a metallic cylindrical "cage" of constant diameter, perforated by thousands of fine funnel-like holes, broadening outward. The single or double worms have continuous helical flights of the same pitch, the shaft diameters increasing to increase the pressure (Fig. 3.4). The process liquor is worked up, as described above. After grinding, the high-protein, high-fat pressed meal is used as animal feed, although the residual oil can be recovered by solvent extraction. An effective washing/scrubbing/burning of the process air used for meal drying is a prerequisite for terrestrial installations. Decanters/centrifuges and/or screw-press based complete plants are produced by most of the well-known large centrifuge and screw-press manufacturers, but also by numerous others (e.g. Stord Bartz).

IV. Edible-fat Processing from Oily Fruit Pulps and Nuts

A. Palm and Palm-kernel Oils

Palm oil (about 20% on the fruit) is obtained from the deep-orange coloured fruit flesh (mesocarp) of the ripe fruits of the palm tree. These also contain nuts (stones) which, after crushing their shells in a separate milling operation, deliver a very different brownish-white coloured palm-kernel oil (5% on the fruit, 50% on the shelled kernel).

When ripe, the palm fruits are collected (this is an all-year-round operation) by cutting the bunches from the trees of the nowadays quite low-growing plantation-cultivated varieties. Ripe means the transformation of monosaccharides to oligosaccharides, the formation of a high (not necessarily the maximum) amount of triacylglycerides from the mono- and diacyl ones, the reduction of chlorophylls, an increase in carotenoid content, and increase in the proportion of loose fruits in the palm bunch to about 10%; all these characteristics are judged by experts. A bunch of palm fruits may weigh 10–20 kg and contain about 1500 fruits. Transport is by road or rail, but intermittent storage is generally deleterious to the quality, and should be kept to a minimum.

1. *Pretreatment*

The palm-fruit bunches, collected from the plantations, are deposited in perforated iron cages on wheels and immediately sterilized in generally batch-wise working horizontal cylindrical vessels (front-loaded channels). The bunches are heat treated for 80–90 min at a steam pressure of 3 bar in two or three consecutive steps. Sterilization not only stops lipase enzyme activity (mainly exogenous, yeast infection caused, but some endogenous), but also loosens the fruit bunches, softens the fruits and coagulates some of the proteins. Stripping (lifting and dropping) of sterilized palm bunches in horizontal rotating drums causes the shaking out of fruit, which is followed by washing on vibrating strainers with boiling water to remove the coarsest impurities. The empty bunches are incinerated and the ashes are used as a fertilizer.

2. *Fruit Digestion or Mastication*

Digestion of the loose, washed palm fruits is done in a (steam-jacketed) vertical kettle (digester) provided with several pairs of horizontal arms fixed on a vertical central shaft. The fruits are heavily mashed (masticated) for about 20 min. This part mechanical, part heat-treatment process detaches the fruits from the nuts at 90–95°C.

3. *Oil Separation from the Fatty Pulp*

The hot pulp is transferred into single- or double-screw horizontal or vertical deoiling/dewatering presses which separate the process liquor from the fibres and the nuts. The "barrels" are perforated with thousands of small holes and the screws have continuous flights of equal pitch. Live steam is admitted in order to aid in "rendering" the oil out of the tissues. The fibres and the unbroken nuts should be removed at the cone of the press. The oil loss on fibres is low (about 6%). The process liquor may contain up to 53% fat and finely dispersed solids and about 40% water. The liquor is strained on a vibrating screen and the coarse fibres and dirt so separated are returned to the digester.

The oil from the strained liquor is recovered in modern factories first by gravity settling for a few hours, and then by centrifugal purification. The separated process liquor is reheated with the fine sludge to almost 100°C and led from the settling tank via a descending cyclone to a solids-ejecting centrifuge which separates the oil, solids and waste water. An alternative method is to use horizontal desludging/decanting centrifuges, immediately after pressing, to separate continuously the three phases.

Oil separated/purified in centrifuges can be post-settled or post-separated by a second centrifuge (finisher), and the remaining moisture can be removed by drying under vacuum (80 mbar residual pressure). The residual moisture content in crude palm oils should be around 0.1%, as this level has been found to be fairly protective against autoxidation during overseas transport and storage, without increasing the free fatty acid content.

4. *Palm-nut Processing*

Morphologically, palm nuts are the endocarp of the palm fruit transformed to a woody encasement, also called the stone. This stone protects the seed (the kernel) which contains about 50% whitish-brown palm-kernel oil, having a distinctly different fatty acid composition to that of palm oil.

The production of palm kernels requires quite a large processing space, because of the necessary air aspiration and drying of the kernels. The press cake from the screw presses is disintegrated by conveying it through a steam-jacketed screw conveyor, extra paddles on the shaft loosening the dried fibres from the nuts. At the end of the conveyor this mass is directed to a vertical aspirator, where an air stream separates the low-density fibres from the nuts. The nuts are polished from residual fibres in a drum and dried in hot air to a moisture level of 11% in silos over a period of 10 h. Heating the nuts in an autoclave with rapid pressure release is also practised for quick drying. After drying, the nuts are crushed in centrifugal impact breakers, the separation of shells, unbroken nuts and kernels being done either by a winnowing column and hydrocyclones or by a wet two-stage classification process. The shells are screened in order to recover fine kernel particles and are then burnt in the boiler houses; they have a fuel value of about 17 MJ kg^{-1}.

The kernels are strained on vibrating screens and dried to 6–8% residual moisture content on continuous vertical-column driers or batchwise in silos for 10 h. The kernels are either put into sacks and transported to an oil mill or processed on site (if an oil mill is attached to the fruit-oil factory) by methods already discussed.

B. Olive-oil Production

Olive oil is obtained from the greenish fruit flesh (mesocarp) of the purple-skinned ripe fruits of the olive tree, a special variety cultivated only for oil and not for fruit consumption. The oil content is about 15–20%. The stones of olives contain an oil which is somewhat more unsaturated than that of the fruit-flesh oil. This oil is sometimes produced in a separate milling operation, similar to the separate milling of the stones of palm fruits.

1. *Selection*

Handpicked, non-bruised or minimally bruised olives are best used for producing "virgin" olive oils in the first pressing. Abuse during transport/storage causes a large increase in free fatty acids by hydrolysis, i.e. quality (value) losses.

Olive fruits are washed in concrete troughs and foreign material (leaves, stems, stalks, earth, etc.) is removed on strainers; vibrating screens are often used for this purpose.

The fruits are ground in small mills to a pulp either by vertically mounted stone edge-runners, by cylindrical or conical stones gyrating over a circular zone of a horizontal mill stone, or by rollstands containing pairs of fluted cylindrical rolls. The pulp (process liquor) contains the oil partly as an emulsion. Hammer-and-tooth disc mills, called triturators, used previously, are less commonly used now, partly because of the less complete breakdown of the cells and partly because of the possible metal contamination caused by abrasion. More recently, cylindrical drums, in which a slowly rotating inner roller mashes and kneads the fruits to a pulp with little emulsion formation, are often used. The stones from the mashed pulp are sometimes removed by rotating centrifugal screens, with slowly rotating inner reels.

2. *Malaxation*

Malaxation is a post-grinding and oil-coalescing treatment of the already milled olive pulp (paste). The aim is to obtain the maximum amount of free oil and the maximum amount of solids free from an occluded oily micro-gel. The apparatus is a double-walled mixer with paddles, Z-formed or helical stirrer arms; varieties with wall-fixed baffles are also available. Heating by means of lukewarm water to not higher than 25°C decreases the viscosity of the pulp. Higher temperatures, which rapidly lead to increased free fatty acid levels by hydrolysis, should be avoided. The duration of malaxation is about 20–30 min.

3. *Oil Separation from the Fatty Pulp*

The finely ground and malaxated pulp (paste) of olive fruits is processed in two ways: (i) it is pressed by hydraulic disc or cage presses, or (ii) is separated by horizontal decanting centrifuges (Dekanters). For pressing, the paste is filled into special filter discs or pressing sacks. The latter are placed into the special closed presses, the cages of which are built of separate rings of about 100 mm height and 550–580-mm diameter. The pressure is built up slowly to not higher than about 400 bar. The modern presses have a perforated needle sticking out from the middle of the bottom press tray, which reduces the distance which the liquid has to travel to half the original. Process liquor, containing oil and the aqueous phase, is removed and the press residues are then mixed with hot water and pressed a second time. This second pressing delivers an oil which cannot be used for consumption without refining. The oil/water process liquor of the first pressing is settled, the oil decanted and strained. This oil is called "virgin" oil, which not only means a varying, but limited free acidity ($<4\%$), but, as importantly, the exclusion of significant oxidative changes caused by the pressing and/or heating. Instead of sacks, already separated stones layered on preformed paste rings can serve as an inert separating filling in the hydraulic presses (Baglioni method).

In the centrifuge process the ground fruit is mashed in a special peddle mixer (malaxator) liberating the oil from the cells. This pulp (paste) is diluted by water and pumped into a horizontal desludging centrifuge (the Dekanter) which separates the husks (stones + fibrous matter) from the oil and the aqueous phase. These solids are again removed by a scroll inside the bowl. This process is labour saving and quick, reduces the costs, and improves the oil quality and the "virgin" oil yield. Finally, the separated oil (emulsion) is treated in sludge ejecting (nozzle) centrifuges. The oil can also be dewatered in ordinary centrifuges. In a variation of this process the oil-in-water emulsion (aqueous phase; French: *margin*) is sieved, in order to get rid of the solids, prior to separation by the Dekanter and, finally, settling. The emulsion can be returned to the mixer/masher, replacing the added fresh water needed for the "second pressing".

The residues of pressing(s) (French: *grignons*) are either extracted immediately (after drying) or are stored, heaped up in holes dug in the ground. These heaps are covered by earth, until the residue is required for further processing. These extracted "olive residue oils", in contrast with those of the second or even the first pressing, are organoleptically not up to standard and must be fully refined. According to relevant legislation they may be mixed with "virgin" olive oils, which of course cannot be sold as such anymore. If the oil from the stones is not processed, the husks are used as a boiler-house fuel or as inert filling for the presses, already mentioned. The dried pulp (extraction meal freed of stones) can be used as animal feed.

V. Special Quality Aspects

Chemically and microbially/enzymatically induced changes, like hydrolysis, autoxidation of the fats, and decomposition of the proteins and carbohydrates (rotting, fermentation) in the moisture-rich surroundings (also in the animal carcasses) may cause an extremely rapid decrease in the fat quality.

The fat present in freshly slaughtered animals is generally neutral, the free acidity is low and remains so if the separated tissues are cleaned, chilled and/or immediately rendered. Fats in plant/fruit sources are much more endangered, due to bruising of the fruit from falling onto the ground or in loading and also to the duration and distance of transport, all of which cause quick deterioration.

In order to maintain the highest possible quality (i) the time elapsing between slaughter/harvest and processing should be kept short and, (ii) the temperature during transport, before storage/processing should be reduced (refrigeration) under or raised (sterilization) above levels which are far from the optimum for (bio-)catalytically accelerated decomposition processes. The addition of phenolic antioxidants to the raw materials, intermediates and final products of animal origin is a logical measure to increase the oxidative stability of fats not protected by nature against autoxidation.

The control of quality should be executed by standard methods, as described in Herschdoerfer's monograph (1986), the loss of fat (oil) should be minimized by closely controlling the weight (volume) and fat content of all the products and effluents.

VI. Surface-water and Air Pollution

All the processes described above produce huge amounts of process liquors, either wash waters and aqueous glue from animal-fat rendering or wash waters and exhausted aqueous phases from palm and olive fruit processing. Therefore these fluids, which contain fat and dissolved and suspended organic material, must first be treated by a suitable fat recovery process (chemical or biochemical sewage treatment systems) to reduce the total fatty matter and biological and chemical oxygen demand (BOD, COD) to levels at which the wash waters can safely be disposed of to surface waters or city sewers. Separation swamps (ponds) in the open air for the collection of fat, acidification and anaerobic biological digestion of the sludge before transfer to sewers or rivers require a lot of factory ground, but, if possible, this method is the cheapest one. Anaerobic tank digestion (the methane produced can be recovered) combined with mechanical aeration and ponding is an alternative. The activated sludge can either be suspended or attached to filter systems. Fluidized-bed systems are also known. Aerobic biological oxidation seems to be less effective. The use of the already described decanter/desludger centrifuges as part of the process reduces the effluent load on the subsequent anaerobic ponding, as 75% of the solids can be removed and, after drying, used as animal feed or fertilizer.

Air pollution is mainly caused by the malodours of the proteinaceous (animal) tissue degradation during heat treatment. This nuisance can generally be reduced by washing exhausts in water scrubbers, cleaning the air exhausted from the rendering kettles/cookers/digesters, etc., and, for example, using it in the boiler house for combustion.

4

Refining

The aim of refining is to achieve the maximum attainable elimination of all non-triacylglycerol or partial acylglycerol components of crude oils with the maximum saving of unaltered triacylglycerols, natural antioxidants and vitamins. The components to be removed are thus primarily the free fatty acids, the phosphoacylglycerols (the rest of the hydratable ones, if pre-degummed, and the nonhydratable ones), sterols, pigments, glucosides, waxes, hydrocarbons and all those glyceridic and non-glyceridic lipid transformation products which, together with contaminations (metal traces, etc), may be detrimental to the flavour or oxidative stability of the refined oil. The general composition of a crude oil is given in Table 4.1.

The conventional classical (caustic) refining operation consists of four (or three) steps, with three optional cleaning operations. The sequence of steps is:

Pre-cleaning and/or mixing of different batches (optional),
(i) (Post-)degumming, if not included in steps (ii) and/or (iii),
(ii) Deacidification (by caustic neutralization),
(iii) Bleaching or decolourization,
post-cleaning: polishing filtration (optional),
(iv) Deodourization/(deacidification by distillation/heat bleaching),
Winterization (Dewaxing), to be executed preferentially on deslimed/neutralized, but not yet bleached oils (see Chapter 5, Section II).

A discrepancy in terminology should be noted: in Europe the term "refining" means all the main steps (i–iv) of the above sequence including the deodorization, whereas in the Western Hemisphere, influenced by the U.S.A. terminology, the term refining means only one step, the caustic-based neutralization.

The industrial refining operation can be executed in a number of different systems, i.e. batch, semi- or fully-continuous sequences. However, we should make a differentiation here as to the ways in which the different methods are achieved. The conventional method is the "wet" or aqueous caustic-based

EDIBLE OILS AND FATS

Table 4.1 *The types of possible changes caused by chemical processes during refining*

Substrate of change	Chemical processes						
	Hydrolysis	Autoxidation	Isomerization	Conjugation	Polymerization	Pyrolysis	Dehydration
Major components (glyceridic)							
Glycerides (fatty acids)	+	+	+	+	+	+	+
Phosphoglycerides	+	+	+	+	+		
Minor components (non-glyceridic)							
Sterols		+	+				+
Sterol esters	+	+		+			
Sterol glucoside esters	+	+					
Tocopherols		+					
Tocopherol esters	+	+		+			
Tocopherol dimers/adducts	+	+				+	
Waxes	+						
Pigments	+	+	+		+	+	
Hydrocarbons		+			+		

"chemical" process, a less conventional one is the "dry" (steam, adsorbents based) or "physical" process. Most of the different fat-modification processes (discussed later) cannot be executed properly without certain pre- and post-refining operations. Refining in miscella and other possible ways are discussed as special cases at the end of this chapter. The general survey of refining-process sequences is shown schematically in Fig. 4.1.

I. "Chemical" Versus "Physical" Refining

Refining processes which involve sequences of operations, superficially described as "physical" processes, are nowadays preferred to the traditional ones, which are characterized superficially as "chemical". This preference is mainly based on certain economic and environmental benefits, physical processes providing a shortcut in the process sequence and with less material loss. Despite this it should be recognized that the criteria which should be used when deciding on which type of process to use are the accepted quality and keeping property norms of the final products. In other words not all sorts and qualities of oils are suitable for physical refining. Genuine unsaturation, the levels of accompanying substances, the amount and nature of impurities, and the past history, thus the oxidative and hydrolytic damage suffered previously, must influence such decisions imperatively.

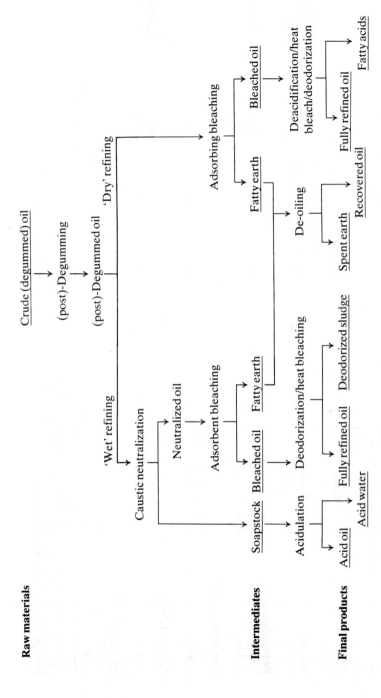

Fig. 4.1 General survey of the refining processes.

Table 4.2 The fate of relevant fat-composition classes during modern refining

Process stage	Components — Major		Minor		Transformation products		Contaminants	
	FFA	Phospholipids	Tocopherol	Pigments	Glyceridic (ex fatty acids)	Non-glyceridic (ex pigments and sterols)	Metal compounds	Toxic agents like pesticides (PAH)
Post-degumming								
Dry	—	Red.	V	Red.	V	Red.	Red.	Red. (Act. carbon)
Wet	Red.	Red.	—	—	—	—	Red.	—
Caustic neutralization								
Weak lye	Red.	Red.	—	—	—	—	—	—
Strong lye	Red.	Red.	Red.	Red.	Red.	Red.	Red.	—
Ads. bleaching								
LT[a]	—	Red.	V	Red.	V	V	Red.	Red. (Act. carbon)
HT[a]	V	Red.	V	Red.	V	V	Red.	Red. (Act. carbon)
Deodorization								
LT[b]	Red.	—	Red.	—	Red.	—	—	—
HT[b]	Red.	—	Red.	Red.	V	V	—	Red.

[a] LT, low temperature, 75–110°C; HT, high temperature, 120–150°C.
[b] LT = 180–200°C; HT = 210–270°C; also neutralization by distillation and heat bleaching.
Red., reduced; —, no influence; V, depending on conditions either reduced or increased or transformed or unchanged; FFA, free fatty acids.

It is inevitable that during the mechanical or solvent extraction of the oils detectable changes occur in the different lipid components, as compared to their "virgin" state in the cells. These so-called lipid transformation products are also generated during all the different further (chemical and/or physical) processing operations, which are aimed at giving some added value (better final quality = better keeping properties) to the purified (refined) or modified (hydrogenated, interesterified, fractionated) fat products. Because of their low levels and, to a lesser extent because of their nature, these transformation products are not at all harmful, at least not those produced in the present-day usual (physical and/or chemical) processing of oils and fats, representing good manufacturing practice. Lipid transformation products are constantly the subject of diverse food-safety assessment programmes, carried out by industrial and government laboratories. They can be detected by different sensitive methods and they give useful analytical clues as to the (final) quality of the oils.

The types of changes caused by different chemical and physical processes acting on different lipid components are summarized in Table 4.1. In general, added adsorbents and applied heating cause as few transformation (side) products or changes, as the diluted acid or alkaline reagents, hydrating phospholipids and neutralizing free fatty acids. Table 4.2 gives the fate of the relevant fat-composition classes during modern refining.

A. (Pre)-cleaning and/or Mixing of Different Batches

Despite the filtration of the miscella and the crude oil and the pre-degumming some solids are present in the oil, mainly because of less careful transport or local storage. It is best to remove these solids in an additional filtration, or at least sedimentation, in separate tanks, the latter process lasting for 1–2 weeks, and so very time consuming. The choice of method for solids removal (filtration on plate, plate and frame, open or closed filter presses, leaf or other types of mechanized filters, by using filter aids, or by centrifuges or "decanters") depends on local circumstances (quality of oil received, the type of processes applied and the local economy). In any case the oil should be heated to the minimum temperature (40–55°C) at which the viscosity is reasonably decreased in order to obtain the best separation effects.

Oils of the same sort but of different quality (i.e. different free fatty acid and phosphoacylglycerol contents) should be either stored separately (batch process) or mixed to a standard-quality crude material in order to be able to cope with the fixed-proportion caustic-requirements of "continuous" processes.

B. (Post-)degumming

Soft oils (soya-bean and rape-seed oils in the first place) are generally delivered to a refinery in the pre-degummed state, i.e. with an actual phosphoacylglycerol level of 150–250 mg kg^{-1}. The remaining phospholipids are mainly of the non-hydratable type, and so it is a question of the locally applied process (also considering the sewage-treatment facilities and local regulations on permissible surface-water pollution levels), whether they are removed first or handled in a later step of the refining process. Their removal by water alone (hydration) is no longer sufficient.

The use of super-degummed and/or Staley-process degummed oils is a big advantage from the point of view of the refining process, as no post-degumming is required. The sole use of either caustics or heating (240–280°C are customary for non-edible oils) or adsorbents in a separate step and only for desliming is less usual with oils processed for edible purposes, as all these means are applied at some step in the total refining process in combination with each other. Considering the options, post-degumming can be executed in two ways: the "wet" and the "dry" methods.

1. *Wet Post-degumming*

"Wet" post-degumming can be executed by adding to the prewarmed (55–85°C) oil acidic additives, e.g. 0.1–0.2% phosphoric acid (85%) which is well dispersable in oil, or a 20–50% aqueous citric acid solution or even dilute brine (sodium chloride solution). The process is done as a batch operation in jacketed (or coil-heated) open cylindrical kettles with conical bottoms and appropriate slow-moving (gate) mixers (30 rpm). After the addition of 0.5–2% water and a few hours of mixing, the flocculated sludges are most efficiently separated by conventional centrifuges of the partitioned-bowl (discs) or tubular type. Sedimentation in the kettle itself is an old-fashioned option.

The dosed addition of the acidic reagent in a knife or static type premixer to the plate-heated oil is followed by a minimum residence time in so-called "dwell mixers", which are holding tanks containing a slow mechanical mixer. The subsequent addition of the hydration water just before centrifugation is another variant, used in the equipment manufactured by Alfa Laval, De Smet and others.

When acids are used the precipitated sludges (if not used directly as animal-feed additives) must be worked-up after separation in the deoiling/waste-water treatment (not necessarily in the soapstock splitting) installations.

As the final phosphoacylglycerol levels arrived at are (except in oils with higher starting levels of "lecithin") often not very satisfactory, this step is very often omitted and the residual "gums" are removed (by previous addition of some phosphoric acid) in the caustic refining step. A very useful measure for raising the starting level of hydratable phospholipids is the mixing of predegummed and non-predegummed oils before desliming. The subsequent addition of the superdegumming or Staley-process reagents to this mixture delivers a reasonable quality lecithin and satisfactorily degummed oils. Acids have the special ability to reduce the chlorophyll (pheophytin) level of "green" soya-bean and rape-seed oils produced from unripe seeds.

2. Dry Post-degumming (*Step 1 of "physical" refining*)

Oils and fats containing lower original levels of phospholipids, like palm and lauric oils, or slimy substances of proteinaceous nature, like fish oils, are "dry" degummed by a combination of inorganic acids and adsorbents. This operation is also the first step in the "physical" refining sequence. The operation involves the thorough (pre)-mixing of, for example, 0.05–0.2% phosphoric acid (85%) of edible grade at 90–105°C with the oil in a kettle or a continuous on-line (knife or similar) mixer and with subsequent admixing of 0.5–2% activated bleaching earth in a dome-covered closed conical-bottomed cylindrical kettle provided with a (gate) mixer, actually a batch bleacher (see below). After 30 min the mixture is filtered on a, preferably, stainless-steel (automated) pressure filter to remove with the acid and earth most of the sludges, trace heavy metals and colour bodies (pigments).

Pellerin Zenith offers the complete semi-continuous Zenith system consisting of "three trays stacked in a cylindrical outer shell". The oil, preheated to 90°C, is mixed with acids in the first tray with an intensive mixer and is held subsequently in a holding tray placed below the first one for about 25 min, before discharging it to the third tray for desludging by decanter-type horizontal centrifuges (Fig. 4.2).

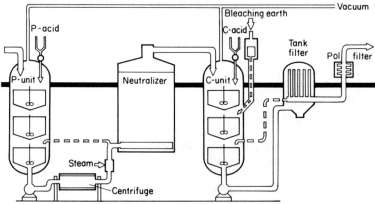

Fig. 4.2 Zenith semi-continuous refining line. [Reproduced, with permission, from Hamilton and Bhati (1980).]

The Simon-Rosedowns "four trays stacked in a cylindrical single shell" design deaerates and heats the oil and adds the phosphoric acid in the first pan, the second serving for the addition of bleaching earth, the third one being the holding tray and the fourth a buffer tray, from which the slurry is pumped to the (automatic) filter(s). Both this and the Zenith design are similar to the compartment (tray) stacking principle of the Girdler/Votator continuous bleachers and deodorizers, described later in this chapter.

Other proprietary systems, like Alfa Laval, HLS, Lurgi, De Smet and others offer these steps in separate units, energy-saving being the leading principle (Forster and Harper, 1983). Very low residual levels (<3 mg kg^{-1} of phosphorus must be achieved before the deodorization step, by any "dry" or "physical" refining sequence, in order to produce oils of acceptable final flavour stability.

3. Intermediate (Wet/Dry) Degumming Processes

Soft oils produced by the Alcon process have been successfully post-deslimed by mixing 1% water and 1% acid-activated bleaching earth with the oil at 65°C, subsequently drying it at 110–115°C at 40–70 mbar residual pressure for 45 min. This method, if applied on common water-predeslimed oils, delivers unsatisfactory results, mainly because the prescribed drying temperature is too high.

In another process dry crude soft oils (mainly low erucic Canadian rape-seed oil) are first wetted with 0.1% water at 25°C for 15 min with stirring. They are then further mixed with 0.1–1.5% phosphoric acid (85%) at 40–70 mbar residual pressure for 15–30 min. Then 1–2.5% highly active bleaching earth is added and the mixture is heated under vacuum to 170°C and then cooled quickly, in order to filter out the acid/earth/mucilage on closed, stainless-steel filters (Mag, 1973). The higher temperatures applied together with the non-earth-absorbed, thus free, phosphoric acid residues might cause alterations, like isomerization and dimerization, of the component polyunsaturated fatty acids. This latter action limits the wider application of this method.

A special procedure used for the continuous post-desliming sequence is claimed by HLS. First soft oils are mixed with edible-grade acid (probably phosphoric), and an aqueous undisclosed flocculating agent (probably water glass) is added with mixing. The heated mixture is allowed to settle for amorphous (probably silica gel) floccules to form. After centrifuging the (co-)precipitated phospholipids, the oil is dried and bleached on a batch (holding/reacting) kettle based system, using modern automated filters.

The added acids have an additional, different role: the sequestering of trace metals. This was discussed in Chapter 1 (Section IV.B.4.b) and everything said there is relevant to the different acid-based simultaneous removal of gums and metal traces (Segers, 1983), the completeness of both in most cases being far from total.

II. Deacidification by Neutralization

The removal of free fatty acids can be executed in different ways, depending on whether the reactivity, the solubility or the (very low) volatility of the fatty acids is employed. The aqueous caustic solution (lye) not only neutralizes the acids, forming salts, called soap (+water), but saponifies (according to the time of contact, concentration of the alkali and the temperature) some ester type compounds, catalyses the formation of, for example, tocopherol–glyceride (fatty acid) adducts, dissolves phenolic compounds (tocopherols) as phenolates, dimerizes gossypol to gossypurpurin, oxidizes fatty acids (keto–enol compounds) to fatty acid (glyceride) dimers, etc. Notwithstanding these slight side effects, alkali deacidification also removes residual phospholipids (post-degumming) and colour bodies (bleaching) in addition to the free fatty acids.

A. Theoretical Background

1. *Emulsions and Emulsifiers*

According to the theory of condensed phases, molecules of a pure liquid are kept together by their mutual attraction caused primarily by van der Waals forces. Depending on the molecular structure, more specific interactions, like hydrogen bonds (water) may also be relevant. The attraction of the molecules is exerted uniformly in the bulk of the liquid by the neighbouring molecules. The molecules at the surface of a single liquid (or at the interface of two immiscible liquids) have no outward (similar) neighbours. This missing attraction is not substantially compensated for by the molecular attraction of the surrounding atmosphere (or by the molecules of the second, different liquid) causing an energy imbalance. If two immiscible liquids were originally dispersed, they try to separate again, because of the higher mutual attaction exerted by their own molecules. In both cases the surface molecules are drawn into the bulk of the liquid(s), promoting a contraction of the surface(s). In this way an equilibrium is restored which lowers the free energies of the system. The force of contraction per unit length along the surface (interface), caused by the imbalance, is called surface (or interfacial) tension.

The surface tension induces a given volume of liquid to take, if possible, the form of the lowest surface area, i.e. a sphere. The interfacial tension against the other liquid at the common surface is generally smaller than the surface tension of the separate liquids against air.

An emulsion is a fine and stable dispersion of one of two immiscible fluids in the other. In order to form an emulsion the two separate liquids must be mixed mechanically. The strong shearing action of mixing produces small droplets (spherical globules), the "internal" phase, in the continuous second phase, the "external" phase. By this dispersion the specific surface of the system (the ratio of the total surface to the total volume of the newly formed droplets) increases strongly. This forced separation also causes an increase in the free energy, so that the system becomes unstable.

As regards the sizes, the droplets in emulsions may be dispersed coarsely (diameters of >20–$5\,\mu m$, finely ($<5\,\mu m$) and very finely ($0.1\,\mu m$), although in practice there may be a wide size distribution.

The stability of a mechanically prepared emulsion can be increased most suitably and effectively by the use of emulsifiers. These agents unify in the same molecule a water-repelling "hydrophobic" or "apolar" tail group (generally a long aliphatic fatty acid chain) and a "hydrophilic" or "polar" head group which has a relatively strong affinity to water (e.g. oxygen based functional groups). The head groups of the emulsifiers can be of anionic, cationic or non-ionic character. Because of their affinity to both phases, but generally somewhat limited solubility in one or the other, the emulsifiers preferentially accumulate (adsorb) at the surface/interface of the dispersions in question. By this local adsorption they reduce the free energy of the system, so that the energy needed to form the new surface is lower than that required to create the same surface area in the absence of adsorption (Gibbs). Because these molecules accumulate at the surface (interface), they are also called "surface active agents" or "surfactants".

If in a dispersion of water and oil an (anionic) emulsifier like the sodium salt of fatty acids (soap) is present, its negatively charged surface-active anions will be adsorbed on the oil, whereas its positively charged cations will remain in the aqueous solution. The separated negative and positive charges produce a diffuse electrical double layer around the dispersed and stabilized droplets with a distinct potential gradient. The surface charge density and the thickness of this double layer at the interface [measured as the so-called zeta (ζ-) potential] regulates the stability of emulsions in aqueous milieus.

As we have already seen, the emulsion formed has a constant tendency to destabilize, with gravity playing a distinct role in most cases. Only if the droplet sizes are very small can thermal motion keep the droplets dispersed. At a lower force of repulsion, the droplets can adhere to each other with only a thin film of the continuous phase separating them. This aggregation of droplets retaining their original size is called flocculation (generally segregation of water). If the droplets, again retaining their size, are brought together by density differences (e.g. in an oil-in-water emulsion, like cream), the process is called "creaming up". If the separate droplets merge to form larger ones and finally coalesce into separate phases, the emulsion undergoes total "breaking" (e.g. segregation of oil in an oil-in-water system).

Beating (churning of butter), freezing–thawing, centrifugation (artificial creaming), the addition of electrolytes (which increase the ionic strength and, thereby, decrease the thickness of the electrolytic double layer) and antagonistically acting emulsifiers may all promote both flocculation, segregation and coalescence of the system.

2. *Mesomorphic Behaviour of Certain Emulsifiers*

When cholesterol benzoate is heated, it melts sharply to form a turbid liquid which at higher temperatures is sharply transformed to a clear one. The turbid liquid is doubly refracting, which is one of the properties of an anisotropic crystalline material. Double refraction means that in passing through the crystal the light splits into two components travelling with different velocities in different directions relative to the crystal axes (e.g. the birefringence of the Icelandic spar in the Nicol prisms of polarimeters). As they are liquid, the term "liquid crystals" expresses their mesomorphic (intermediate form) state. The existence in this form is characteristic of many surface-active organic molecules of flat (e.g. sterol based) or long (fatty acid based) shape (e.g. soaps), forming an oriented liquid pseudo-microcrystalline mass.

In an aqueous environment, close to their melting points the, for example, fatty acid chains of the water (in)soluble molecules melt. The water penetrates the spaces between the polar groups giving liquid crystals which contain water and have a lamellar, hexagonal, or cubic structure. These crystals may form a continuous phase called a "lyotropic mesophase".

B. The Process of Alkali Neutralization

Besides the component triacylglycerols oil also contains dissolved free fatty acids, partial acylglycerols, residual phospholipids and sterol (glucoside), not all of which are inert against caustics. Neutralization of oils with the aqueous caustic solution is thus a more complex process than that expressed by the simple equation:

$$\text{acid} + \text{base} = \text{salt (in our case called soap)} + \text{water}$$

The neutralization process is primarily a mass-transfer based extraction of the free fatty acids from the oil into the water phase, strongly accelerated by the above chemical reaction at the sites, where the two immiscible carrier mediums, water and oil (triacylglycerols), meet. This process might also be called deacidification.

The contact area between the fatty acids and the caustic is increased by mechanical mixing, producing a temporary and not too stable dispersion of the two phases. From the point of view of mass transfer it is not indifferent whether the (more or less dilute) caustic solution is dispersed in the continuous mass of oil or *vice versa*, considering the different distances the fatty acids have to travel to the interface. Besides, the fatty acid concentration gradients (the driving force) and the size, size distribution, composition and structure of the dispersed droplets, not forgetting the temperature, all play a distinctive role in the whole process.

Fatty acids, further phosphoacylglycerols and sterol glucosides accumulate preferentially at the interface of the mechanically formed droplets, because of their more hydrophilic (or polar) character, as compared with that of the oil. In this way they lower the interfacial tension of oils. If, in addition, another surfactant is present in the oil/water dispersion, a coarse emulsion can be produced and (temporarily) stabilized. The extra "surfactant" of the neutralization process is the alkali soap, formed at the interface. The hydrophobic (anionic) acyl chain of the soap has an affinity for the oil, whereas the sodium containing cationic head group has a high affinity for water, in which sodium soaps are soluble. This solubility is strongly increased by the formation of polymolecular soap aggregates (micelles) above a "critical micelle concentration" (CMC). The CMC is higher for short-chain fatty acid soaps, than for longer chain ones. It is decreased by the cations of the unused, completely ionized caustic lye. Thus soaps may be "salted out" by electrolytes as solid or liquid crystalline precipitates.

Sodium soaps preferentially stabilize oil-in-water type emulsions (Bancroft's empirical rule: the phase, in which an emulsifier is more soluble, becomes the external one), if the volume proportions of the phases make this locally possible (Ostwald's empirical rule: the phase, which is present in the higher volume proportion becomes the external phase). The soap micelles may also include (solubilize) organic (oil) molecules in the centre of the aggregate, this solubilization being one reason for the inevitable neutralization losses.

In batch neutralization the lye is sprinkled onto the surface of the slowly stirred oil. Because of the unfavourable volume proportions of caustic to oil, the soap immediately formed at the interface this time favours and sustains a water-in-oil type emulsion of more or less coarse dispersion. Local high concentrations of caustic, e.g. because of high excesses used, or stable emulsions might cause the saponification of neutral oil. Fortunately, the transport of fatty acids to the interface is initially quick, because of the initially high concentration difference of fatty acids in the bulk oil and at the interface. The quick neutralization reaction reduces the caustic concentration. The soap formed should be removed (or diluted) as quickly as possible, in order to keep the reaction rate high. This can be done if the finely dispersed aqueous phase, containing the (flocculated) soap, can be forced to coalesce to a continuous phase. This is done in a batch kettle by stopping the stirrer and settling the soapstock. In continuous systems, centrifugation achieves this task easily and quickly.

In batch systems there are other problems. The depletion of the fatty acid concentration in the thin oil layers separating the finely dispersed water (aqueous soap) droplets creates a local increase in the interfacial tension. The surface between the water droplets contracts, thickening the separating oil film, and the coalescence of the water droplets, because of the increasing distances between them, is hampered. The small diameter of some water droplets and the increased distance between neighbours is the reason for the sometimes high residual soap levels found even in water-washed batch-refined oils.

As was stated earlier, phospholipds, oxidized fatty acids and sterol glucosides act as emulsifiers. Their presence during caustic neutralization is thus a nuisance. Phosphatidyl ethanolamine (PE), magnesium and calcium, salts of phosphatidic acids (PA) and non-alkali bound oxidized fatty acids also inhibit the coalescence of water-phase droplets by stabilizing the water-in-oil emulsion.

However, the initial water-in-oil type emulsion/dispersion becomes less stable at the end of the process and gives way to the inverted oil-in-water type emulsion/dispersion. The latter is stabilized further by the partially hydrolysed phospholipids and sterol glucosides, generally found as a more-or-less transparent interlayer ("lyotropic mesophase" of liquid "emulsifier" crystals) between the oil layer and the soap stock. It is the most difficult task of a batch-kettle operator, when separating the neutralized oil from the soapstock, to detect this interlayer in the sight glass of the bottom outlet of the kettle. Unobserved, it may cause costly neutral fat losses if abundant.

In the inverse system the oil to be neutralized is dispersed as a continuous stream of droplets through a static (or slowly renewed) mass of very dilute caustic, which reduces the danger of saponification by strong lyes used in the batch process. Here the neutralized oil droplets collecting at the top are prevented from coalescing by the adsorbed soap.

1. *Caustic-based Batch Neutralization*

Batch neutralization is executed in open, or preferably in dome-topped closed, vertical cylindrical kettles with conical bottoms, in order to be able to remove easily the separated aqueous phases after the completed process steps. The kettles are generally made of mild steel with capacities between 1 and 60 tons, 4–30 tons being customary. The kettles contain (gate) mixers, and the heating (by steam) and cooling (by water) is done through jackets or coils. The oils should be washed and dried after neutralization. As the removal of the last traces of water is easier under vacuum, the closed (domed) kettles should be able to withstand 40–70 mbar residual pressure. In this case they can be used directly for the bleaching operation which is also carried out under vacuum (Fig. 4.3). For all appliances the use of any copper or copper containing alloys should be avoided, the pipes supplying the water also being made of copper free material. The easy handling by the (experienced) operator, their versatility (different oils can be refined in quick sequence) and the low installation costs are the advantages of this type of equipment. The sometimes high refining losses and the problems with the subsequent large volumes for soapstock splitting and sewage treatment (reduction of possible surface-water pollution) are the drawbacks, not considering the fact that some oils of very high free fatty acid content cannot be neutralized economically by this process.

In broad outline, the caustic solution is sprinkled on top of the slowly (30–50 rpm) mixed oil previously heated to 50 95°C. The fatty acids react with the lye to form the soap, enclosing the non-reacted lye, phospholipids, oxidized glycerides, colour bodies, etc. If mechanical mixing is stopped, this aqueous mixture, the soapstock, settles to the bottom of the kettle. The rate of settling depends on the diamater of the oil or water droplets, the density differences between the oil and the soapstock [influenced by electrolytes (salts) added to the water phase], and the viscosity of the oil and the soap (stock) are components of Stokes' law and are in balance with the force of gravitation (or, when using centrifuges, with the centrifugal force):

$$u = \frac{g D^2 (\rho^s - \rho)(\rho^s - \rho)}{18 \eta}$$

where u is the settling (gravitational) velocity, g the acceleration due to gravity, D is the droplet diameter, ρ and ρ^s are the densities of the droplet and the surrounding phase, respectively, and η is the viscosity.

The first task, when applying this neutralization process, is to select the appropriate concentration of the caustic soda solution (lye). In making this selection the main aspects to be considered are:

(i)   the free fatty acid (FFA) content of the oil to be neutralized;
(ii) the quality of the soapstock formed by the neutralization;
(iii) the amount of neutral oil (generally co-emulsified and so lost) degraded;
(iv) the sedimentation rate of the soapstock formed (Stokes' law); and
(v)   the required colour of the finished product.

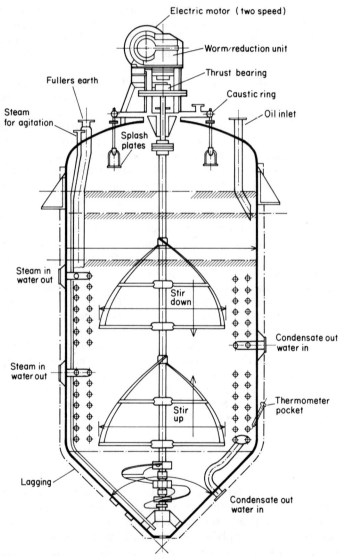

Fig. 4.3 Batch neutralizer/bleacher. [Reproduced courtesy of Unilever N.V.]

The free fatty acid content of the oil is the primary factor (besides the type of fatty acid) which determines the amount and concentration of the caustic soda (and to what excess, generally 10–25%) to be added to the oil. The envisaged amount of residual free fatty acids after neutralization (0.05–0.1%) will determine the necessary level of excess caustic used. As a rule, more concentrated aqueous lye is sprinkled on the top of oils with high initial free fatty acid contents, than the lye sprinkled on those with a lower one:

Free fatty acid content (Molecular weight of fatty acids is c. 280)	Lye concentration (N)	NaOH solution (%)
<1%	0.8	3.2
1–2%	4	16
2–3%	5	20
>3%	9	36

More saturated (lower iodine value) oils, like palm, palm kernel and coconut oils, but also cotton seed and sunflower seed oils, give soapstocks of relatively solid consistency, and so the lye should be (more) dilute(d) in order that the soap formed remains liquid and is not salted-out by the high residual ion concentration (ionic strength) of the solution. Because the CMC of lauric soaps is higher, their refining is less demanding as compared with that of other oils. A too high dilution is again adverse, as this enhances emulsification of the neutral oil. The concentration and excess of lye and the temperature also determine the speed of settling of the soapstock (viscosity factor). Once again stronger lyes favour the removal of colour bodies (tocored, gossypol, gossypurpurin, oxidized dimers of tocopherols and of glycerides, etc.). The addition of the necessary amount of lye and excess of it can be done in three steps, or just water might be first mixed with the oil, before adding stronger lyes which are thus diluted *in situ*. The temperature of the system can be increased during lye addition from 50°C to the final value. All the above-described factors can be used as measures to fulfil the requirement of the most economic free fatty acid removal.

After the removal of the soapstock, and before the washings, an intermediate treatment might be necessary. The oil, like soya-bean oil or a tank residue, even after neutralization may contain high residual amounts of non-hydratable phospholipids. In this case an additional "re-refining" can be applied. Boiling with some alkaline reagents, such as waterglass solution neutralized with a slight amount of soda ash, can reduce the phosphatide content to an acceptable level. Less difficult oils are immediately washed free of the soap by sprinkling hot water on top of them. If problems are encountered during refining, a useful measure is to use, instead of the first water washing, dilute salt solutions to increase the separation speed of soapstocks, and thereby reduce the oil losses. The normal sedimentation time in a kettle is 1–2 h. The draining of the soapstock from the refined oil is an operation that requires a skilled operator who knows the characteristic flow properties of the soapstock in question and that of the kettle. The sight glass in the outlet line is also very important in assessing the onset of drainage of the emulsion layer and, even more importantly, the oil.

The neutralized oil, although washed 2–4 times with water after the separation and removal of the soapstock (the aqueous phase to be treated sometimes amounts to 50–100% of the oil refined before it can be discarded to the sewer), will contain very fine droplets of soap, formed during the primary steps. This soap cannot be removed below a certain level; in ordinary batch kettles this level is 300–500 mg kg^{-1}.

(a) *Refining losses.* The economics of every refining operation should be calculated by accounting for the losses. The practical losses (all in weight) can be calculated most exactly by using the refining factor (RF):

$$RF = \frac{(\text{crude oil} - \text{water} - \text{dirt}) - (\text{neutralized oil} - \text{water} - \text{dirt} - \text{soap})}{(\text{free fatty acids in crude oil} - \text{free fatty acids in neutralized oil})}$$

Another parameter which can be used is the fatty acid factor (FAF), also expressed as a weight:

$$FAF = \frac{\text{total fatty matter (TFM) in the soapstock}}{(\text{free fatty acids in crude oil} - \text{free fatty acids in neutralized oil})}$$

The fatty acid factors are measured with a better accuracy than are the refining factors. A value of FAF = 1.8 means that for every 1 kg of fatty acids, 0.8 kg of oil + gums + other oil-soluble material has been lost.

The refining factors of batch refining are 1.8–2.0 for the neutralization of oils with normal free fatty acid contents, whereas high free fatty acid oils ($\geqslant 2\%$) have factors of 3–5, values which are considered to be disastrous.

The minimum losses can be estimated by, for example, the chromatographic test method "neutral oil and loss" of the American Oil Chemists Society (AOCS). This method shows, at least for soya-bean oils, a good correlation with the officially accepted minimum refining losses based on the so-called "cup" methods of the AOCS. The practical losses can be measured in the batch method by tank-volume gauging. For the continuous methods, described below, there are now automated loss-monitoring devices available (see Section II.B.6, this chapter).

2. Dilute Caustic-based (Semi-)continuous Neutralization

The dilute caustic-based (semi-)continuous neutralization process is based on the already mentioned inverse system, where the continuous phase is a dilute (0.2–0.8 N, 1–3%) caustic soda solution, the oil being dispersed within it in small uniform droplets of 1-mm diameter from the bottom of the vertical cylindrical kettle (e.g. the Pellerin Zenith, now Zenith, system; see Fig. 4.2). The collected and coalesced top oil layer is continuously removed, whereas the lye can be used as a static layer (exhaustion, semi-continuous cross-current system) or constantly renewed (fully continuous counter-current system). No mechanical energy (except pumping) is needed. Because of the weak lyes used, there are no losses caused by saponification, and oil losses only occur if the droplet size is not uniform, the too-small droplets having a longer residence time as they rise up slowly. A disadvantage is that non-hydratable phospholipids, other polar lipids and colour bodies cannot be removed because of the weakness of the lye, so most of the soft oils containing them must be pre- or post-deslimed. In oil-in-water type emulsions stabilized by soaps, the main problem is the coalescence of the oil droplets at the top. This can be counteracted by, for example, applying much higher temperatures than the customary 95°C, thus working under pressure. The refining factors are low (1.2–1.3) for "easy oils" (e.g. laurics) if the initial free fatty acid content is not too high. The high effluent volumes (dilute soap solutions) may cause sewage problems.

3. *Strong Caustic-based Continuous Neutralization*

The problems of batch neutralization already mentioned above can at least partly be circumvented by continuous, in-line or off-line contacting of strong caustic lyes (4 N) with the oil to be neutralized, followed by the separation of the soapstock by centrifuging. A second treatment with alkali lyes may also be made, followed by one water washing step. The advantages of the latter are the complete exclusion of air contact, the low cost of labour, lower space and steam requirements, reduced oil losses caused by saponification, no or low levels of emulsion formation, lower amounts of effluent (mainly wash) water and the easy separation of the soapstock, by centrifugal forces (instead of gravity) which act against the factors represented by Stokes' law. The disadvantages are the low flexibility, the high investment costs and high refining factors, which are, however, still generally better than those of batch neutralization.

The centrifuge-aided methods differ from others in the type of pretreatment before neutralization, the time of contact with any reagent applied and the number of centrifuges (generally two or three) used for a certain oil of a given quality.

The oils to be treated are pumped in metered amounts of a few tens to a few hundreds of litres per hour by either positive displacement pumps with pulsation dampeners on the suction side or by centrifugal pumps. The pretreatment is mainly the addition of some post-degumming agent of the "dry" type (0.05–0.1% of 85% phosphoric acid, 0.05% citric or tartaric acid, as a concentrated solution). These reagents are administered to the oil by metering pumps of displacement or diaphragm type, the amounts being measured by variable-area flow meters (rotamers). Variations of this method involve the addition of the reagents to the oil heated to 90°C either in a static on-line mixer (see below) with 2 s contact times, or by adding them to the oil in a separate holding tank at ambient temperature and mixing slowly for 4–12 h before heating and contacting the prepared mixture with the lye (U.S.A. and Canada practice). Staley-, super-degummed and Alcon-processed soft oils do not require this treatment.

The time of contact with the strong (4 N) lye might be short (1–15 s), in which case the mixing is effected by dynamic on-line mixers, such as a centripetal-force based mixer bowl (Westfalia) or a disc-type rotor/stator mixer (Alfa Laval), or by static mixers (Sharpless). A static mixer is a simple tube with no moving parts, the mixing elements dividing the flow of the components at their leading edges and producing, by their radial mixing, high turbulence in low-viscosity fluids. The mixing time in the paddle-mixed contactors can be quite long, e.g. 1–15 min. These times are more than sufficient for completion of the ionic reaction of lyes with the free fatty acids. The temperature is again 90°C. Really long mixing of the oil with caustic soda in holding vessels (dwell mixers) before heating the mixture to 75°C, in order to "break" the soap before centrifugation, is a variant used in the U.S.A. and Canada. The use of soda ash, the combination of caustic soda and soda ash are steps in the evolution of centrifuge-aided refining, introduced in 1932.

The soapstock and all other phases contacted are removed by disc-bowl (Alfa Laval, Westphalia) or by tubular-bowl (Sharpless) type centrifuges, rotating at speeds of 6000 and 17.000 rpm and producing 7000 and 15.000 g, respectively. The capacities are in the range 2–20 ton h^{-1}. The centrifuges are hermetically sealed, the fluid entering either from the top or from the bottom. Some of the modern ones are self desludging (cleaning), any dirt collecting at the periphery of the bowl being continuously discharged through "nozzles" [Fig. 4.4(a)]. In another variant the collected solids are removed by opening the bowl hydraulically at timed intervals [Fig. 4.4(b)]. They are equipped with back-pressure valves at the outlet, by which the boundary zone between soapstock and oil in the bowl can be regulated, the higher pressure moving the zone outwards. The larger the relative oil volume, the longer is its residence time in the centrifuge and the better the removal of soap particles from the oil. Too high back-pressure may cause oil losses by entrainment into soap stock and, therefore, the periodic disruption of the zone balances should be monitored by breakover warning systems and/or automatic sealing devices, besides the old-fashioned lighted sighting glasses watched by the good operator. The preheated caustic and wash water are metered by previously set proportioning pumps. These are generally positive displacement pumps, of which the stroke is adjustable. However, pumping with centrifugal pumps is also possible, if the flow is controlled by accurate volumetric devices (rotameters or rotating positive displacement flow meters). The wash water obtained before the washing centrifuge is admixed with the oil in special knife mixers; this intensive mixing done before separation gives savings in the amount of wash water needed.

Difficult problems, like the removal of red gossypurpurin from cotton seed oil from glanded seeds, are solved by re-refining the oil with strong lye in excess, as the second step of the operation. Re-refining with lyes in order

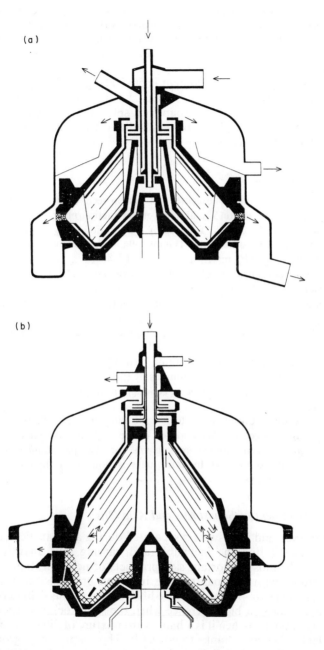

Fig. 4.4 Centrifugal separators: (a) nozzle type; (b) periodically opening. [Reproduced, with permission, from Düpjohann and Hemfort Jr. (1975).]

to remove residual phospholipids is also one of the options applied to soyabean oils of inferior quality. One water wash with 5–(10)% water (4% claimed by Alfa Laval, if the wash-water is used in a two-stage countercurrent contactor) is customary. The washed oil is dried, if because of intermittent storage this is necessary, at 75°C and under 70 mbar residual pressure. The residual soap content is 50–200 mg kg^{-1}.

The use of ammonia instead of caustic soda has been proposed, because its use excludes acid-mediated soapstock splitting (heating of the soapstock liberates ammonia which can be recycled and used again), but it has never found practical application.

4. Miscella "Refining"

Miscella refining is mainly executed on cotton-seed oil from directly extracted glanded meals, obtained on site at a solvent extraction plant. The miscella concentration, preferably 40–60%, is adjusted by evaporation or by mixing of the miscella with oils bought from other mills. The miscella is placed in a closed, jacketed kettle provided with a (gate) mixer and heated to 40°C, a conditioning agent (e.g. phosphoric or citric acid) is added, if necessary. The aqueous 4-N caustic solution with 30% excess on the free fatty acids is also added here. The soap formed is melted by heating the contents to 50–65°C and the slurry, after rapid cooling to 45°C, is centrifuged. The separated miscella is evaporated, the soap stock freed from the solvent by evaporation/drying. If the refining is done not later than 6 h after extraction, the oil obtained is of an excellent light-yellow colour, this being the big advantage of miscella refining on cotton seed oil. The same batch process can be made semi-continuous or fully continuous by the means described for neutralization without solvents, the line mixers being colloid mills or pressure-valve operated (milk) homogenizers.

Alfa Laval offer continuous systems with additional miscella winterizing of the neutralized cotton-seed oil. The addition of the acid and alkaline reagents through metering devices, their intimate mixing via, for example, homogenizers and the separation of the soapstocks by centrifuges is done in a similar way as in the continuous refining systems.

5. The Treatment of Soapstocks and Acid Oils

The soapstock and wash-waters separated during any caustic neutralization process, constituting the inevitable and varying, but anyway annoying, losses, are the most important side products of neutralization. The neutral oil, discarded with the soapstock as inclusion or emulsified products, can never be regained in the form of an edible oil and is thus definitely degraded. There is of course a big difference when the soapstock is derived from the same sort of oil and when it is a haphazard mixture of different sorts because of the lack of separate tanks to collect it. The soapstock of ground-nut oil will, even in the degraded form, bring a higher price than will that of fish oils. However, the amount of soapstock (generally 3–20% of the oil, with

10–40% total fatty matter (content) strongly determines the total economics of caustic neutralizations (this being the reason for using even expensive control/warning systems in the continuous operation). Soapstock splitting is a special part of the total waste-water treatment. A more complete sewage/refinery waste treatment is discussed at the end of this chapter (Section V).

The wash-waters from the centrifugal separations can be added to the separated soapstock, because the amount is much less than that from the batch systems and the regular soapstocks. Soapstocks above 30% total fatty matter are quite viscous and are sometimes solid. Therefore they must be (kept) heated to 60°C for pumping from the tanks to the splitting or drying installations. They ought to be diluted before further handling with water. The huge amounts of second to fourth wash waters of batch refining can be treated separately.

Soapstock splitting can be done in either batch or continuous installations, depending on the amount to be treated. Splitting is basically a very simple operation: sulphuric or hydrochloric acid (diluted to 10%, if corrosion-resistant materials are used) is added in 10–15% excess (above the necessary amount measured by analytical titration) to the soapstock, thereby bringing the pH of the system after completion of the reaction to a value of 2–3.5. If the two immiscible phases can be well contacted, the splitting reaction itself is virtually instantaneous:

soap + sulphuric/hydrochloric acid =

"acid oil" + sodium sulphate, chloride

In the batch variant the mass is heated by live steam to 90°C in an acid-resistant (e.g. Kynar-coated plastic) mild-steel tank for 2–4 h. The previously used lead clad iron or red-pine wooden troughs are out-dated. After mixing (with steam or, preferably, an inert gas, but generally air; the wearing of protective clothing and goggles during this stage is absolutely necessary), the split fatty acids and the oil (constituting analytically the total fatty matter) are left over for settling, in order to separate to acid waste water (bottom) and the fatty "acid oil", as split soapstocks are called. The acid oil can be washed after the acid waste water has been run off with 20–40% water, boiled (mixed) and, after drainage, pumped for storage or transport. Because of the extra effluent problems, this washing is generally omitted.

In continuous systems the necessary acid may be mixed by proportioning pumps to a mechanically stirred prereactor, from where the mixture is transferred to tanks providing substantial residence times for complete separation of the phases. The Alfa-Laval system removes the bulk of acid water from the settlers by decantation. A stainless-steel centrifuge is then used to separate the split and washed bulk. De Smet uses three parallel reactors each of which is charged with a portion of soapstock and treated in the sequence of acidification, water washing and decanting *in situ*. In other automated continuous systems using premixing reactors of small volume, the pH of the as yet unseparated mixture can be used without any time lag for the control of the acid dosing, although practically the acid water outlet is monitored. In any case the main aim is the reduction of the acid excess, which leads to further surface-water polluting loads by forming extra sulphates or chlorides during the neutralization.

A problem in splitting, for example, soya bean soapstocks, is the presence of an oil-in-water type emulsion interlayer (a lyotropic mesophase) which is formed in both batch and continuous systems. The layer is situated between the already split oil and the acid waste water and is mainly composed of sterol glucosides which form liquid crystals.

A possible chemical solution to the problem is the addition of water-in-oil type emulsifiers, like casein or 50–100 mg kg^{-1} ethylhydroxyethyl cellulose (trade name Modocoll) to the acid water before splitting of the soapstock. These agents can counteract the oil-in-water emulsifying action of the sterol glucosides which are not destroyed by the acids.

Instead of a later, postponed measure, Alfa Laval propose the addition of a hydrotropic surface-active agent, called Halvopon (sodium xylene sulphate) to the crude oil before neutralization. Such action should counteract the emulsification of the oil in the soap stock and thereby allow easy release of the neutral oil after the refining, the separation being executed by centrifuges. Alfa Laval claim refining factors of 1.2. Some alkaline-earth metals of alkyl- and aryl-sulphonates are also used as hydrotropes. The process is not used regularly for common types of oil because of the expense of the additives and the need for centrifuges.

Another possibile method is to dry the soapstock in heated rotating drums, whereas the USDA proposes the exact neutralization of the soapstock with mineral acid to pH 7. After drying under vacuum in scraped surface film (Votator) or circulation evaporators, the final (neutral) product can be used as animal (poultry) feed. Sullivan has designed a completely closed neutralizing and drying installation for the soapstock derived from miscella refining with due recovery of hexane from the soapstock in a circulated evaporator/dried (Woerfel, 1983).

The acid oils can be used either directly for soap production or totally hydrolysed to fatty acids and used as such or distilled into separate constituents. The distillation residues (pitch) can sometimes be sold as an additive for road-surfacing (construction) purposes, but it is often burned as a 10% additive in fuel oils.

6. *Automation of Oil Blending and Reagent Addition*

The smooth functioning of a continuous refining line depends strongly on the constant quality of the crude oil entering it. To this end the different deliveries of crude oils of the same type, but of different qualities, must be blended by mixing and homogenizing them in a day tank(s). As the goal is neutralization so their free fatty acid level is necessarily the first practical (quality) parameter on which the blending is based. After having been blended to a suitable (low) average of free fatty acids the "impurities" as expressed by "(minimum) refining losses" and/or by "acetone insoluble phosphatides" must be determined immediately in the laboratory. Free fatty acids and "impurities" produce the so-called "target data" for refining yields on the finished blend, the "pre-mix". This preliminary knowledge makes it possible to create a steady state for at least 12–24 h, during which time the predetermined amounts of reagents like the acids and the alkali lyes can be continuously admixed with the oil in the necessary concentrations and excesses.

In this light the blending of the different deliveries of crude oils to a "premix" with given starting free acidity (free fatty acids) is an operation which could be automated. This means the pumping of the necessary amounts of oils from different tanks in a given sequence into the day tank, measured volumetrically in units of temperature-compensated litres. Weigh scales on load cells can also be used, although because of the large volumes involved they are less practical to operate.

The alkali lye, generally obtained at 50°C in a Be concentration of 49% (w/w) (16–19 N) in tank cars is transferred to caustic dilution tanks, where it is diluted to the required concentration with (preferably condensed or deminerilized) water. The dilution can be easily done in the traditional way, i.e. by float gauging or manual dipping of tank volumes. However, (automatic) ratio control before the tank could also be done by a caustic/water proportionating pump system and sufficient cooling (as heat of dissolution is evolved). Proportionating pump systems with pulsation dampeners are regularly used for adding phosphoric acid and alkali lye in the different steps of the neutralization process.

Automated devices developed for continuous refining lines are degumming and/or neutralization loss monitoring systems. These devices measure to 0.1% accuracy on a constant basis via volumetric flow meters and automatic oil-temperature and moisture-content compensators the weight differences of an oil stream entering and, after washing (and drying), leaving the line. By setting minimum (target) yield or "impurity" values (e.g. based on the "neutral oil and loss" method results and an empirical allowance) the computerized devices compare this present value with the actually measured losses and give constant information, so there is sufficient warning if the line produces oils with losses outside the limits. The control device also makes calibration checks on the flow meters without process interruption at any time and feeds the necessary correction factors to the percentage-yield determining formula. Manual or automatic correction of the set points is optional. Despite automation, one should not forget to keep a check on the unaccounted losses, like water-soluble substances and the amounts of oil which really may be lost. Automatic hydraulic centrifuge sealing devices and the boundary zone balance breakover warning systems have been mentioned already as simple examples of further refinery process equipment automation.

Another place where automation can possibly be applied is in the separation of the soapstock phase after the completed lye treatment in a batch kettle. The objective is to detect the correct cut-off of the emulsion interlayer by means of an, for example, ultrasonic sensor system, after the soapstock has been drained. Turbidity based monitoring has also been described. The pumping of the soapstock is slowed down and stopped at the moment the sensor detects the onset of the pumping of neutral oil instead of the soapstock and/or the emulsion layer. The same sensor systems can be applied on the batch acid splitting of soapstocks, as regards the sharp separation of drained acid oil and the acid water and/or emulsion interlayers.

III. Dry Deacidification Methods

A. Solvent-aided Fatty Acid Extraction

Although it may sound somewhat ridiculous to speak about dryness when using a solvent, the absence or near absence of water allows us to call the process a "dry" one. In general, solvents which are selective for fatty acids should be used, but most solvents also dissolve polar accompanying components, like phospholipids, etc., as well as the free fatty acids. The principle "polar likes polar" is duly expressed when using 98% methanol, which is one of the most preferred solvents. The process is a normal liquid–liquid extraction, and has a well-defined distribution coefficient ($m = 0.024$), as neutral oils (triacylglycerols) are only very limitedly soluble in methanol of this concentration. Contact may be done in different devices: rotating-disc contactors, spray columns, packed columns, in the continuous, or at least multistage countercurrent, way. The process has low refining factors (1.05–1.3), but the equipment required is quite expensive and so is used only for expensive speciality fats having high initial acidity. The fatty acids obtained are of good quality.

B. Deacidification by Distillation of the Free Fatty Acids, (Step II of "Physical Refining")

Deacidification of the free fatty acids by distillation is also the second step in the "physical" refining sequence. Considering the many physical similarities with the removal of volatiles from neutralized and/or bleached oil, this method of deacidification is discussed in more detail in the Section V of this Chapter. It should be mentioned at this stage that, because of the corrosive nature of fatty acid vapours at the high distillation temperatures, all the equipment which comes into contact with them must be constructed of stainless steel, this being a cost factor of considerable importance. There are some distinct advantages to this method. One advantage is the already mentioned environmental benefit because there is no need for soapstock splitting. Another advantage is the low refining factor of the method, being close to 1.1 if other volatile matter (tocopherols, sterols, off-flavour volatiles, hydrocarbons) is simultaneously removed; these other compounds are (partly) removed in any case during the deodorization step.

C. Re-esterification of the Free Fatty Acids

The fatty acids present in crude oils and fats are partly the products of hydrolysis of the triacylglycerols already biochemically synthesized, and partly initial fatty acids which were never esterified due to the incomplete synthesis during the ripening process. Free fatty acids are always accompanied by partial acylglycerols most probably generated for similar reasons and in the same processes as the free fatty acids. The idea of reversing the process of hydrolytic splitting of esters by adding glycerol and removing the water from the reaction products was the basis of this chemical type of "dry" deacidification. Thus merely heating the oil/fat to high temperatures (240–280°C) with or without esterification catalysts (in an inert atmosphere, to oppose oxidation) should deliver an oil of low free acidity, the reaction water being constantly removed. The process (as will be shown in Chapter 5, Section III) results in a mixture of triacyl-, diacyl- and monoacyl-glycerols, their relative amounts being determined by the solubility of glycerol in the reaction mixture. The fatty acid distribution of the acylglycerols is due merely to the laws of chance, resulting in the so-called "random fatty acid distribution". The re-esterified oil has physically and chemically (analytically) quite different characteristics to the starting one. Olive oils have been treated in this way in the past in order to regain their lost "virgin" quality as regards the free fatty acids, notwithstanding the easy detectability of the re-esterified character by the changed fatty acid distribution.

IV. Bleaching or Decolorization

As we have seen, the colour of the crude oil is changed during the processes of desliming and neutralization. Desliming might co-precipitate some less soluble, slightly coloured components and the mineral acids used might decompose some chlorophyll, but the main colour improvement is caused by the removal of the colloidally dissolved lecithin. Caustic neutralization improves the colour of oils, partly by reacting with, for example, the phenolic components (tocopherols, gossypol, sesamol, oxidized hydroxylic fatty acids and their condensates, also their acylglycerol forms, etc.), and partly by co-emulsifying and co-precipitating with them. Despite this the more sophisticated markets demand oils which have an even lighter colour than that of the neutralized ones, so for this and for other reasons given below, decolorization or bleaching by different means is a standard part of both wet and dry (physical) refining operations.

Bleaching or decolorization of oils can be achieved in many ways:

(i) by adsorption on solids;
(ii) by heating;
(iii) by catalytic hydrogenation;
(iv) by chemical bleaching agents.

A. Adsorptive Bleaching

Adsorptive bleaching is a most important method which is used daily in refining practice, and aims not only at the removal of colouring bodies, pigments, etc., but also at the removal of residual amounts of phospholipids, mucilage, oxidized tri- or partial acyl-glycerols, metal traces in ionizable and non-ionizable (complexed) forms, and soap traces which survived the washing of the neutralized oil. As we shall see, some solid adsorptive bleaching agents have the ability to fulfil these and even more objectives. Before going into details of the types and various uses of decolorizing processes, it is interesting to look at their nature and mechanism of action.

Adsorbents are solid materials, generally in granular form, having varying sizes from diameters of a few millimetres down to some micrometres, produced generally by grinding or by aggregation. On the one hand adsorbents must have substantial strength and hardness to withstand these operations and also to support their own weight, if used, for example, in fixed beds. On the other hand they must possess a large surface area per unit weight, meaning a large surface within the internal pores of the particles. Above all that they must exhibit a distinct specificity when adsorbing certain solutes or gases.

As we have already seen, when discussing moisture binding in seed material, there are two types of adsorption that occur between the molecules of the material to be adsorbed (gaseous or dissolved in a carrier medium) and the adsorbent. The first is physical adsorption, which is reversible and is based on intermolecular forces of low strength. The second one is the activated adsorption, or chemisorption, which is often irreversible: the compound once adsorbed might not be reconverted to its original form because of the strong interactions which cause (catalysed) chemical reactions. The temperature and the physical conditions (e.g. the polarity of the milieu) often determine which type of adsorption occurs, although the simultaneous occurrence of both types is also quite common.

1. *Physical Aspects of Bleaching: The Freundlich Equation*

When speaking about a physical model, the Freundlich equation is often used to describe the physical phenomena. This equation takes the form:

$$Y^* = Y_1 = m X_1^n = m \left[\frac{L(Y_o - Y_1)}{S}\right]^n$$

where Y_o and Y_1 (Y^*) are the initial and final (equilibrium) amounts of (colour) substance adsorbed (mass substance/mass oil), respectively, L is the mass of oil, S is the mass of adsorbent added, $(Y_o - Y_1) L/S$ is the apparent adsorption of (colour) substance (mass substance absorbed/mass adsorbent), m is proportional to the distribution coefficient, and n is characteristic of the affinity between the substance and the adsorbent, determining the shape (concave, linear or convex) of the

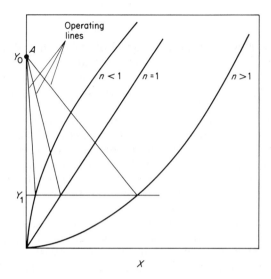

Fig. 4.5 Typical adsorption (Freundlich) isotherms from dilute solutions. [Reproduced, with permission, from Treybal (1981).]

adsorption isotherm (Fig. 4.5): $n = 2\text{--}10$ represents good, $n = 1\text{--}2$ moderately difficult and $n < 1$ poor adsorption characteristics. The equilibrium (colour) substance concentration (Y_1) can be plotted as the ordinate versus the concentration of adsorbate on the adsorbent (X_1) as a colour change caused, for example, by a certain percentage of adsorbent used on the oil (the last member of the equation) on the abscissa using logarithmic coordinates. This plot is a straight line with m as the intercept and n as the slope.

The colour of the oil can be measured by any, but preferably by standardized, type of colorimeter or spectrophotometer, including automated ones. The Freundlich equation can be used for the calculation only if the units, in which colour or absorbance changes are expressed, are additive in nature and proportional to the actual colouring agent concentration. As not only carotenoids and chlorophylls, but also other unknown colour substances are removed during the bleaching of oils, then measurements at one or two prescribed characteristic absorption wavelengths cannot guarantee the fulfillment of above prerequisite. In addition, colour changes caused by simultaneous catalysed reactions are included in the data obtained.

The factors determining the effect of physical adsorption are the type of the adsorbent, the temperature, the medium (solvent) and the concentration of the solute to be adsorbed. In general, but not exclusively: the higher the temperature the smaller the amount of material physically adsorbed; the more dilute the solution the better the absorption; and the lower the solubility in the solvent the more complete the adsorption. All these occur because of the more favourable conditions for attaining equilibrium.

2. *Adsorbents*

(a) *Bleaching earths.* Natural clays like Fuller's earth and bentonite, are aluminium silicates containing magnesium, calcium, sodium and iron. The chief components are the minerals montmorrillonit and attapulgit. When heated (by steam) and dried, these clays disclose their layered molecular-lattice structure, giving rise to macropores in which many active spots for the adsorption of water, oil, phosphatides, soap, colour bodies and metals also become available. Their activity is modest because of the occupation of many more active spots (negative silicate hydroxyl valences) by the metals already mentioned. They are used mainly in physical adsorption, causing no alteration to the substances adsorbed at normal bleaching temperatures (90–110°C) or to the oils treated. They retain from the oil about 30% of their own weight; this is a straight loss.

Activated earths are bentonites treated by steam and sulphuric or hydrochloric acid in order to remove the already mentioned metals occupying the active hydroxylic sites of the silicate lattice. Washed, although still retaining some protons as titratable free acidity, and dried they are the most frequently used bleaching agents in the oil industry. They retain not only all the above-mentioned oil impurities and colour bodies, but also exhibit ion exchange (of metal traces) and diverse catalytic activities (see Section IV.A.3 in this chapter). They can adsorb an amount of oil equal to their own weight and so, despite their usefulness, their use means neutral oil losses, even if the earth is later de-oiled. Oil treated with acid activated earth has a much lower residual colour tan than do those treated in the same amount with an earth of natural activity. However, if more than 2% activated earth is used, the bleached oil may show some increased acidity. This is due to splitting of the adsorbed soap to free fatty acids above 95°C, or less probably to the release of mineral acid traces into the oil. Use of 3–5% activated earth can improve the flavour stability of soya-bean oil, if this can be discounted in the price.

The earths used in a refinery generally belong to one of three types: (i) Natural (non-activated) clay of low activity. A 10% aqueous slurry of pH 8, the apparent bulk density being 0.7–0.9 g ml^{-1} is good for removing soap and for bleaching slightly coloured oils, such as coconut, lard, tallow, and hardened oils, their main advantage being their cheapness and their inertness to causing chemical changes in the oil treated with them. (ii) Acid activated clay of medium activity (e.g. a 10% aqueous slurry of pH 4, the apparent bulk density being 0.5–0.7 g ml^{-1}) is most generally used for routine bleaching. (iii) For special tasks a very active earth. A 10% aqueous slurry of pH 3, the apparent bulk density being 0.4 g ml^{-1} can be kept in storage for oils, like unripe soya-bean or rape-seed oils. The particle-size distribution is important: particles with diameters above 40 μm have less active (available) surfaces, whilst those with diameters below 20 μm not only cause filtration problems but also lead to a lower stability of the finished oil if not adequately filtered out. The minimum dosage should not be less than 0.2%, this amount being necessary to form a thin cake on any filter device. The average dosage used is 0.5–1.5%, depending on the activity and the final colour required. The moisture content of earth [depending on its equilibrium relative humidity (ERH)] is generally 10% in moderate climates, but is much higher under tropical conditions. Some (too dry) earths function better if the oil to be bleached is not dried before bleaching, as is the usual case. The moisture is also adsorbed by the earth, restoring its original layered-lattice structure and thus making the active surfaces in the macropores available for absorption.

(b) *Activated carbons.* Activated carbons are prepared by partially combusting and carbonizing sawdust, coal dust, lignite, wood, coconut shells, sugar or any other carbonaceous vegetable matter with or without the addition of acidic additives, like calcium chloride or sulphuric or phosphoric acid, with the exclusion of air. This step is followed by blowing steam (or hot air for non-oil, e.g. solvent recovery, uses) through the porous mass obtained. Washing, drying and grinding gives various external sizes, internal pore sizes, activities, which serve to remove some specific colours from oils, like coconut and palm-kernel oils. The removal of 5–6-membered ring polycyclic aromatic hydrocarbons, sometimes present in coconut and/or sunflower seed oils is done by using special types of activated carbons (the smaller 3–4-membered rings can be volatilized during high-temperature deodorization). Activated carbons adsorb up to 150% of their own weight in oil; i.e. this oil is lost. However, activated carbon is generally used in lower dosages. The activated carbons are more expensive than bleaching earth. Combinations of 6–9 parts of bleaching earth and 1–4 parts of carbon are sometimes more effective for the removal of certain colour bodies than are their separate uses.

Activated carbons have apparent bulk densities of 0.4 g ml^{-1}, a 10% solution having a pH of 6–10 (some acid-activated types have a pH of 2.2), the particle-size distribution showing more fines under 20 μm diameter than earths.

3. *Chemical Aspects of Bleaching*

Caustic neutralization generally does not decompose any primary products of autoxidation. This means that the amount of hydroperoxides remains at a constant level or even increases if air access if too abundant during this step. The oil to be bleached enters the bleaching step with a certain amount of primary and secondary oxidation products. Hydroperoxides are catalytically decomposed by the protons of acid-activated earths, delivering volatile (e.g. aldehydes, hydrocarbons) and non-volatile secondary oxidation products, such as monomeric oxygenated triacyl-glycerols, containing, for example, hydroxy acids, and dimeric acylglycerols. After a bleaching operation with activated earth, the peroxide level ought to be zero or close to zero. Volatile carbonyls formed in this or previous steps are not removed by the earth, thereby increasing the levels of the *p*-anisidine-(AV) or any other carbonyl-value.

Dimers are formed, although only above 150°C, under the catalytic action of activated earth with glycerides containing oleic acid, and in even higher levels with glycerides containing polyunsaturated fatty acids. The dimers will be partly adsorbed by the earth, together with some polycondensates formed by the reaction between keto- and normal triacylglycerols. Keto compounds are formed by the dehydration of hydroperoxides. Acylglycerides with (poly-)unsaturated fatty acids might be isomerized in different ways by earth catalysis: both positional (conjugation) and geometric (*cis/trans*) isomerization are observed. The $E_{1\,cm}^{1\%}$ values are generally increased during bleaching in the dienoic (232 nm) and the trienoic (268 nm) range. This is due to the partial dehydration of hydroxymonoenes and hydroxydienes by the earth. The triene level forms the balance between the dehydration and subsequent dimerization, the latter diminishing the number of conjugated double bonds.

Sterols may be dehydrated to steroid hydrocarbons and tocopherol dimers might be cleaved to monomers; however, the level of tocopherols might also be slightly reduced by adsorption/chemisorption. Carotenoids might be first protonated by the mobile protons of the acid-activated earth, catalysing the decomposition of the highly conjugated molecules during the mixed earth/heat bleaching process at intermediate (150°C) temperature.

The ion exchange properties of the acid-activated earth are demonstrated by some reduction in the original trace heavy-metal content of the oils, despite the sometimes appreciable amounts (up to 2%) of non-ionic iron present as part of the contaminating minerals not sufficiently washed out during the activation process. Traces of bleaching earth remaining in the oil after filtration might, because of their ionic iron content, serve as pro-oxidation catalysts, causing flavour defects in the deodorized oils if stored.

4. Adsorbent Bleaching Operations

(a) *Calculations on adsorbent use.* The oil to be bleached can be treated in batch, semicontinuous or fully continuous fashions, the process being theoretically similar to that of liquid–liquid extraction, the oil being one solvent, the colour the substance to be extracted, and the adsorbent the extraction solvent. The operation, if executed in stages, might be of either cocurrent or cross-current mode. The countercurrent mode, even in the only feasible two-stage version, is never applied. The method most frequently used in the oil industry is the stage-wise contact filtration of the oil in batches.

The basic material balance can be determined from the equation given below. If a mass of oil (L) contains unabsorbed colour substance(s) the concentration of which are reduced from Y_0 to Y_1 mass colour substance (or colour unit)/mass oil, when adding a mass of S adsorbate-free adsorbent, then the colour substance concentration on the adsorbent will increase from X_0 to X_1 mass colour substance (or unit)/mass adsorbent. If fresh adsorbent is used, $X_0 = 0$ and the material balance is

$$L(Y_0 - Y_1) = SX_1$$

This equation is based on the assumption that sufficient time of contact was allowed for the system to reach equilibrium. $-L/S$ is the slope of the so-called operating line shown in Fig. 4.5, determined by the coordinate points X_0, Y_0 and X_1, Y_1.

The equilibrium conditions can be described by using the equilibrium equation of Freundlich, already given, in the form

$$Y^* = m X^n$$

and for the final equilibrium conditions

$$X_1 = \left(\frac{Y_1}{m}\right)^{1/n}$$

The necessary adsorbent/oil ratio can be calculated for a given change in colour substance concentration using:

$$\frac{S}{L} = \frac{Y_0 - Y_1}{(Y_1/m)^{1/n}}$$

In the two-stage cross-current case, the above equation should be used for the second step starting (Y_1) and final (Y_2) colour substance concentrations. The minimum amount of total adsorbent required is given by:

$$d[(S_1 + S_2/L]/dY_1 = 0$$

and can be calculated using the (reduced) solution of the above differential equation:

$$\left(\frac{Y_1}{Y_2}\right)^{1/n} = \frac{1}{n} \times \frac{Y_0}{Y_1} = 1 - \frac{1}{n}$$

5. Batch-wise Operations

The equipment used for batch-wise bleaching is generally the same type of vertical, cylindrical, conically bottomed, domed, pressure-resistant kettle or autoclave, as used for neutralization. The kettle should be equipped with heating/cooling, adsorbent adding devices, oil charging and discharging lines, vacuum connection, thermometer, etc., and a slow-rotating (40 rpm) gate-type mixer. The more modern variants have a two-speed (slow and very slow) mixer. The oil is first dried under vacuum (40–70 mbar residual pressure) to a moisture level of 0.1–0.3% and heated to and kept at 90–110°C, according to the type of oil bleached. The earth should be sucked (or dropped) in by means of vacuum from a closed, vacuum-resistant container, where it was, preferably, deaerated of the interstitial air occluded in it. The formation of an evacuated bleaching earth/oil slurry (1:3) is also a preferred solution. Evacuation at low temperatures is necessary because of the oxidative damage caused by air and suffered by the hot oil when earth is sucked, dropped or pressed in with little or no precaution as occurs in the present-day bad general practice. The earth acts both as the catalyst and the carrier of this oxidative attack. The bleaching equilibrium is theoretically arrived at after about 5 min, but for the sake of completing other processes (e.g. the cleavage of hydroperoxides) about 30 min is allowed for the adsorption process taking place. There are different views as to whether the admission of air during the rest of the bleaching operation is detrimental or not. In daily practice the vacuum is broken (again with air, instead of with an inert gas such as nitrogen) before starting earth filtration. Ideally the temperature should be lowered before this step to 70°C in order to minimize the oxidative damage. Using closed filters and cooling the oil directly after filtration, the filtration temperature might be kept at 90°C. During filtration the mixer is only used incidentally for some 30 s, giving a burst of movement to the earth settling in the bottom in an attempt to prevent the flow of a thick slurry to the filter; alternatively a very slow speed mixer is constantly in action.

Instead of adding a certain amount of bleaching earth in one step, the addition of a lesser amount in two steps, but not necessarily in equal dosage (split-feed treatment) is sometimes done (10 min interval between doses). Split-feed treatment requires less earth, but is practically more difficult because of the need to break the vacuum twice if addition cannot be managed under vacuum. Two-stage cross-current operation, in which a smaller amount of earth is added than in the single stage method, in two not necessarily equal doses with intermittent filtration is not a better choice: the adsorption of colour bodies and soap are irreversible and so the extra filtration is not needed. Theoretically the counter-current two-stage process with intermittent filtration of the earth gives the highest savings in earth consumption, if the two stages can be built as compartments in the same hull and the process is made continuous.

Fixed or partially fixed-bed filtration might use the residual activity of earth better, because of the increased contact time of the earth with the oil: the oil needs about 1–2 min to pass the normal layer thickness of the press cake formed in the filter. There are refineries where the used filtered earth, left to accumulate in the cells of the plate or plate and frame filter presses or other filtering devices, is used for a longer period than normal. This is done by decreasing the amount of bleaching earth, after having bleached the first two/three batches of oil with the regular amount of earth, down to 50% of the original dosage and yet obtaining satisfactory residual colour and soap results. These data are not relevent for reversible soft oils with high flavour stability demands. This fixed bed "press bleaching" process is a variant of the split-feed treatment, ingeniously circumventing the need for intermittent addition of earth during the course of the same treatment in the bleaching kettle.

6. *Semi- and Full-continuous Bleaching*

Full-continuous cocurrent vacuum bleaching was described by King and Wharton (1949). In this method the oil and earth are mixed and the slurry is sprayed by a pump into the headspace of the upper compartment of a stacked two-tray "single-hull" column kept under vacuum. At 45°C the water and air are flashed off and the slurry withdrawn and heated outside the column to the bleaching temperature. The slurry is then sprayed by a pump into the headspace of the bottom tray in the column where the bound moisture in the earth is released and the oil is bleached. The slurry is led to closed filter presses and, after cooling, to storage.

The Girdler/Votator (1955) apparatus applies in a stacked two-tray "single-hull" column the same "flash off" spraying principle under vacuum, but the counter-current two-stage method with intermittent filtration also involves fixed-bed filtration on partially used earth in the first stage.

In later (semi-)continuous systems (e.g. Pellerin-Zenith or Simon-Rose-downs) concentrated citric acid solution is first added to the oil in the first drying tray of a column composed of three or four stacked trays (Fig. 4.2). Citric acid decomposes most of the soap residues which inactivate bleaching earth. In the second tray the earth is added, the third (and fourth) tray(s) serving as stirred holding units during filtration, all this done under vacuum. Similarly, the addition of, if necessary, some dilute phosphoric acid and the bleaching adsorbent in the evacuated premix kettles of the, for example, Alfa Laval, HLS or De Smet continuous systems, but also in those attached to any other closed (old batch or modernized) system, is generally a big advantage because of the exclusion of (interstitial) air. Air damage is reduced not only during earth (acid) addition and treatment (brisk agitation under vacuum), but also during the closed circuit filtration of the treated oil by any (automated) filter device. The use of stainless steel as construction material and the heat-energy saving aspects are of course additional important extras of the Votator and other modern installations.

Because of the periodicity of the different consecutive steps, the automation of these systems can be quite straightforward. The weighing/degassing/mixing/treating/holding/filtering of the additives and the oil can be controlled individually, because, apart from the material streams, the time and the temperature are the most controlled variables. A dedicated microprocessor can collect, store and retrieve all the data of the controlled loops.

(a) *De-oiling of earths.* As we have seen, the earths (carbons) retain quite a substantial amount of oil, this being one of the most important reasons for trying to minimize the amount of earth used. Striving to recover the oil in good quality is a reasonable, but economically not always feasible, part of the adsorbent bleaching processes. The methods of recovery are different. The most effective method, but not always practical in terms of cost, is the solvent extraction of the oily earth by hexane in a solvent-tight filtering device, i.e. like the method used in modern tank filters. Among others perchloroethylene can also be used as a de-oiling solvent. The high investment and operating costs and hazards are only compensated for in those cases where more expensive and yet oxidatively more stable oils should be regained, like cocoa butter, peanut and olive oils. These latter types of oil are therefore generally handled at a separate site and not in the refinery itself. If solvent-aided extraction is not done within a reasonably short time interval (say 24 h), the regained oil can deteriorate strongly in quality. Low-in-oil solvent-extracted earths have better disposal possibilities than do the only partially de-oiled ones.

Instead of solvent(s), blowing with steam, or better, rapid circulation of hot water at 95°C through the oily filter cake in the filtering device itself at about 5 bar is the most generally used method, delivering about 75% of the oil. Nearly 90% of this oil is obtained in the first 10 min, but in degraded quality, by which it is unacceptable for edible use. This oil is generally used in mixed animal feed, etc.

The Thomson process involves boiling the earth removed from the presses in an autoclave (under pressure) or a kettle with hot water containing, for example, 3% sodium carbonate and 3% salt. This method is of interest, not because it produces a better quality of oil, but because of the problems with the disposal of oily or partially deoiled bleaching earth.

The adsorptive power and catalytic activity of the spent bleaching earth during all these de-oiling (definitely not regeneration) processes should not be forgotten, particularly earth which has been applied on cakes covered with highly unsaturated oils like soya-bean, rape-seed or sunflower-seed oils. Self ingnition by induced polymerization in the air may cause a serious fire hazard, even during temporary storage, transport or when discarding these spent earths.

Soda boiled montmorillonites are transformed back to bentonites. They withhold much water which can only be released if the pH is raised to 4 or less. The regenerated bentonites can be used in cement (concrete) manufacture and the more oily variants as adjuvants for sandy soils. Partially de-oiled bleaching earths containing oil residues constitute an energy source for, e.g. certain animal feeds. This mixing (discarding) of bentonites with extracted meals for animal feed should only be done strictly according to government regulations and they should not be fed to young animals even if diluted.

7. *Filtration/Separation*

(a) *Filter media.* When filtering oils through a bed of solids (layer of adsorbent and/or filter aids), the adsorbent must first be retained by either filter cloths, paper or wire gauzes (screens) which serve as filter media.

In filter presses, an all polyester cloth of twill weave staple fibre of 680 g m^{-2} with a water permeability of 3 $m^3 m^{-2} min^{-1}$ at 25 mm WG is usually used due to its five-fold longer life span as compared with that of cotton or mixed cotton/polyester cloths. When using paper overlayers, a much lighter over and a somewhat lighter under cloth of polyester can be used in order to arrive at an oil filtration rate of at least 200 kg $m^{-2} h^{-1}$. No synthetic fibres can be used if the spent earth is extracted directly by hexane in filter presses especially devised for this purpose because of the danger of static-electricity caused explosions.

The media for leaf and self-cleaning filters are stainless-steel wire gauzes of plain Dutch or Hollander weave of 24 × 110 warp (weft per inch) or the very similar Panzertressengewebe with wire diameters which can produce an aperture of 60–80 μm. In this case the filtration rate should be 350 kg $m^{-2} h^{-1}$.

Filter cartridges used for fine filtration are made of different materials, like cellulose (cotton, paper) or synthetic fibres. They are woven or pressed in different shapes and structures, their aim being the on-line retainment of solid particles of 1–100 μm diameter from oils and fats which have already been filtered.

The structure of the filter media used must provide the highest selectivity for given particle-size ranges with the minimum pressure drop (decreasing pressure involves a decreasing filtration rate).

The rate of filtration can be expressed simply by the common engineering rate relationship:

$$\text{rate} = \text{driving/force resistance}$$

The driving force is the presure applied upstream of the filter medium or the reduced pressure (vacuum) applied downstream of it. The resistance is caused by the amount and nature of the solids formed as a cake and by the filter medium. The variation in filtration (flow) rate and pressure with time depends on the types of earth/oil slurry transport pumps used. The pumps most often used as slurry transporters, centrifugal pumps, give conditions of variable rate and variable pressure, because of the increasing cake resistance. The discharge rate decreases with increasing (back) pressure. Reciprocating (piston, plunger) or rotary (gear or screw type) positive displacement pumps, producing a constant rate, or slurry supply by constant gas pressure (montejus, acid egg) are less often applied.

The general rate of filtration can be roughly characterized by a modification of the D'Arcy equation:

$$v = \frac{dV}{dt} = \frac{\Delta P \times A}{\eta(R_c + R)}$$

where v is the rate of filtration (volume/time), V is the volume of filtrate, t is time, A is the area of the filter (cake) surface, η is the viscosity, ΔP is the constant pressure difference, R_c the resistance due to the filter cake, and R the resistance due to the filter medium covered with the initial cake. The weakness of this formula lies in its non-representation of the compressibility of the cake and the change in R_c during the filtration.

At constant-rate filtration (with a positive displacement pump) we obtain, as a (simplified) characteristic equation:

$$P - p = \eta \, mr \, v^2 \, t$$

where m is the mass of earth deposited per unit volume of the cake, r is the specific cake resistance, P is the applied filtration pressure, and p is the pressure at the filter medium. In this case v and p are constant.

(b) *Filtering devices.* There are four types of filtering devices: (i) filter presses; (ii) leaf filters; (iii) disc filters; and (iv) leaf filters with filter cloth.

The most often used filter presses are the (corrugated or pyramidal) multiple plate-and-frame (Fig. 4.6a) and the recessed-plate (chamber) type (Fig. 4.6b) pressure-operated ones, made of cast iron and originally

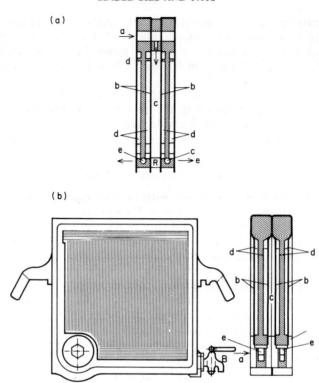

Fig. 4.6 Filter presses: A, Plate and frame type; B, chamber type press; R, frame; a, slurry inflow; b, filter cloth; c, space for filter cake; d, filtrate; e, run-off drain for filtrate. [Reproduced courtesy of Unilever N.V.]

manually operated. These presses are simple, robust and cheap to maintain. The plate-and-frame filters serve for filtering earth/oil slurries of higher solid volumes. The recessed plates of the chamber-type substitute for the frames. The cavities can easily house up to medium amounts of filtered solids. Filter cloths (and if necessary also filter paper) are placed on each side of the plates. The plates have feed and discharge holes at the corners of the plates which form continuous longitudinal channels. The plates and the frames are hung on a pair of horizontal support bars and pressed together during the filtration cycle between two end plates. Electrohydraulic sealing, and automatically thrown out filter cakes ease the requirement for manual labour. A disadvantage is the mostly open-to-air collection of hot filtered oils (via individual taps or cocks) in the side troughs. The modern presses may have a closed manifold, collecting the hot oil with air-exclusion. The filter-presses (at least the chambered ones) must be properly filled with earth: the weight of the final cake must be enough to overcome its adhesion to the filter material. Bottom/top feeding and top/bottom discharge of the filtrate are well-known variants. With larger presses flexibility problems might be

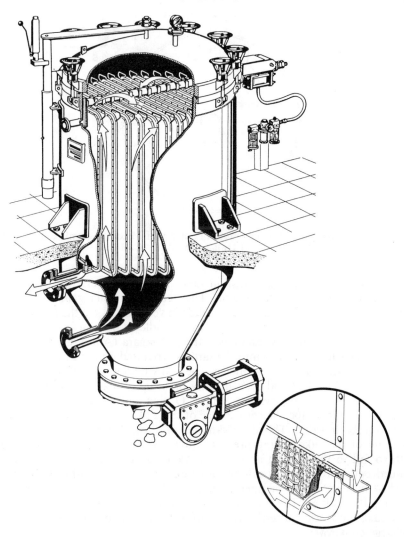

Fig. 4.7 Vertical leaf filter and part of an element. [Reproduced, courtesy of Amafilter B.V.]

encountered if a quick change of feed stock is required, e.g. in a smaller refinery working with smaller batches of different raw materials.

Leaf filters are mounted in vertically or horizontally placed pressure-resistant closed cyclindrical shells (tanks). They contain vertically or horizontally placed round or square flat metallic filter elements (Fig. 4.7). The elements are constructed of stainless-steel wire gauzes attached to stainless-steel frames or plates. One type of element, a filter leaf, consists of three to five layers of wire gauze, the two outer layers are of very fine material in

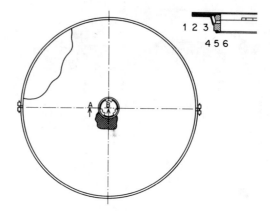

Fig. 4.8 Disc-filter element. [Reproduced courtesy of M & M Schenk Filters Ltd. and Unilever N.V.]

order to perform the actual filtration task. The inner layers function as a drainage chamber for the filtered oil. The package of filter gauzes is fitted into a tubular frame. The leaves are applied in vertical rows in the horizontal or vertical tanks and supported at the top, middle or bottom.

These filters are built by, amongst others, Niagara (Ama), Chemap and Lochem. The slurry pumped into the tank is separated on the gauze and the filtrate is drained out via an outlet nozzle. This nozzle is positioned at the lowest point of the leaf and prevents any build up of material. The nozzles are plugged individually with a simple O ring onto a common filtrate manifold discharge pipe. The advantage of these filters is the usability of the screen on both sides. The disadvantage of these filters is the possible loosening of the cake during pressure fluctuations which might cause the earth to fall to the bottom and some darkening of the filtrate. The cakes retained between the leaves can be loosened and removed without opening the shell by mechanical or pneumatic leaf vibrators and, for example, by reversing the oil flow. The cake is removed via conical bottoms containing a large cake door or butterfly valve and/or by horizontal bottom cake discharge conveyors.

Disc filters consist of a flat-bottomed solid disc on which a coarse screen, serving also as drainage area, supports the fine wire-gauze layers which perform the filtration. The disc and the screen are held together by a clamp ring (Fig. 4.8). Spacer rings are welded to the bottom in the middle around a central hole in the disc and spacer pins are welded to the clamp ring at the outer radius of the plate. Many of these elements with sealing rings around the spacer rings are pressed together by a compression device to a compact stack and placed in a vertical tank. A discharge shaft through the central hole collects the liquid filtrate and a part of the shaft also serves for driving the discs during discharge of the filter cake collected on the gauze (Fig. 4.9). All these components are housed in a cylindrical pressure-resistant vertical

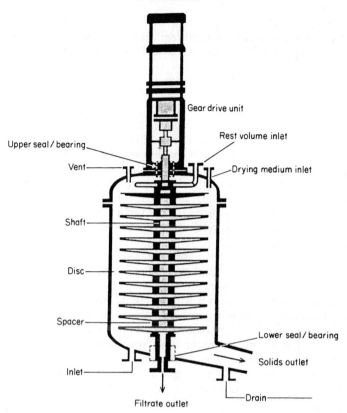

Fig. 4.9 Vertical disc (Funda) filter. [Reproduced, courtesy of Alfa Laval A.B.]

tank, fitted with flanges for charging the slurry, a deaeration valve, nitrogen and steam (solvent) inlets, a pressure gauge and a safety valve.

Vertical tank filters with horizontally mounted metal-gauze disc elements are produced by, among others, Funda/Alfa Laval (top driven shaft) and Schenk/Chemap (bottom driven shaft). Depending on the type of solids to be filtered, the slurry is charged at the top or the bottom of the tank by pumps or other pressure sources in order to minimize the settling of solids in places other than on the gauze between the individual leaves or elements. The filtering layer is formed by the substances present in the oil; i.e. in this case the bleaching earth and/or carbon. Also these filters work batch wise and so, after the gaps between the elements have been filled with earth, the cakes formed must also be discharged. The residual oil (the rest volume) before emptying a full filter is filtered under nitrogen pressure through the two lowest discs. Turbid oil is recirculated through these two discs until clear. The removal of earth cakes is effected by the centrifugal force of rotating the discs, (when the earth is "dry") or by backflushing the cake formed on the upper side of the discs by oil ("wet").

All these filters can be used with or without additional precoating by filter aids. The cake produced from all leaf and disc filters is generally less dry than that produced from filter presses. During all charging, filtering and cleaning steps these filters are kept closed. This means that little or no manual labour is required and the quality of the oils filtered is less impaired by oxidative damage. Fouling is sometimes a real danger. The leaves and elements are expensive, so care in handling them is necessary. If used under up to 4–5 bar pressure, their throughput rates are high.

Some older types of leaf filters used filter cloth outer covers as the filter medium, calking the outer edge of the fabric into the rim of the, for example, U-shaped frame. Cloth-based types in horizontal tanks with vertically standing rows of leaves are built by, amongst others, Niagara, and those with horizontal leaves in vertical tanks by, for example, Sparkler. Horizontally placed elements can be turned into a vertical position inside the tank for easy cleaning, whereas with vertical leaves the cakes formed can be removed by rotating cutting knives.

All the filters described above are best used in pairs, the rise of pressure in one being taken as the signal to change over to a cleaned one. The automation of filters is based on flow controllers, filter openers, turbidimeters and pressure build-up warning signals, all of which can be used to operate batteries of filters by just one operator. The cycle might be fully automatically stopped if the predetermined pressure build up is reached. The other operations are preset by the times required for their completion.

With mainly relatively low amounts of adsorbents, sedimentation or Dekanter type horizontal centrifuges are sometimes used. Hydrocyclones which separate suspended solids into different classes are not applied as fines are not retained by them.

(c) *Cake filtration*. Besides the removal of the solid particles loaded with adsorbed material, the other main task of cake filtration is to obtain sufficient clarity of the filtrate, also called depth filtration. Fine earth particles in the filtered oil with sizes below 20 µm might pass the cake if channels or cracks have been formed in it. (N.B. An increase in fines in the earth is considered to occur because of its transport by pneumatic lines during its handling, which causes some mechanical degradation (attrition) of the particles.)

Cake filtration always starts with circulating the oil/earth slurry out of the bleacher through the filter device, until it is seen to be clear enough to be diverted to the filtered oil tank. However, even after this recirculation a lot of invisible fines could pass the filter, the chances for being trapped being better when the thickness of the cake layer increases: the minimum thickness is about 3 mm, above this it is not the cloth or wire gauze but the cake itself that serves as the filter medium. This works effectively for fine particles, if coarse-sized and large-pored particles form bridges, well wetted by the adhering oil. If the particles give up the oil wetting them, they may clog together and close the passage of oil through the filter medium.

Filter aids, like diatomaceous earth (silicaceous shells of once-living microscopic sea molluscs of the type *Foraminifera* and *Diatomacea*), added in a small amount (0.1–0.15%) can help in forming a better primary medium for the rest of the filtration. The maximum recommended pressure during filtration is about 4–5 bar; if achieved rapidly, this signifies fouling of the filters which should then be cleaned. The metal gauzes are susceptible to corrosion, mainly because of the action of the citric and phosphoric acids which are occasionally added or the hydrochloric acid residues used for earth activation (look for chloride quotations in the supplier's specifications).

(d) *Polishing filtration of fines.* The fines present in (mainly metal gauze) filtered oils, cause because of their iron content, not only a stability hazard above certain levels (10–100 mg kg^{-1}), but also unwanted sediments in the strippers/deodorizers. Fines can be well retained by the use of electrostatic filters: high voltage static fields are developed between the electrodes, trapping all these earth and other inorganic (nickel catalysts of hydrogenation) fines. However, in practical terms small filters, placed in the filtered oil line with, for example, filter paper or woven cotton on their metal-gauze elements (AMA, Seitz), special filter cartridges (Cuno, AMA, Ronningen Petter) sometimes covered by a polyester outer filter cloth sleeve (reasonable savings!), or simple perforated filter tubes at the end of a manifold with the single pipes covered by a cloth bag, and all housed in a shell (Gaflo) are effective in "polishing" out the fines from an already filtered oil. The filtered oils should be stored only for short periods and be well protected against oxidation.

B. Heat Bleaching

Palm oil contains quite high amounts of carotenoids (2000–4000 mg kg^{-1}), which are responsible for its deep-orange colour. The colour carriers of the molecule (chromophor) contain up to 11 conjugated double bonds. These conjugated double bonds are quite sensitive towards acids and even more so towards heat, their decomposition starting above 150°C. The heat-cleaved volatile compounds are β-ionon, toluene, xylene and naphthalene derivatives, and the non-volatile residues (about 80% of the total) are cyclic, non-aromatic compounds with double bonds; in addition there are some dimeric products. Carotenoids are also sensitive to oxidation, again because of the presence of the many double bonds. If oxidized, their cleavage delivers more intensely coloured products than those of the non-oxidized ones. This will lead to technological problems in the heat-bleaching process. It is established that all the cleavage products, at least at the levels at which they occur, are completely harmless.

The heat sensitivity of carotenoids is thus the basis of removing the bulk of the non-oxidized colour bodies by:

(i) in a separate earth pre-bleaching and subsequent high-temperature heat bleaching; or
(ii) a simultaneous activated-earth based medium-heat bleaching process.

The real heat bleaching is executed on either a neutralized or on a crude, filtered (in this case of standardized quality, see below) palm oil. If neutralized, the pretreatment will be a bleaching with about 1.5% acid-activated earth at 105°C under the normal conditions already discussed in Section A.

1. *Palm-oil Qualities*

Because of the huge amounts produced and the wide applicability of palm oil, there have been attempts to standardize crude palm oils by three main aspects: the level of free fatty acids, the amounts of pro-oxidizing trace heavy metals, and the bleachability, the last being determined by using different more-or-less standardized "bleachability tests".

Besides these three measures other analytical values are also used for quality characterization, all of which are aimed at revealing the past "oxidation history" of the oil. Such values are the spectrophotometric $E_{1\,cm}^{1\%}$ values of dienoic and trienoic conjugation, these two values being parallelled by two chemical ones: the amount of hydroperoxides (PV) and of certain (mainly saturated) oil carbonyls, expressed as the p-anisidine value (AV). The combination of the latter two as the total oxidation value (TOV) or Totox ($1 \times AV + 2 \times PV = TOV$) is considered to be of the same quality-describing power as are the three aforementioned values. However, the Totox is a value chosen as the basis for the premium quality oils produced (Lotox = low total oxidation value < 10).

Particularly in the case of palm oils, the final colour to be attained determines the quality of crude palm oil which can be subjected either to a heat-bleaching operation only or to a combination of procedures [total caustic neutralization before or after the heat bleaching/(partial) fatty acid stripping step]. If palm-oil based "white fats" are required, one has to start with an oil which has the qualities called special quality (SQ), super prime bleachable (SPB) Lotox. These most probably deliver after the dry (physical) refining sequence an oil having a final, e.g. Lovibond, colour of 20 Y(ellow), 2 R(ed) or lower if measured in a 5 1/4-inch cell. Standard palm oil qualities, if treated in the dry refining sequence will deliver products with final colour 50Y, 5R or lower measured in the 5 1/4-inch cell. The already mentioned combination of dry and wet processes gives better final colour, because of some chemical decolorizing action of the strong caustic lyes.

2. *Separate Pre-bleaching by Earth*

In this method the crude oils (with 2–6% free fatty acids) are first treated with the dry post-degumming sequence, described previously. The acid used is somewhat more dilute (about 0.2% of a 20% phosphoric or 33% citric acid solution) than that used in the dry post-gumming process and the activated bleaching earth used is about 0.5–1.0%. The neutralization of the added acid in the vacuum mixed slurry after 30 min treatment might be done with chalk, but this causes some difficulties during filtration due to clogging of the filter devices. No air should be admitted either during treatment, or during filtration, because it causes damage to the carotenoids (colour fixation) which is not balanced by the subsequent heat bleaching. If the acid is well dispersed in the oil, e.g. by means of static mixers, the amount of it can be reduced. This makes the use of chalk superfluous, the earth simply adsorbing the lower amounts of finely dispersed acid + hydrated mucilage.

3. *Separate Heat Bleaching*

The actual heat bleaching of this process is executed in practically the same equipment as that used for the stripping of free fatty acids and the removal of volatiles. Therefore the bleaching is actually an integral part of the fourth official step of the classical refining operation sequence: the deodorization/ (deacidification by distillation). Because of the high temperatures ($>210°C$) applied, the construction material should be a highly acid-resistant variant of stainless steel. The apparatus used will be discussed when dealing with that used for the final deodorization step.

The rate of carotenoid decomposition by heat is a time/temperature dependent pyrolytic process, the reaction time needed being hours at 210°C and a few minutes at 270°C. Higher temperatures are not permitted because of the dangers of the pyrolytic decomposition of the triacylglycerols themselves. German Federal Republic rules prescribe that during batch heat bleaching a temperature limit of 220°C for a maximum of 3 h and during (semi-)continuous processes a limit of 270°C for a maximum of 20 min should not be exceeded. Heat bleaching is generally executed in the common vacuum of the oil-stripping processes (2–6 mbar residual pressure) the continuous or semi-continuous equipment permitting rapid heating and cooling of the oil, in contrast to the batch ones. The oil heat bleached in the batch processes should be mixed; this could be done by using an inert gas, but is generally done by the stripping steam of the deodorization.

4. *Simultaneous Earth and Heat Bleaching*

Simultaneous earth and heat bleaching is only applicable in those cases where, despite the lack of high temperature steam or other heating media and/or the lack of stainless-steel strippers/deodorizers, the refiner is still required to produce light-coloured palm oils. Such oils can be produced by using 5% acid-activated earth of high activity, added to the caustic neutralized palm oil under vacuum in a mild-steel batch bleacher (neutralizer/ bleacher), stirring it for 1 h at a maximum temperature of 150°C. In spite of the high earth use and oil losses this method can be used to provide from special quality crudes or the better standard ones quite lightly coloured, even "white" fats. The free fatty acid content, being increased by hydrolysis, can be reduced again by the subsequent low-temperature (180–190°C) batch deodorization.

The use of less-active earth, previously mixed with some concentrated sulphuric acid (sulphuric acid/Fuller's earth = SAFE bleaching) is generally excluded when producing oils for edible purposes. By the free mineral acid catalysed hydrolysis of the oil its free fatty acid content is increased much more than in the previous case in which only acid activated earth was used. The presence of non-volatile mineral acids not absorbed by the earth, together with the free fatty acids produced, urge the producer to neutralize (re-refine) the oil by using dilute lyes. The soap formed must be washed out and the oil must be treated again with small amounts of bleaching earth, making an extra filtration step absolutely necessary. As all these steps involve extra costs, so SAFE bleached palm oils are only used for technical purposes such as soap boiling.

C. Hydrobleaching

Hydrobleaching is only applicable to palm oils, the basis of the operation being an aimed selective saturation of the middle double bond on the carotenoid molecule, causing the destruction of its chromophore. The interruption of the long conjugated chain into two groups of much shorter conjugation, shifts the colour from orange-red to yellow or where there was even more saturation to colourless. This theoretically effective idea cannot be fully realized during a nickel catalyst based hydrogenation of palm oil, because unsaturated fatty acids are also generally cohydrogenated causing slight or sometimes severe changes in the melting and solidification characteristics of the carrier palm oil. If for any reason the latter cohydrogenation is not considered to be a disadvantage hydrogenation is a quite powerful means of decreasing the colour of the final product in this elegant way. Some more details on this will be given in the discussion on hydrogenation.

D. Chemical Bleaching

Many chemical oxidizing agents, like hydrogen or benzoyl peroxides, sodium hypochlorite (whether acidified or in alkaline milieu), dichloromonoxide, if added under intensive mixing to warmed palm oil, destroy the oxidation-sensitive carotenoids (the main colour bodies). This can be achieved even more easily by blowing air through the warmed palm oil, air also decolorizing palm oils quite effectively (but not completely). However, chemical or air oxidation which are often used in, for example, the soap industry, attack quite unselectively not only the carotenoids, but also the unsaturated fatty acids of palm oil, causing severe degradation of the refined quality. For this reason chemical bleaching is out of the question for oils intended for edible use.

V. Deodorization/Deacidification by Distillation/Heat Bleaching

The last regular step of the refining sequence (European terminology) is stripping of the oil with live steam at elevated temperatures and under reduced pressure, called deodorization and/or deacidification by distillation. Because of the combined nature of the process, this operation, which is carried out at higher than regular batch deodorization temperatures, is also step II of the physical or dry refining sequence. The objectives of this process are of physical and/or chemical nature, such as:

(i) the removal of odoriferous volatile compounds;
(ii) the removal of residual amounts of free fatty acids;
(iii) the heat bleaching of carotenoids, if present in large amounts; and
(iv) the rendering of the oil, by means of some (chemical) change, as more flavour-stable during its shelf life.

Because of these different aims, the theoretical backgrounds to them are discussed separately, the physical removal being treated first.

A. Theoretical Background

1. Mechanism of Steam Stripping

Volatile substances dissolved in a non-volatile liquid obey ideally the laws of Dalton and Raoult, these forming the basis of the equilibrium equation of the mass-transfer process.

$$p^* = p_0 x$$

Raoult's law

$$y = p^*/(p_s + p) = p^*/P = p_0 (x/P)$$

Dalton's law

$$p_s + p = P \text{ and } p_s = P, \text{ if } p \ll P$$

where y is the molar fraction of volatile component in the gas phase, x is the molar fraction of the gas phase in the liquid phase (both in kmol kmol^{-1}), p_s is the partial pressure of steam, p^* the equilibrium partial pressure, and p the actual partial vapour pressure of the volatile, p_0 the vapour pressure of the pure volatile component, and P the system pressure composed of p_s and p.

Instead of molar fractions we can use mass/volume concentration units of the volatile compound in the liquid and gas phases, C and c, respectively:

$$C = \rho(\text{oil}) \, x \, \frac{M(\text{vol})}{M(\text{oil})}$$

where $\rho(\text{oil})$ is the specific density of the oil, and M is the molecular weight (in mass kmol^{-1}).

2. Material Balance

If steam is passed at a volume rate of S through a volume of oil V for a time t in a closed (batch) vessel, the concentration of the volatile compound will decrease from the initial concentration, C_0, to some other concentration C. The decrease in the oil concentration is equivalent to the amount of oil carried away by steam, c. Thus the mass transfer rate is

$$-V\frac{dC}{dt} = -KA(C_0 - C) = Sc = mKA\Delta c$$

where A is the interfacial area between the oil and the steam in (m^2), and K and k are the liquid and gas phase side mass-transfer coefficients, respectively [$k = mK$; the distribution coefficient of the volatile between the two phases being $m = \rho(\text{oil})RT/p_0M(\text{oil})$]. High m values represent low vapour pressures and are associated with long-chain fatty-acid-like substances, whereas lower values are due to odiferous volatile compounds with high vapour pressures which are to be removed.

The steam leaving the oil does not actually carry the amount of volatiles (p) it could carry if a substantial contact time and optimal mixing provided the necessary equilibrium conditions (p^*) according to Raoult's law. This fact is discounted by the steam efficiency, E, being expressed, instead of as the partial pressures quotient, as a gas-concentration quotient

$$E = \frac{c}{c^*} \text{ and } 1 - E = \frac{\Delta c}{c^*}$$

where c^* is the equilibrium concentration of the volatile compounds in the gas phase. Dividing this concentration by the third and fourth parts of the mass-transfer equation we obtain

$$SE = mKA(1 - E)$$

$$E = \frac{mKA}{S + mKA} = \frac{1}{(S/mKA + 1)}$$

If the rate equation is solved with the efficiency included, then if the same distribution coefficient (m) is valid for both the gas and the liquid concentrations of the volatile compounds ($c^* = C/m$; $c^* = c/E$; $c = EC/m$), we obtain

$$V\frac{dC}{C} = \frac{SE}{m}dt$$

and, after integration

$$\ln\frac{C_0}{C} = \frac{St}{mV}E$$

and rearranging

$$\frac{St}{V} = \frac{M}{E}\ln\frac{C_0}{C}$$

This equation, first proposed in its simple form by Adriani in 1920 and later independently developed by Bailey, is relevant not only to batch deodorizers, but also to intermittent flow semi-batch deodorizers (Girdler principle). The equation states that the amount of volatiles removed is directly, but logarithmically (i.e. the levels decrease steeply at the beginning), proportional to:

(i) the duration of deodorization;
(ii) the relative steam rate on the oil volume (this at low residual pressure);
(iii) the (high) temperature, because of the logarithmic increase in vapour pressures;
(iv) the steam efficiency (if S/mkA is $\ll 1$, then the efficiency is high, i.e. close to 1); and
(v) is directly proportional to the vapour pressure of the volatiles present, because of the inverse relationship with their distribution coefficients.

B. Process Conditions

1. *Stripping Agent*

Any inert gas could be used as a stripping agent, but, because of its cheapness and inert nature, steam is the most frequently used stripping agent. As steam can be condensed under practical conditions (ambient temperature and even low pressures), this makes its use possible under vacuum. No oxygen should be present in the steam used for deodorization and, therefore, modern boiler-house technology is required.

2. *Amount of Stripping Steam*

When removing volatiles, it is not the mass but the volume of live steam which determines the result (see the above equation). Therefore, higher temperatures and low residual pressures are beneficial for increasing the steam volume. High temperatures, besides increasing the vapour pressure of the volatiles, also reduce the amount of steam needed. However, because of the low concentrations in which off-flavour volatiles are perceptible, their different vapour pressures in their pure state at the deodorization temperatures and the lower steam efficiencies necessitate the use of high volumes of steam in order to reduce their concentrations to below their threshold values. A practical value for the stripping steam demand is, for example, 8000 m^3 of steam per 1 m^3 of oil at 180°C with a condenser pressure of 30 mbar. The live-steam consumption at 10 mbar and 180°C is decreased to 10% of the original value at a temperature of 220°C and 5 mbar. The steam demand of the booster ejector remains almost unchanged in both cases, all these data being relevant to both batch and semi-continuous deodorization. The extra costs of heating the oil to 220°C or above can be easily, although by extra initial investment, compensated for by efficient hot/cold oil heat exchange.

3. *Temperatures*

For reasons already given the temperature should be above 180°C but not much higher than 240°C. Undoubtedly, the increase in the vapour pressure of volatiles also favours higher temperatures. For heat bleaching and fatty acid stripping, 270°C is an admissible temperature with short residence times at the top. Above 210°C special stainless steel should be used as the construction material instead of mild steel. These high temperatures can be achieved by 12–15 and 20–70 bar gauge steam pressures, achieved by a fire-tube boiler and a special closed-circuit high-temperature water-tube boiler, respectively. The use of various low-pressure synthetic solid/liquid (Dowtherm, Marlotherm, Thermex and others) heat-transfer media halves the investment costs of the use of high-pressure water boilers. Although most media are fully volatile under the conditions of deodorization, because of some accidents with burst heating coils in the past, the use of toxic media should be discouraged.

4. *Time*

The operation is not only temperature but also time dependent. Therefore, if the same volume of steam could be passed through the volume of oil in a shorter time, this could physically shorten the time required to arrive at a certain concentration reduction of volatiles. (Chemically determined rate processes are not considered for the moment.) On the other hand, this time reduction would lead to steam velocities which would give rise to unacceptable frothing and severe mechanical oil entrainments. Therefore in batch vessels with a layer thickness of at least 2 m and substantial froth and head space (about 45% of the total volume) no more than 30–60 kg m^{-2} live steam should be passed through, depending on the residual pressure.

The use of improved live-steam admission designs and entrainment-reducing devices makes it possible to apply higher steam rates in some semi-continuous deodorizers (see later), and thus achieve shorter treatment times. Even with these devices, to obtain a completely bland taste at least 3–4 h total residence time is advised in Europe for "difficult oils". Continuous cross- and counter-current designs, very often applied in both the U.S.A. and Canada, are claimed to achieve the desired result in half this time, partly because of the different pattern of residual flavour acceptance by the local consumers. The treatment time required in film deodorization is claimed to be reduced to the theoretically quite possible interval of a few minutes instead of hours (but not for "difficult" oils).

5. *Pressure*

The necessary and, for the whole process, favourable low pressure is nowadays mainly produced by a set of steam ejectors (see later). A combination of mechanical and steam-jet vacuum devices, with direct cooling (barometric) or indirect (surface) steam condensers is also a possibility, if they are well dimensioned.

C. **Physical Changes**

Substances which have a substantial volatility will be removed more or less completely by the stripping during deodorization, depending on the factors already mentioned. The physically removed substances are: free fatty acids, monoacylglycerols (and very little diacylglycerols), alcohols and phenolic compounds, like sterols and tocopherols (α-tocopherol being vitamin E), sesamol, pesticides, hydrocarbons (aliphatic, cyclic and polycyclic aromatic ones) and carotenoids, the volatile scission products of hydroperoxides (the off-flavour carriers, like aldehydes, etc).

Tocopherols are very important volatile components, being the natural antioxidants of oils. The total losses of tocopherols during the whole wet-refining process with a final batch deodorization at medium temperature but longer treatment times is mostly less than 20%. High-temperature stripping may remove 30–50% of the tocopherols which gives rise to a serious decrease in their protective power against future autoxidation. The recovery of tocopherols from the fat catchers is only feasible if the natural product can compete with the purity and (low) price of synthetic (α)-tocopherols. The latter can be recovered from the deodorizer catchpots and scrubber/surface condenser effluents, together with the sterols (see below).

Free sterols, if not dehydrated during the bleaching operation to sterol hydrocarbons, show lower losses on deodorization of about 10% at low (180–190°C) and 20% at medium (210–220°C) or high (240–270°C) temperatures. With the same reservations as for tocopherols, the sterols also can be collected from the deodorization distillates for pharmaceutical purposes.

Organochlorine pesticides are generally not removed by either neutralization or bleaching and so the only stage at which they can be removed is during the deodorization. This is easily performed by the medium- or high-temperature stripping, the low-temperature stripping leaving more of the less-volatile pesticides in the finished oil. The volatility of pesticides increases in the sequence: DDT < DDE < dieldrin < hexachlorhexane < aldrin < lindane.

Polycyclic aromatic hydrocarbons, possible contaminants introduced primarily by the unintended condensation of fuel-gas components on some directly dried seeds and tissues (copra and sunflower seed if they are indeed contaminated) must be removed in two operations: the penta-, hexa- and hepta-cyclic ones by special carbon in the adsorptive bleaching step, and the tri- and tetracyclic ones, which are the more volatile ones, during deodorization.

D. Chemical Changes

The main aim of deodorization, not considering the (residual) fatty acid stripping, is to produce a good quality oil or fat with a long shelf life. This not only means a finished product which is initially odourless and tasteless, and thus flavourless, but also one which keeps this acquired quality for the longest possible time. Removal of off-flavours below their threshold perception level determines the minimum treatment time. The extent to which mainly oxidative chemical reactions, catalysed by traces of non-decomposed or newly formed hydroperoxides, influence the minimum treatment time or, in other words, how far they are rate determining, has not yet been satisfactorily determined. This problem arises immediately when producing oils in the most modern devices in which rising or falling films of oil (thin layers) are contacted with the carrier gas streaming along, instead of steam bubbles, to be saturated with the volatiles when rising in the oil. This type of operation considerably diminishes the physical treatment time. However, despite not being able to name the relevant chemical reaction, there is evidence that the optimum temperature needed to produce a better-keeping "difficult" oil is bound at least to a certain minimum treatment time. Low- and medium-quality soya-bean oils, for example, are better keeping or slower flavour "reverting" if deodorized at 220–240°C for some hours. Too prolonged deodorization at much higher temperatures should be discouraged not only because it impairs flavour quality, but also because of the analytically detectable chemical changes discussed below.

First of all the changes based on the primary and secondary reactions of autoxidation should be mentioned (see Figs 1.5 and 1.6). In spite of their destruction during previous steps, some newly formed hydroperoxides decompose during the heat treatment, partly to volatile compounds (aldehydes, the flavour carriers and hydrocarbons) and partly to other products. The non-volatile decomposition products may remain monomeric or might dimerize under the action of heat. Therefore, proportionally more dimers (alkyl/alkyl or alkyl/oxygen bridged) besides polycondensates (practically also dimers) will be found above 240°C according to the time and temperature applied. The heat-induced geometric (*cis/trans*) and positional isomerizations (conjugation) will also be increased above temperatures of 220–240°C and when applying longer treatment times. Thus more *trans* isomers and also more dienoic conjugation will be detectable, e.g. in oils containing linolenic acid. The trienoic conjugation will be less, not because of the minimized extent of isomerization, but because of the formation of (cyclic) dimers which decrease the conjugation. No harmful levels of (apolar and polar) dimers (and/or polymers, if any) are produced during the normal high-temperature treatment of oils with higher linolenic acid content. Which of these reactions (or others) stabilizes the oil against "flavour reversion" and to what extent is not yet established.

The (non)volatile decomposition products of carotenoids have been discussed amply in heat bleaching in the previous chapter. Whether volatile or not, decomposition products have been found to be harmless in the amounts generally produced.

It is an established fact that steam does not act as a hydrolysing agent under the conditions of deodorization/stripping/heat bleaching, and so no increase in free fatty acids should occur during this treatment. The fatty acids may be mobilized by the inter- and intra-molecular change in their position in the triacylglycerol molecule by the process of interesterification. Without discussing this phenomenon in detail, the possibility at higher temperatures (240°C and higher) of a slight "randomization" of the oil is real. This not only means a chemical (positional) change of the component fatty acids, but also a very annoying change in different physical characteristics, like melting and solidification. Liquid table oils expected to stay free of sediments (solid fat crystals) might suffer under this phenomenon, whereas palm oil might be somewhat improved in its crystallization behaviour if undergoing this spontaneous partial randomization.

Pyrolysis of triacylglycerols begins far above 300°C. Although decreasing amounts of volatile decomposition products are collected during deodorization of oils at, for example, 175°C, new volatile products begin to be produced in constant or even increasing amounts if the oil is heated to higher temperatures. These products are thus generated by pyrolytic decomposition at much lower temperatures than the >300°C mentioned previously. Whether these products are derived from (e.g. oxidized) triacylglycerols or from the accompanying components has not yet been established.

Metal ions, surviving the bleaching step or released to the oil during it, should be inactivated together with those which might enter the oil in the deodorizer/stripper since they might be detrimental to the future oxidative and flavour stability of the finished products. This can easily happen, if the deodorizer is made of mild steel and used at temperatures above 210°C. For this reason the addition of about 100 mg kg^{-1} citric acid (as a 30% solution), preferably in the cooling-down period of the deodorization cycle (below 120°C), is highly recommended. Addition before deodorization is pointless as citric acid is partly decomposed by heat and partly stripped off during the cycle.

The removal of bleaching-earth fines, which are possible autoxidation catalysts because of their iron content, after the main filtration by additional fine filters has already been discussed.

There is no doubt that a heavy pre-oxidation of the crude or partially refined oils before deodorization (thus bad starting quality, bad processing and bad housekeeping) will generate final products which are less flavour- and oxidation-stable and even more so if sensible measures are not taken to protect the oil during deodorization. These measures are:

(i) removal of residual amounts of bleaching earth;
(ii) deaeration of steam and the oil to be treated;
(iii) prevention of air leakage during and after deodorization; and
(iv) use of metal complexing (sequestering) agents.

E. Deodorization/Stripping Equipment

The most frequently used, and the relatively most cheap and most flexible equipment for straight deodorization is the batch-type equipment. The fully continuous and semi-continuous types of equipment are somewhat less flexible, but because of the better economy of the surrounding equipment (heat exchange), the shorter cycle times and also their materials of construction which allow higher working temperatures, they are often favoured by bigger factories.

As there are many different designs, it is not possible to describe them all. Instead, brief descriptions are given here only of those which are essential to the understanding of the working principle or the advantages, where appropriate.

1. *Batch Deodorizers*

Batch deodorizers are generally vertically standing cyclindrical vessels of 5–20 ton volume, with all the necessary appliances for live-steam and oil inlet,

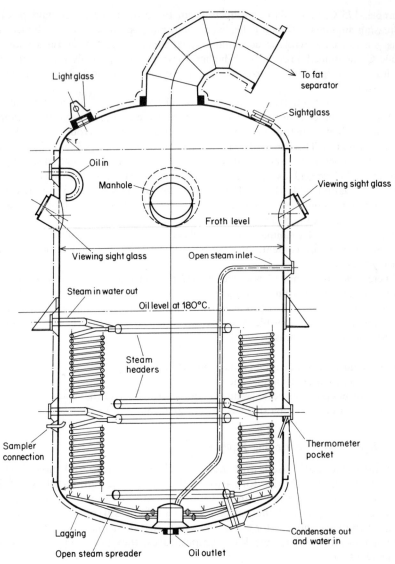

Fig. 4.10 Scheme of a batch deodorizer. (Reproduced with permission of Unilever N.V.)

a shallow bottom and domed top with a broad duct for the vapours and the mechanically entrained oil (Fig. 4.10). They work under a residual pressure of 6–20 mbar, so they have to withstand pressure and are subject to safety controls similar to those of autoclaves. They are mostly heated and cooled by internal helical coil bundles connected to distribution headers (steam, water in the same coil). About 5–6 m^2 heating surface per ton of oil is needed. The deodorizers are provided with thermometers, sighting glasses, manholes, light, and sample and oil inlet and outlet valves. The live steam is admitted at a steady flow rate via a steam-reduction assembly consisting of a pressure gauge, a regulation valve and an orifice plate. The reduced steam enters the vessel via a central box in the bottom and is divided through radial pipes (spiders) with small holes. These similar holes, of about 3 mm diameter and directed upward, are drilled in the radial pipes with 60° alternation in sequence. Considering the expansion of the oil bed by the live steam passing through it, the total area of the holes should be in a certain optimal proportion to the cross-section of the vessel (the total oil surface area) at a given pressure (steam load or steam velocity), in order to obtain small bubbles and not much entrainment of oil.

The material of the batch vessels might be mild steel, but if used at higher temperatures (generally $>210°C$), the material should be special stainless steel (V2A, 316). Welded construction is customary even with batch equipment. Batch vessels are not generally used for fatty acid stripping and if they are only for oils of low free acidity.

The other surrounding pieces of equipment are the (tangential) fat separators with catchpot. Distillate recovery towers (hot vapour scrubbers, absorbing the volatile products and entrained oil by spraying and circulating cooled oil or other medium) or surface condensers are less customary with batch installations.

The vacuum is produced by mechanical vacuum pumps and/or by three or four stage steam ejectors (one or two of them being booster—pressure augmenting—ejectors) annexed to the necessary barometric condensers with hot wells. The fat traps for the cooling water of the barometric condensers and waste-fat collectors are standard to all other installations.

Centrifugal pumps transfer the pre-cooled oil (120°C in the deodorizer) to post-cooling (down to 85°C and finally to 40–50°C) in a dropping tank or through plate, shell and tube or spiral coolers/heat exchangers to the deodorized oil storage tank.

2. *Continuous and Semi-continuous Deodorizers/Strippers*

(a) *General features.* There is some similarity between the most frequently applied continuous cross-current and the semi-continuous deodorizers in terms of the internal arrangement of the different sections. In general both these deodorizer designs are of the multi-chambered compartment/tray types, stacked one above the other, or side by side, the sections representing combined heating/deaerating, heating/pre-stripping, heating/stripping, holding/stripping and cooling/stripping regions.

Medium-temperature heating is done by steam, and high-temperature heating by (very) high-pressure steam or (less preferably, but cheaper) synthetic transfer media, as described previously. As the movement of a synthetic heat transfer medium is by the thermo-syphon principle, so the positioning of the heating unit should be carefully considered.

The high volumes of live steam for stirring and stripping passed through

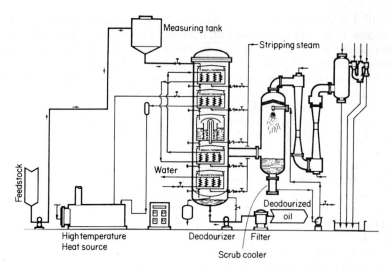

Fig. 4.11 Semi-continuous deodorizing installation with fat scrubber. [Reproduced, with permission, from Kramer *et al.* (1983).]

the system require well-dimensioned spiders and microbubble steam distributor boxes in nearly all sections. In particular, spiders must be designed carefully in order to avoid vibrations caused by the action of live steam. In some of the designs the live steam is added by means of the so-called gas-lift or mammoth pumps, devices having a well-constructed baffle situated above them which is an effective mixer, making possible a three-fold increase in the steam velocities across the tray area (Girdler/Chemetron's third tray; Fig. 4.11). In general these constructions bear much resemblance to the batch system, as the steam has only a single passage and needs the characteristic bubble contact with the oil. Therefore, the live-steam consumption in these two systems is nearly equal. However, the total steam consumption (heating, live, booster and ejectors) of the continuous system can be much lower than that of the batch one, mainly because of heat recovery/exchange achieved by built-in or surrounding equipment (see later).

A difference in the construction and investment costs lies in whether the mostly vertical cylindrical columns are of the so-called "single-" or "double-shell" type. In the single-shell type the side wall of each compartment/tray is that of the column, whereas in the double-shell type the compartments are built in as separate trays (individual kettles or elements) within the shell (Fig. 4.12).

Because of the $>210°C$ working temperatures, the construction material of the compartments/trays is always special stainless steel. The shell can be made of mild steel if the double shell deodorizer is only used as such and for heat bleaching. If used as a fatty acid stripper, the shell should also be made

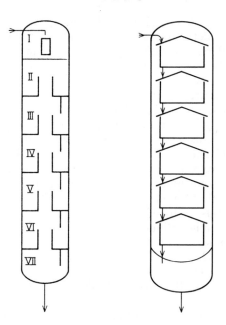

Fig. 4.12 Schematic view of single- and double-shell (semi-)continuous deodorizers. [Reproduced courtesy of Unilever N.V.]

of special stainless steel or normal steel clad by stainless steel. The compartment-type single shell deodorizers/strippers are all made of this latter material. In theory the trays in a shell give better protection against air leakages.

An important aspect is the use of special centrifugal pumps of glandless construction (e.g. magnetic drive), as no air may enter the system when very hot oils are pumped from the system under vacuum through heat exchangers to the buffer or other vessels.

All systems, if connected before the barometric condensers to a (tangential) catchpot annexed to well-dimensioned fatty acid (surface) condensers or distillate recovery towers (fat scrubbers), can also be used for deacidification by distillation, i.e. for step II of physical refining.

(b) *Cross-current continuous deodorizers/strippers.* In cross-current continuous systems the oil flows in a constant stream through vertically stacked compartments and/or compartments positioned laterally/circularly side by side. Within a compartment the oil may pass along vertical baffles placed concentrically, spirally or in parallel in a zig-zag, labyrinth or other pattern. There may also be sections with perforated trays, metal-gauze packings and, most importantly, internal holding vessels in order to arrive at the minimum treatment time (hold-up) necessary for an acceptable residual flavour level (blandness). Stripping steam is admitted in (nearly) all sections via the devices described above. The oil layer thickness (sometimes down to 250–350 mm) and the residence time are controlled by overflow weirs, pipes, etc., connecting the different sections. The residence time is short and varies between 20 and 120 min. They work at a very low residual pressure of 1–8 mbar. The systems can be easily automated, as the working conditions in the different sections are stationary.

Deaeration of oil is obligatory because of the high maximum temperatures applied. The working temperatures are gradually raised to a maximum of around 230–270°C. Oils with a tendency to polymerize might cause problems of clogging. Gradual cooling to 60°C can occur inside or outside the system, with intensive heat exchange (see later). Changing from one sort of oil to another takes some time (a disadvantage), although 30 min changeover times are claimed with the modern systems. By using automated drain valves to connect the sections, many of the cross-current continuous types can also be operated in the semi-continuous mode.

Among many others Blaw Knox, EMI, Cherry Burrel, Krupp, Olier, Extraktionstechnik, De Smet, Lurgi, Gianazza (Deorapid), Pintsch-Bamag, Sullivan-Alfa Laval, HLS, Kirchfeld, Mazzoni (for fatty acids) are well-known manufacturers of this type of deodorizer.

(c)   *Semi-continuous deodorizers/strippers.* Semi-continuous deodorizers can be considered to be intermittent-flow semi-batch ones, and thus the theoretical background is very similar to that of batch ones. Many of these deodorizers use the original Bailey design of 3–7 stacked trays Girdler (today Chemetron's or Cherry Burrel's Votator Division) construction or some derivative of it (Fig. 4.11). Again there are both "double"-shell construction types with trays within a shell (amongst others the original by Girdler, Extraktionstechnik, Simon-Rosedowns, Gianazza-Dovert-Timerized, partly built under licence), and "single"-shell types (amongst others Simon-Rosedowns, Lurgi, De Smet), where the shell forms the wall of the tray.

The working principle is that each portion of the oil is completely subjected to each and the same condition before proceeding to the next step. Thus the oil is either pre-deaerated or deaeration occurs in the first heating tray with prior stirring by live steam. The actual deodorization by live steam takes place in the subsequent trays and the pre-cooling in the last tray, also stirred by live steam. The layer thickness per tray is about 700 mm, which is required to reach the necessary gas density at the spider. Live-steam admittance was described under the general features, given above.

The residual pressure is again between 3 and 8 mbar and is produced by the surrounding equipment, as described for the batch system. The residence time in each tray is predetermined and equal, this being 10–50 min. The system is controlled automatically by the valving system between the trays. The temperature of the oil is gradually raised to a maximum of around 230–270°C, the latter temperature only being used in one of the trays. Gradual cooling to 60°C may take place in the last two trays, or with less trays in external coolers, effective heat exchangers again saving energy.

(d) *Continuous counter-current deodorizers/strippers.* The true counter-current continuous deodorizers are the perforated plate, bubble-cap tray or packed-column types placed in high cylindrical towers. The steam and oil have an obligatory counter-current passage through the columns, the residence time varying between 10 and (with internal holding vessels) 120 min. Known problems which can be solved by the right design are the pressure drop, the uniformity of liquid distribution, the decreasing steam efficiency and, therefore, increasing live-steam consumption, which is, however, lower than that of the batch system, and the high investment costs. Frequently used systems are those built by Foster-Wheeler and Wurster and Sanger.

(e) *Continuous-film deodorizers/strippers.* There are a few systems on the market offering continuous-film deodorizers/strippers which are claimed to have very short residence times, excellent versatility and energy savings, easy cleaning facilities and flavourless and flavour-stable finished products. For example, Cambrian is offering a single internal tray with horizontal baffles in a horizontal cylindrical shell (called Campro Compact). The shell can be slid back to expose the tray for cleaning. The oil enters the system for deaeration and heating. The oil then overflows in plug flow into the deodorizer section consisting of the vertical baffles. Here it is compelled by cross-currently added sparging steam introduced via spiders on the bottom to surpass the baffles in a thin film over a vertical serpentine-like route. The total residence time in the deodorizer (including cooling) is 30 min, the residence time in the deodorizer section is 1 min. Its use for deacidification by distillation is also claimed. A vertically placed variant is also available.

Parkson Corporation offers a low hold-up, short residence time strip deodorizer unit, in which oil and steam are contacted in a tortous pass into a two-phase system, which is then discharged (sprayed) into a cyclone-type separator, with the vapours leaving it overhead to the vacuum booster and/or a distillate scrubber, the treated oil leaving the system at the bottom.

Extraktionstechnik offers a so-called spherical liquid-film (SLF) column for physical refining and deodorization, in which counter-current streams of steam and oil flow in a column around special metal packings, developing spherical falling film strings of oil.

F. Energy Savings

Energy savings are possible, but also necessary because of the high heat-energy levels accumulated in the large amounts of oil treated at the high working temperatures characteristic of modern deodorizing practice. The heat-energy consumption can be reduced by exchanging heat during the heating up and cooling down (e.g. to 120°C) of the oils entering and leaving the installations. The heat-exchanger installations can be built either in the top and/or in the lowest two trays of the deodorizers or placed outside of it. The heat exchangers are of the shell-and-tube, plate, spiral, etc., type. Besides oil-to-oil heat exchange, oil to softened water heat exchange has also been devised, the soft water from the boiler house being used to prevent scale formation on the inner surfaces. The steam developed by the heat exchanger in the cooling section can be used as motive steam, e.g. for vacuum generation. The final cooling is done mainly by water, which can be reused via cooling towers without heat recovery, or, later, by softened water which is used as boiler feed.

G. Steam-jet Ejectors

A steam-jet ejector (Fig. 4.13) is in fact a compressor (like the mechanical vacuum pumps) consisting of three main parts: (i) the steam nozzle, (ii) the mixing chamber, and (iii) the diffuser (a venturi-type device). It converts the kinetic energy of the medium-pressure (e.g. 3–12 bar) driving steam to gas-pressure energy.

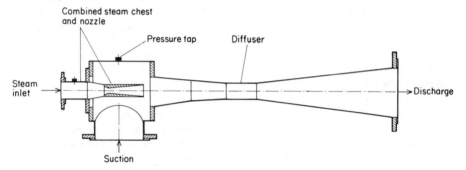

Fig. 4.13 Steam-jet ejector. [Reproduced, with permission, from Gavin (1983).]

In the low-pressure (1–8 mbar) initial range, to be maintained in the deodorizer headspace, the pressure of the condensable gases leaving it are first compressed ("boosted" or augmented) to the pressure of the cooling water leaving the barometric condenser. As in tropical countries this temperature might be 35–40°C, the corresponding equilibrium pressure is 55–76 mbar. In this, for example, using a double (two-stage) steam-jet ejector "boosted" first stage, the compression ratio for each step is between 5 and 7 and for the two thus between 25 and 49, depending on the construction. In countries where low-temperature cooling water is available, just a one-stage first step booster may suffice. The booster(s) is first connected with the fat trap/separators or the scrubbers, before the rest of condensables and non-condensables are transferred to the first barometric condenser. After this the third- and the fourth-stage ejectors (with a second barometric intercondenser between the two) must compress the non-condensable gases to the 1 bar total of the atmosphere.

The system is designed to a given fixed residual pressure in the deodorizer and to a fixed suction capacity, which cannot be varied with changing load. The daily maximum of the cooling-water temperature determines whether all the steam is needed for driving the first-stage ejector nozzle and how much of it is wasted. In the evenings and at night, in cooler seasons, and if cooling towers are used to decrease the hotwell water temperatures to a few degrees centigrade, the boooster-steam consumption of the stage immediately before the barometric condenser can be reduced. This can be done by manual or (better) automatic throttling devices, resulting in 10–15% steam savings. The highest steam consumption (50–75% of the total vacuum system) is in this stage, because it is here where the non-condensed driving steam of the first ejector, if present, is handled (Gavin, 1983).

Artificial indirect cooling of the barometric water by refrigerators, replacing the last stage of the steam-ejector unit by mechanical compressors (vacuum pumps) or water-ring pumps, are further possibilities of economizing on usage. The second stage of the steam-ejector units is the higher steam consumer, so the last measure can be very useful, if for any reason steam generation is a restricting factor.

H. Deodorization Losses

The losses encountered during deodorization are of different origin. First of all, the fatty acids as such cannot be considered as a loss, even less so if collected for later use as a "non-acidified acid oil". Their removal is important, as the residual amount of free fatty acids in the treated oils should be reduced according to the process or to the locally prescribed levels, for fully refined oils this level is mostly 0.05–0.1%. Secondly there are those compounds, besides fatty acids, which regularly distil off because of their volatility under the given conditions. Finally, there are those mechanically entrained oil droplets which are carried away by the mechanical energy of the expanding steam.

The refining factors, as a measure of losses, are at any rate low and in the range 1.1–1.5. The higher refining factors arise because of possibly increased entrainment of oil at high steam velocities and the increased volatility of the aforementioned components at higher temperatures.

1. *Retrieval of Deodorization Distillates*

The main sources of external air and water pollution originating from deodorization/deacidification/heat bleaching are the incompletely condensed/collected volatile distillates and the oily entrainments leaving the deodorizer vessels. In particular the high temperature/low residual pressure units are quite active in producing these pollutants. For this reason the primary booster ejectors/augmentors generally direct these volatile compounds through, for example, (tangential) fat traps/separators, also called catchpots. The use of catchpots is imperative for reducing the fat contamination of barometric condenser water.

In modern installations one- or two-stage direct distillate absorption towers (fat scrubbers) as well as fat separators are used to safeguard the barometric condenser and thereby the hot-well and the cooling tower from any excessive (500–1500 mg kg^{-1} total fatty matter) organic material separation/deposition. In these scrubbers (Fig. 4.11) fatty acids or vegetable oils are sprayed and recirculated as absorbing and cooling liquids for the fatty acids and other volatile compounds. For shorter chain fatty acids ex laurics water is the best absorbent/cooler. The working principle is that the scrubbing liquid is recirculated in the system at a temperature which is not below the "dew-point" of steam at the given pressure. The absorbent liquid is externally cooled, recirculated by a pump and distributed by spraying it either in the inner space of the scrubber counter-currently to or in the high-pressure side of the booster cocurrently with the incoming vapours. As a mechanical trap against entrainment a de-mister/arrester filling of "Knit-mesh" type gauze sheet in the tower is the preferred type, because of the relatively low pressure drop experienced by the system. The scrubber/absorber may reduce the total fatty matter content of the effluent water from the barometric condensers to low (15–50 mg kg^{-1}) or even very low (2 mg kg^{-1}) levels depending on the deodorization temperature and method (batch or semi-continuous) used and the molecular weight of the fatty acids or other volatile compounds. This device greatly reduces the frequency of the necessary maintenance (cleaning) of the filling elements in the water-cooling tower (see below). In the case of laurics, for exmaple, 10% of the scrubbing water can be "bled-off" to the hot-well before replenishment with the same amount of fresh water. By scrubbing a part instead of the whole amount of water, only this 10% has to be deoiled/treated separately because its high total fatty matter level.

The final fat separation and steam condensation can also be achieved by the use of surface condensers, in which case the barometric condensers are completely eliminated.

The barometric water produced is usually 10–50 m^3 per 1 ton of oil depending on the temperature of the incoming water and the type of deodorizer used. For saving fresh cooling water the good quality (low total fatty matter content) warm barometric water can be pumped directly from the hot-wells to the top of the cooling towers, where it is rinsed through appropriate cooling elements (e.g. slat-type wooden grids). The water is forcibly and cross- or counter-currently cooled by air sucked into the tower by motor-driven fans. Any residual fatty material separating from these effluents on the cooling elements of the tower blinds the air access and clogs the pumps, so that the elements must be removed and cleaned from time to time anyway.

In order to reduce the air pollution caused by volatiles in the open-to-air towers, the most objectionable odorous substances present in the recirculated cooling water can be partly destroyed/removed by the addition of oxidizing agents (acidified potassium permanganate) to the system and/or by pumping it through activated carbon beds. The use of the odiferous cooling air for combustion in the boiler house is also an option.

VI. Refinery Effluent Treatment/Fat Recovery

In the absence of scrubbers/absorption installations or surface condensers, quite significant amounts of separated/condensed fatty material may be present in the hot-wells below the barometaric condensers. These and any other type of condensable fatty material must be separated from the refinery/factory effluents via quiescent or other fat settling/separating installations.

The fat (grease) trap (Fig. 4.14) is the traditional device used where 95% of the free fat can separate from the water phase during quiescent gravity settling. This can be accomplished in well-dimensioned settlers because of the genuinely more hydrophobic character of fatty matter and their density differences versus water. The fat trap should be designed for a specific volume of fat to be collected over specific time periods. Easily separating fatty matter ($\leqslant 1$ min) can be settled quiescently in traps dimensioned on the basis of $0.3 \, m^2$ of surface per $1 \, m^3$ of effluent per hour. The trap should be carefully and regularly cleaned (skimmed), as neglected or poorly designed settlers are worse than no settler at all.

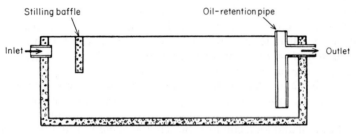

Fig. 4.14 Simple fat trap for removing oil from effluent waters. [Reproduced, with permission, from Watson and Meierhofer (1976).]

It is feasible to have delocalized fat traps near to the larger effluent sources where the water first comes into direct contact with fatty material. Separate fat traps should serve those effluent waters which, for any reason, were contacted with crude unrefined oils, these containing gums, like phosphoacylglycerols, and sterol glucosides, both native or partly decomposed ones. Water used for cleaning the production vessels, floors, internal areas of the production buildings, holding and storage tanks, tank cars, drums, etc., should be passed through fat traps. It is always best to minimize the mixing of different quality collected fatty material while it is in the water phase.

If the total fatty matter content and/or the BOD are still above the accepted norms, all the waste waters must be directed to the central wastewater treatment facilities. Here the treatment consists of pH adjustment, dissolved air flotation and/or chemical flocculation. The flocculating chemicals used are, e.g. aluminium sulphate, ferric chloride and lime (calcium hydroxide), besides polymeric flocculation aids. The precipitate formed adsorbs large molecular organic substances except for water-soluble organic material, like some amino acids, glycerol and sugars. These can be decomposed after filtration through rotating filters by subsequent (anaerobic) biodegradation. This biological treatment can be executed much better outside the plant, i.e. the municipal installations.

The acid oil of certain vegetable oils may contain interlayers (emulsions) which are difficult to separate because of the presence of sterolglucosides or other emulsifiers. This situation requires quiescent gravity settling at pH 2 for better oil recovery. The waste water from the soap stock splitting/acid oil installations may thus be very acidic, which necessitates the use of acid resistant, plastic-clad fat traps. After having skimmed off the oil, the water layer is neutralized with, for example, sodium hydroxide.

The oily/fatty matter collected from the separate primary traps can be used according to their origin and their freshness, the latter being determined by the regularity of the recovery.

VII. Quality Control of the Refining Operation

The incoming crude (pre-degummed or post-degummed) oils should be analysed as a minimum measure, for their colour, free fatty acid content, peroxide value, dienoic and trienoic conjugation (E values), moisture and volatile matter, insoluble impurities (dirt), and trace heavy metals. Of course in this and in any subsequent stages any additional analyses (like softening-slip-point, solid fat content, chlorophyll, erucic acid, phosphorus, tocopherols, wax content, *trans* isomer content, sulphur, mineral oil, solvents, etc.) and tests (acetone insolubility-turbidity, of residual amounts of phospholipids, neutral oil and loss, cup loss, bleachability, etc.) should be executed, if appropriate. Neutralized oils should be tested for their free fatty acid and residual soap content. Bleached oils are controlled on colour, peroxide value, dienoic and trienoic conjugation, and deodorized ones on colour, free fatty acid, and trace heavy metal contents, and last, but not least, their taste when fresh and their taste after some weeks of storage in the dark, at 15°C. A smoke point determination at this stage is an extra test, like the control analyses on chlorophyll, tocopherols, wax content and sulphur. The necessary analyses and tests on side products (the total fatty matter in soap stocks and acid oils, fatty matter recovered from the bleaching earth, deodorizer distillates and fat traps) and on the quality of the auxiliary materials (caustic lyes, acids, bleaching earths, carbons, etc.) should also be executed according to schemes and frequencies being appropriate to the operation.

5

Fat-modification Processes

The first industrially produced light-fat product was margarine, the butter substitute invented by Méges-Mouries in 1869, which has been made since 1872 according to a detailed prescription. The margarine had more-or-less constant physical characteristics due to the fact that, at least in the very beginning, it was based on a single fatty raw material, the oleomargarine. This semi-solid fraction of beef tallow was made by fractionation and so this was the first fat-modification process producing a "tailor-made" fat.

Although by mixing non-fractionated (beef) tallow and liquid oils (mostly rape seed) it was possible to replace the oleomargarine (the substitute for butterfat!), towards the end of the 19th century a growing shortage of tallows stimulated the scientific community to search for a new source of the semi-solid/solid fats. The most exciting solution to the problem (from an industrial chemical viewpoint) was provided by Normann, who proposed, applied and patented the hydrogenation of liquid triacylglycerol oils by nickel catalysts (in Germany in 1902 and in the United Kingdom in 1903). This discovery was a sort of *"deus ex machina"* solution for the problem at the right moment.

The first plant was erected by Crosfield's at Warrington in the U.K. in 1906. Cheap liquid oils, like whale and fish oils became available through the industrialization of the whaling and fishery businesses between 1910 and 1930, and these oils could only be made edible by the process of hydrogenation. Hydrogenated cotton-seed oil was one of the first stable shortenings, produced under the name of Crisco oil by Procter and Gamble in the United States. Hydrogenation is the most versatile fat-modification process, offering new opportunities for the edible and technical fat industry.

The abundance of liquid soya-bean oil since the 1950s, that of liquid sunflower-seed oil since the 1960s, and of semi-solid palm oil since the 1970s all heralded new periods in oil-processing history. This led to the large-scale application of the third modification process—the interesterification of solid and liquid fats. The inter- and intra-molecular exchange of fatty acid/acyl chains in the component triacylglycerol molecules gave rise to the production of a new range of fatty products. This process, combined with

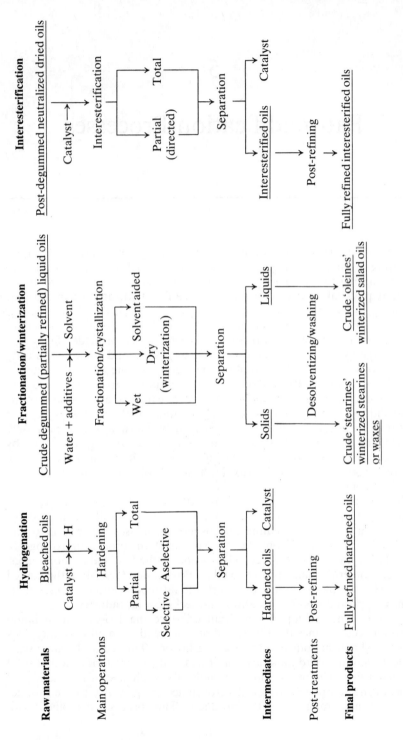

Fig. 5.1 General survey of the hydrogenation processes.

fractionation, not only saves biologically important n-6,9 *cis/cis* (*Z/Z*) polyunsaturated fatty acids, but also provides tailor-made semi-solid fats, needed in many fat products, without resorting to partial hydrogenation. This gives a reasonable processing flexibility, as regards final consistency, if hydrogenation cannot be employed for any reason.

The three modification processes (singly and/or in combination) produce the full spectrum of the modern fatty intermediates used in many high- or low-fat containing margarines, confectionery and other products. These processes give the producer the flexibility to interchange the available fatty raw materials when some are unavailable or in times of shortage due to a bad harvest or political/economic troubles, in other words in an ever-changing market situation. This interchangeability is also an advantage for the consumer, because it ensures products of constant quality at reasonable prices, in certain countries the latter often being regulated by government price control.

A survey of the technological sequences followed by these three modification processes is given in Fig. 5.1.

I. Hydrogenation of Edible Oils

By means of dissolved hydrogen in the presence of a catalyst, hydrogenation saturates some (or occasionally all) of the double bonds originally present in a liquid (poly)unsaturated fatty acid. These acids are in the form of triacylglycerols, the constituents of liquid oils or semisolid fats. This process changes the melting and solidification characteristics of the oils treated and, therefore, it is also called fat hardening. By reducing the original unsaturation of the oils, it also increases their stability with regard to oxidation and thereby in general, their flavour stability. Colour improvements obtained by simultaneous saturation of some carotenoids or reduction of some coloured quinoid compounds are side reactions which are sometimes applied on purpose, like hydrobleaching.

Normann's patent on hydrogenation of fatty oils in the liquid phase was the first industrial application of the discovery made by Sabatier, Senderens and co-workers, between 1895 and 1905, that finely dispersed nickel particles, among other transition metals, could be used for the gas-phase hydrogenation of organic compounds. The use of other metals, like copper, platinum and palladium, as catalysts is an important extension of the industrial possibilities, but in practice less often applied.

Hydrogenation requires that the three reactants of the system are effectively contacted whilst in three different (liquid, gas, solid) states at a suitable working temperature. In the light of this the process is a special example of heterogeneous catalysis, based on the chemisorption of the participants on the catalyst, accompanied by the chemical reaction.

Catalysts are formally unchanged by the reaction and favour special reaction mechanisms, depending on the prevailing conditions. The favouring of a certain mechanism represents the selective nature of a catalyst. In the present case the favoured reactions catalysed by (predominantly nickel) catalysts, are: (i) the saturation of different double bonds present in the fatty acids or other unsaturated compounds; and (ii) the simultaneously possible geometric (*cis*/*trans*: *Z*/*E*) and positional isomerization of double bonds (conjugation, shifting). The possible intra- or inter-molecular cyclization of unsaturated acyl chains is considered undesirable and is the result of careless processing.

A. Theoretical Background

1. *Activated Adsorption*

During the course of heterogeneous catalysis the participating molecules are first adsorbed onto the surface of the catalyst. The generally used nickel catalyst has a high specific surface/volume (weight) ratio. Nickel catalysts may have a specific surface of $50–100 \, m^2 \, g^{-1}$. The catalyst attracts the substrate (oil and hydrogen) molecules not only by physical (van der Waals) forces, but also by forces which, under certain circumstances, involve energies equivalent to those of chemical reactions, i.e. tens of $kcal \, mol^{-1}$. This type of adsorption (chemisorption) is called activated adsorption and, in order to make it practical, a certain amount of external energy in the form of heat or pressure must be added to the system.

In order to explain the micromechanism of saturation and/or isomerization reactions the actual binding of reactants (triacylglycerol molecules and hydrogen gas) on the active sites of the nickel catalyst during chemisorption can be hypothesized in many ways. One of the possibilities is the following (Fig. 5.2).

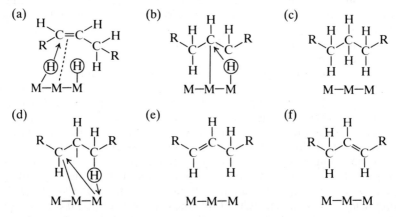

Fig. 5.2 A possible mechanism of the hydrogenation and geometric/positional isomerization processes. M–M–M, metal catalyst. (a) π Complexed; (b) half hydrogenated; (c) fully hydrogenated; (d) σ-bound allylic complex; (e) geometrically isomerized; (f) positionally isomerized.

The electron deficiency of the stable d orbitals in transition metals (M) together with the π-electrons of the double bonds of an unsaturated fatty acid (the subtrate), make possible the formation of a temporary π complex, M $\cdots$ C=C). This complex is based on a loose electron donor–acceptor relationship, consisting of two bonds. On the one hand an electron from a filled bonding π orbital is donated to (overlapped by) a vacant d orbital of the metal, forming a σ bond. On the other hand, an electron from a filled d orbital of the metal atom is (in this case less completely) donated to (overlapped by) a vacant antibonding orbital of the olefin, forming a π bond. In this way the π bond electrons of the double bond obtain a delocalized character, circulating in the whole system.

After the catalyst-induced dissociation, a hydrogen molecule (H—H) is bound to two active sites of the metal catalyst forming (very simplistic) two σ-like (polarized) M—H bonds with it. One carbon atom of the π complexed substrate's double bond may also be temporarily σ bound to the catalyst metal atom, M—C.

From this point on there are a few possibilities (Fig. 5.2).

(i) In the close neighbourhood of the catalyst-bound hydrogen atoms the available free valence of the other carbon in the partially bound double bond can be occupied by one of these hydrogen atoms. In this way the substrate is transformed into a so-called "half hydrogenated state". If subsequently the temporarily σ bound first carbon atom in the half hydrogenated substrate becomes free, it can accept a second hydrogen atom from the neighbourhood, by which its "full saturation" is accomplished. The molecule can then leave the catalyst surface.

(ii) If there is, for example, a shortage of hydrogen atoms, then the π-complexed substrate might lose an α-methylenic hydrogen atom to the catalyst. Such hydrogen atoms reside on the two neighbouring methylene groups next to the double bond. At this moment the catalyst is confronted with an allylic group of three adjoining methyne carbons: CH—CH—CH with strongly delocalized electrons. At least one of these three carbons is also temporarily σ bound to the catalyst. The carbon atom having a high probability of forming this bond is the one at the far end of the original double bond counted from the methylene group originally involved. If the temporarily abstracted hydrogen atom then returns to this methyne (—CH—) carbon atom and the substrate leaves the catalyst, a double bond is restored. However, the double bond is not the original one, but it has slid off sideways in the direction of the originally neighbouring met(y,ile)enic carbon. This "shifting" therefore leads to "positional isomerization".

(iii) With no or no close neighbourhood bound hydrogen atoms on the catalyst surface, the C—C bond of the one σ-bond fixed substrate molecule (M—C) can make a 180° rotation about its axis before the bond is broken. The slowly moving hydrogen atoms have no chance to react and so the double bond is reconstituted, but not in the original configuration. A geometrical (*cis/trans* or *Z/E*) isomerization of this bond thus occurs, by which the substrate leaves the catalyst in the *trans* configuration. Shifting is nearly always accompanied by a change in configuration to the *trans* form.

In the case of serious hydrogen atom shortage, the residence time of the unsaturated substrate on the catalyst surface can be long. Consequently there is ample chance for the repetition of this hydrogen attraction/rejection (Horiuti–Polányi) mechanism by possibility (ii). Therefore, some wandering of the existing double bond(s) along the carbon chain can occur. Although the chances for shifting are equal at both sides of the double bond, in the presence of a second double bond and/or of a carboxyl group in the same (fatty acid) molecule the shift will be performed preferably into their direction. These groups exert some attraction on the similarly negatively charged double bonds. Positional isomerization of two or more double bonds leading to no methylene group interruption, e.g. 1,3-tetradienoic structures, is called conjugation.

Extreme reaction conditions not only cause isomerization, but also cyclization of some very unsaturated fatty acids. This may happen if in the initial phase there is a relative shortage of hydrogen and/or other stronger adsorbed substances, like catalyst deactivating "poisons" prevent the substantial coverage of catalyst active sites by hydrogen.

2. The Heat of Reaction

The saturation reaction, accomplished after the activated adsorption, is exothermic, having an enthalpy of 0.92–0.95 kcal kg^{-1} per iodine value unit (or 27–28 kcal mol^{-1} per double bond present). This quantity of heat is sufficient to raise the temperature of the hydrogenated mass of oil by 1.9–2.1°C per iodine value unit saturated, if the oil is already heated to 130–200°C as usual. However, because the reactor walls lose some of the evolved heat, the net rise will be 1.6–1.7°C per iodine value unit saturated. Besides an exact metering of the hydrogen gas uptake, an efficiently designed cooling system in or outside the hydrogenation vessel is a very effective means of controlling the extent of hydrogenation.

3. Activity and Selectivity

In order to illustrate the chemical concepts of activity and selectivity of the hydrogenation reaction, we can discuss the mechanism of the hydrogenation process from the viewpoint of the three reactants: (i) the liquid substrates (the triacylglycerol molecules); (ii) the solid catalysts; and (iii) the hydrogen gas.

(a) *Selective adsorption of liquid substrates.* It has been found in model experiments that oleic, linoleic and linolenic acid methylesters show different rates of hydrogenation. A close look at their structure shows that linolenic and linoleic acids both contain a typical 1,4-pentadienoic isolated double-bond system in their molecule. These one methylene interrupted pentadienoic groups participate much more readily in complex formation with the catalyst than does oleic acid which contains a single double bond. Linolenic acid has two of these 1,4-pentadienoic carbon systems in a superimposed mode, and so it is not surprising that the relative rate of hydrogenation of linolenic acid is approximately twice that of linoleic acid. The relative rates for hydrogenation of linoleic acid esters, as compared to those of the oleic ones, vary by 50–20:1, depending on the reaction conditions.

A very often encountered isomeric or iso-linoleic acid, derived by the partial hydrogenation of linolenic acid is, for example, 9,15-octadecadienoic acid. This and similar isomeric octadecadienoic acids, containing double bonds separated by more than one methylenic group are also called non-(directly-)conjugatable, in contrast to the easily conjugatable 1,4-pentadienoic structures. These iso-linoleic acids behave according to their affinity to the catalyst like oleic acid and so their rate of hydrogenation is like that of oleic acid.

(b) *Selectivity and saturation sequence.* Although there is competition for the active sites of the catalyst by the different unsaturated fatty acids, their structure differences, described above, co-determine a certain saturation sequence of these fatty acids. This sequence influences fairly strongly the final fatty acid composition of the hydrogenated oil. The phenomenon is called the "fatty acid saturation selectivity" or simply the "selectivity". There are different definitions of selectivity but one of the most often used expresses it as the proportion of the amount of linoleic acid transformed to oleic acid, as compared to the amount of oleic acid transformed to stearic acid. According to this definition, with perfect selectivity no stearic acid will be formed from oleic acid, until most of the polyenoic acids present have not been transformed to oleic acid.

It will be shown later that if hydrogen is present in excess, the (sequential) saturation reaction (from linolenic to stearic acid) can be described as an irreversible, pseudo-first-order one. The step-wise consecutive reaction model, mathematically elaborated by Rakowski (1906) in the form of a series of linear differential equations, was first applied to the hydrogenation of fatty acids in a very extensive form (isomers too) by Bailey (1949). This model was later simplified by Albright (1965) as follows: a unit (basic) rate constant was assigned to the saturation of linoleic acid to oleic acid, all the isomeric dienoic and monoenoic fatty acids formed were considered to be similar in their reactivity to hydrogen and the low chance of direct saturation of linolenic acid to oleic by two molecules of hydrogen was disregarded.

The sequence is as follows:

$$\text{(linolenic)} \xrightarrow{k3} \text{linoleic (isolinoleic)} \xrightarrow{k2} \text{oleic (isooleic)} \xrightarrow{k1} \text{stearic}$$

where $k1$, $k2$ and $k3$ are the pseudo-first-order rate constants. The constants are only averages of the many possible pathways and may change during the reaction, as may the order.

A more detailed terminology expresses this saturation selectivity (S) as the ratio of the relevant rate constants, i.e.

$$S32 = \frac{k3}{k2} \qquad S21 = \frac{k2}{k1}$$

These ratios should be as high as possible, i.e. reaction conditions such that the preference of linoleic acid (L) over oleic acid (O) hydrogenation (linoleic selectivity, $S21$) or that of linolenic acid (Le) over linoleic acid (linolenic selectivity, $S32$) is absolute ($S = \infty$). In real situations the established linolenic selectivity with nickel or partially poisoned nickel catalysts is 2–3, a value close to the theoretical one. The linoleic selectivity is much more variable, being 5–100, this depending not only on the type of the catalyst but also on other factors.

The computer program of Albright based on the above model depicts a family of curves, giving as parameters the different linoleic ($S21$) selectivities (ratios from 2 to 50) as functions of given stearate gains against given starting and final linoleic acid levels. By means of these parameters the selectivity of different hydrogenation reactions or of different catalysts can be compared. It should be remembered that the use of this and other simplified calculations is merely comparative.

High linoleic selectivity gives an oil with the lowest melting point for a given unsaturation. High linolenic selectivity increases the oxidative stability of an oil without changing its liquidity; e.g. when salad oils with no more than 1–2% residual linolenic acid content, starting with soya bean or low erucic rape seed oils, are needed. For reasons to be explained later, copper catalysts improve the linolenic selectiviy to (ratios) 8–15.

The second most important selectivity definition is that for the possible parallel geometric (*cis/trans*, Z/E) isomerization reaction. The *trans* isomerization selectivity

$$Si = \frac{85.6 \ trans \ \text{content}}{IV}$$

represents the ratio of all the *trans* E double bonds formed to all the double bonds remaining, the latter being expressed as the iodine value (IV).

If during the hydrogenation of triacylglycerol molecules the newly formed saturated fatty acids remain randomly or close to randomly distributed, the mole fraction (s^3) of tristearates (SSS) with respect to the mole fraction of stearic acid present (s) would be small and the so-called "triacylglycerol selectivity" (ST) will be high (or close to 1). If all newly formed saturated fatty acids accumulate in the form of tristearates, their mole fraction will be equal to the mole fraction of saturated fatty acids (s) formed during the hydrogenation and this selectivity will be very low (or zero).

Colour selectivity is the ratio of the efficiency of colour-bodies removal (Δ colour) to the removal of double bonds (Δ IV) during hydrogenation.

(c) *Selectivity and structure of the catalyst.* Following Coenen (1976), selectivity can also be discussed from the point of view of a supported catalyst, focussing on its structure as a starting point. The catalyst metal is generally precipitated onto a porous supporting inorganic carrier (natural diatomaceous earth or silica gel) which has particle sizes of 1–50 μm (preferably 10–15 μm) diameter and internal pore widths of 0.5–70 nm. (Particle diameters below 10 μm are difficult to filter.) The precipitated and reduced nickel catalyst itself has an average diameter of 3–5 nm and a high specific surface area per weight. There are some 1 million nickel "crystallites" on a 10 μm diameter porous supported catalyst. However, not only the surface area of the catalyst, but also its accessibility to the reactants plays a decisive role in the selectivity and activity of a catalyst.

The size of a triacylglycerol molecule, containing polyunsaturated fatty acids, is about 1.5 nm. These large molecules must be able to enter the sometimes narrow pores and channels (medium width $c.$ 2.5 nm) of the supporting silicaceous structure. After having been selectively saturated at the active sites of the nickel "crystallites", the molecules should be able to leave them. Besides the easy one-way traffic (mass transfer) of small hydrogen molecules (if available) there is the two-way traffic of the triacylglycerol molecules. The latter can be slow due to the long residence time of partially saturated triacylglycerol molecules on the catalyst surface caused, for example, by the slow transport of new, more unsaturated substrate molecules from the depleted bulk of the oil. This longer residence time leads to inevitable "over" hydrogenation [reduced $S21$ and triglyceride (ST) selectivities].

The mass transport in the pores is regulated by molecular diffusion, the driving forces being the substrate concentration differences, e.g. along a longitudinally extended pore (channel). If the pores are too narrow, the effective diffusion coefficient will be zero and there will be no reaction. If the pores are wide, the pore walls will exert no influence on the diffusion and the bulk diffusion coefficient prevailing outside the pores will characterize the mass transfer. Under steady-state conditions, as we shall see, the mass transfer is equivalent to the chemical reaction rate.

The activity and selectivity of a catalyst can thus only be exploited in supporting structures where the active nickel surface is present in pores wider than 2.5 nm, this being strictly related to the average particle size (average diameter) of the support. Considering this restriction, a catalyst with small average particle size will have a slow filtration rate, but if its pores are wide and shallow it will have high activity and selectivity. If the particles are large, they are easy to filter out, but if the pores are too narrow the catalyst may be inactive. An optimum situation must be determined by the catalyst producer.

4. Mass-transfer Effects

Activity and selectivity should be considered not only on the basis of the affinity of the substrate for the catalyst and the nature of the catalyst (the more "chemical" aspects), but also from the point of view of the mass transfer of the reactants (triacylglycerols and hydrogen, Fig. 5.3) to and from the active sites of the catalyst (the "physical" aspect).

A constant high presence of hydrogen (the reactant least adsorbed by the catalyst) is highly important for the success of the hydrogenation reaction. The solubility (Sol) of hydrogen (vol/vol) in 1 m^3 of oil at temperature t (°C) and 1 bar pressure (abs) is approximately given by

$$Sol = 0.0295 + 0.000497t$$

The hydrogenation is accomplished in a heterogeneous system, i.e. a long and

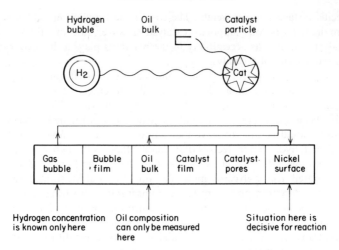

Fig. 5.3 Mass transport in fatty oil hydrogenation. [Reproduced, with permission, from Coenen (1976); and courtesy of Unilever N.V.]

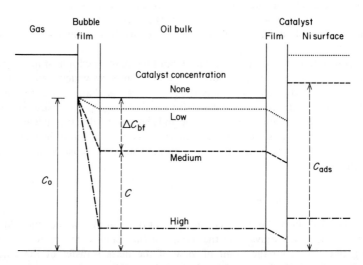

Fig. 5.4 Concentrations of dissolved hydrogen in oil (gas bubbling through stirred oil). [Reproduced, with permission, from Coenen (1976); and courtesy of Unilever N.V.]

complex logistic route for reactants and products. According to Bailey (1949) the supply of hydrogen is from the gas phase (head space or gas bubble) via a boundary film layer to the bulk of the oil, where it is dissolved, this part having a mass-transfer resistance, called bubble resistance. This process will be the same under all physically comparable conditions. From the bulk of the oil the hydrogen passes through an oil-film boundary layer around the interstices of the pores to the catalyst "crystallites". The hydrogen is then adsorbed at the active sites. The driving force of all the above steps is their own concentration gradient (Fig. 5.4). The second resistance is only partly physical (the oil/solid interface), the chemical reaction rate not being separable from the calculated values. Increasing amounts (activity) of the catalyst, absorbing more hydrogen, are quite decisive in establishing the hydrogen concentration near the catalyst surface: with high amounts of catalyst the concentration can be as low as 20% of the saturation concentration or even zero. As regards the transport of triacylglycerols, we have already discussed their two-way traffic to the catalyst active spots where, after chemosorption, the chemical reactions occur.

B. Effect of Process Variables on the Reaction Rate

In the light of the foregoing it is clear that the process has both physically and chemically controlled aspects. The physical aspect is the gas/liquid and less clearly the liquid/solid mass transfer. If we assume that at the start of the process the unsaturated fatty acids are in an overwhelming excess and hydrogen transport (bubble resistance) is problematic, the rate of the reaction can be considered as being hydrogen supply controlled. If in a later stage the hydrogen supply is steady and all the dissolved hydrogen is transferred to the catalyst and consumed in the chemical reaction, then no accumulation of hydrogen will occur in any of the phases. Under these steady-state conditions the process is kinetically controlled and the relevant rates in the different regions are equal. One can write, according to Puri (1980):

$$\underset{\text{Transfer to oil}}{R = kA(c^* - c_1)} = \underset{\text{Diffusion to catalyst}}{k'A'(c_1 - c_s)} = \underset{\text{Reaction at catalyst}}{k''A''c_s U}$$

where R is the rate of transfer and/or chemical reaction, c^* is the saturation hydrogen concentration in oil, c_1 and c_s are the hydrogen concentrations at the liquid and catalyst interphases, respectively, U (Le, L, O) is the unsaturated fatty acid concentration at the catalyst surface, k and k' are the mass-transfer coefficients, k'' is the chemical rate constant, K is the overall transfer coefficient, A and A' are the specific interface (bubble to reactor volume and particle surface to reactor volume, respectively) areas of the mass transfer, and A'' is the specific active nickel (active surface to volume) area.

If

$$\text{Rate} = \frac{\text{all driving forces}}{\text{all resistances}}$$

then

$$\text{Rate} = \frac{c^* - c_1 + c_1 - c_s + c_s}{(1/kA) + (1/k'A') + (1/k''A''U)} = \frac{c^*}{(1/K)}$$

$$\text{Rate} = Kc^*$$

According to Henry's law

$$c^* = bp \text{ (hydrogen)}$$

where b is Henry's constant and p (hydrogen) is the partial hydrogen pressure in oil.

The reaction rate is thus always proportional to the saturation concentration or partial pressure of hydrogen in the oil.

In the case of high bubble resistances (low values of kA) the rate of reaction can be represented by:

$$\text{Rate} = kAc^* = kAbp$$

i.e. by changing the hydrogen pressure and the intensity of mixing at the start of the hydrogenation some control can be exerted on the mass-transfer process.

If resistances to the hydrogen mass transfer are low (mainly kA, but also $k'A'$ are high), then the rate is described by the kinetic equation, which has already been met in discussing unsaturated fatty acid selectivity

$$\text{Rate} = k'' \, U \, (A'' \, c^*)$$

A high hydrogen demand at the start of hydrogenations can be explained by the selective adsorption of triacylglycerol molecules containing different polyunsaturated fatty acids on the catalyst. Besides the degree and type of unsaturation of the fatty acids, the hydrogen demand also depends on the amount and activity of the catalyst and the temperature. Temperature has an overwhelmingly strong effect on chemical reactions and the material-transport phenomena (Albright, 1973). If the hydrogen demand cannot be satisfied for any reason, such as a low capacity of hydrogen production (supply), the nickel catalysts will promote reactions causing geometrical and some positional isomerization (see below).

In the more advanced stages of practical hydrogenations, the situation is somewhat different. After the most unsaturated fatty acids have been (more-or-less selectively) saturated, there is an ample supply of hydrogen to satisfy the hydrogen demand of the remaining unsaturates. The hydrogen demand will depend less explicitly on the amount of catalyst or the degree of residual unsaturation. The rate will be determined by the (reduced) activity of the catalyst (poisoning) and thereby the retarded desorption of intermediate products, having special structures. Such intermediates are well- or non-conjugated *trans,cis*-dienoic fatty acids and *trans*- or *cis*-monoenoic acids formed by partial saturation of genuinely *cis,cis,cis*-trienoic and *cis,cis*-dienoic acids which are involved in the positional isomerization of these double bonds by, for example, the Horiuti–Polanyi mechanism. The most important findings of the practice are summarized in Table 5.1.

C. Isomerizations with Different Catalysts

As has been shown above, at the start of hydrogenation any factor which lowers the hydrogen concentration on the catalyst surface will increase the

Table 5.1 *Effect of factors on the physical and chemical aspects of practical hydrogenations*

	Initial stage				Advanced stage			
Factor increased	Rate	c_s	Selective	trans isomerization	Rate	c_s	Selective	trans isomerization
Physical aspects								
Pressure	+	+	−	−	+	+	−	−
Mixing intensity	+	+	−	−	+	0	0	0
Chemical aspects								
Temperature	+	−	+	+	+	−	+	+
Catalyst amount	+	−	+	+	+	0	0	0
Catalyst activity	+	−	+	+	+	0	−	−
Unsaturation of oil	+	−	+	+	+	0	0	0

+, Increase; −, decrease; 0, little or no influence.

trans isomerization (Si). In the more advanced stages of hydrogenation both geometric and positional isomerization are promoted by poisoning of the catalyst by deactivating substances.

The main substrates for the initial isomerization are the fatty acids with 1,4-pentadienoic unsaturation, which possess a near monopoly for the catalyst active sites. The adsorbed 1,4-pentadienoic group can be transformed into a 1,3-tetradienoic one. This new "conjugated" group is preferentially hydrogenated by a poisoned nickel catalyst via a 1,4-hydrogen addition to monoenoic isomers. This saturation mechanism also produces from linolenic acid the positional isomers of linoleic acid. Many of the double bonds newly formed or shifted are in the *trans* configuration. At equilibrium the maximum level of *trans* (*E*) double bonds is 70% of all the double bonds present.

In nickel catalysed hydrogenations, the genuine oleic acid molecules, because of their lesser affinity for the catalyst, are more or less excluded from the competition for adsorption in the initial stage. These molecules do not participate in any *trans* forming isomerization as long as the bulk of the linoleic acid molecules has not been saturated by the normal 1,2-addition (or, after conjugation, sometimes 1,4-addition) of hydrogen atoms to monounsaturated acids, i.e. oleic or iso-oleic acids.

Copper has a strong tendency to stop the saturation of polyunsaturated fatty acids at the monoenoic level. The 1,4-pentadienoic double bonds are first transformed to 1,3-tetradienoic, conjugated ones. However, after a 1,4-addition of hydrogen, the remaining double bond (which in this case is not necessarily *trans*) cannot be further saturated by hydrogen. The different non-conjugatable isolinoleic acids, their double bonds being separated by more than one methylene group, behave like oleic acid. They cannot be hydrogenated by copper catalysts.

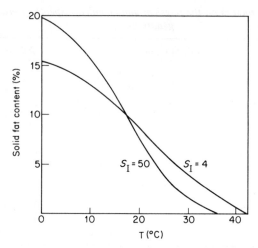

Fig. 5.5 Solid fat content *vs.* temperature curves for soya-bean oil hydrogenated to IV = 95 under conditions of high selectivity ($S_I = 50$) and of low selectivity ($S_I = 4$). [Reproduced, with permission, from Coenen (1976)]

D. Effect of Selectivity Differences on the Properties of Hydrogenated Fats

The amount of solids present in a fat during warming up and/or cooling down can be measured by different means [dilatometry, pulsed nuclear magnetic resonance (NMR) spectroscopy] and can be expressed as the solid fat content (SFC) at the given temperature. The SFC percentages, if plotted versus temperature, can show a steep or a flat curve. Within a narrow temperature range the steep curve shows a rather drastic solid content decrease, which is characteristic of a selectively hardened fat. Conversely, a flat curve, showing within a wide temperature range only small solid content changes, is characteristic of a less selectively hardened fat (Fig. 5.5).

The SFCs measured at 20 and 30°C thus characterize the applicability of hydrogenated fats for different purposes. In Fig. 5.6 the SFC values of soya-bean oil at 20°C are plotted against the SFC values at 30°C. This figure clearly shows which products can be made from this important oil with the available hydrogenation techniques: the useful products are to be found in the narrowly striated area. The more selective the hydrogenation, the further to the right the SFC 20/30 point is located. Cocoa butter, a quite ideally melting fat, is included for comparison. The straight line on the left signifies the complete lack of selectivity: all the solids produced are trisaturated triacylglycerols. A horizontal line drawn from any point to the straight line gives the difference of SFC $\Delta(20–30)$, this being characteristic of the steepness between these two temperatures.

It is evident that hydrogenation raises the melting point of fats. It is less evident that *trans* isomerization can also raise the melting point of hardened

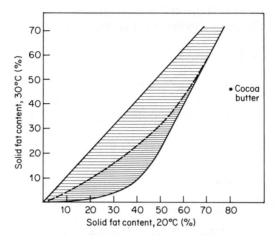

Fig. 5.6 Hydrogenated products from soya-bean oil, accessible by present-day methods: solid fat contents at 20 and 30°C. The solid fat content of cocoa butter at 20 and 30°C is also given for comparison. [Reproduced, with permission, from Coenen (1976)]

fats. The latter effect can be easily understood by the presence of monoenoic *trans* isomers in triacylglycerol molecules, mainly formed in the more advanced stage of the hydrogenation process. Trielaidinylglycerol, if formed from trioleylglycerol, has a melting point of 42°C, compared to 73°C for tristearylglycerol and 5.5°C for the starting trioleylglycerol. It should be realized that trielaidinglycerol retaining its original level of unsaturation melts uniformly and quite quickly at body temperature, if eaten. The *trans* isomer formation in monoenoic acids thus leads to high SFC values (based on *trans* isomers and the inevitable saturates) at a certain temperature, but at a relatively high level of unsaturation.

Natural fats composed of only a few types of saturated (C_{12} and longer chain) and unsaturated (also iso-oleic) fatty acids, which are relatively uniformly distributed within the molecules have a shorter melting range than do most of the very complex, multicomponent ones. The above-mentioned cocoa butter, consisting of large amounts of symmetric palmitoyl–oleyl–stearoyl–glycerol, melts within the narrow temperature range of 20–35°C and has a steep SFC curve. Similarly, a selectively hydrogenated and/or *trans* isomerized fat may contain a lot of triacylglycerols with elaidinic or other *trans*-octadecenoic acids such as disaturated monoelaidinate, monosaturated dielaidinate and saturated elaidinate-oleate, as well as only low levels of tristearate. This type of fat will thus also have a narrow melting range and a steep SFC curve. It is this latter type of fat which is often used in the production of classical, wrapped, non-refrigerator stored block (stick) margarines because, besides having good room-temperature spreadability, it also has good oral melting properties. A disturbing feature is still the liability of mono- and di-stearyl elaidinates to crystallize in the highest melting β-modification (see Chapter 6, Section I.C.5) thereby causing some margarine defects.

A non-selectively hardened fat containing high levels of trisaturated triacylglycerols besides the polyunsaturated ones will have a very wide melting range (up to 50°C) and will show a flat SFC curve. Because of this, this type of fat is very good for use in, for example, shortenings used in the baking industry where the strong β-crystallizing tendency of trisaturates is a real asset.

E. Hydrogenation Techniques

1. *Off-flavour Reducing Techniques*

The reduction of unsaturation generally increases the stability of the fatty acids against autoxidation and polymerization. Reduced unsaturation is also generally advantageous with respect to the flavour stability of the oils. The off-flavour aldehydes which are formed during the decomposition of the autoxidized (poly-)unsaturated fatty acid and which sometimes have very low threshold values of perception, can be very effectively controlled by the partial elimination of the unsaturation of the parent acids.

However, even hydrogenation does not always eliminate off-flavours. Some isomeric fatty acids, developed during the hydrogenation of e.g. linolenic acid, like 8,15-*trans*- and 9,15-*trans*-octadecadienoic acids, if autoxidized and decomposed may give rise to the formation of 6-*trans*-nonenal. This latter compound is the flavour carrier aldehyde of a very well-known hydrogenation off-flavour, called linolenic acid hardening flavour isolated by Keppler *et al.* (1965). It can be perceived by the nose at a concentration of 1 mg per 1 ton of fat. [Hydrogenated soya-bean, rape-seed, fish and whale oils should have an improved flavour stability, because of the (selective) saturation of linolenic and higher unsaturated fatty acids.] In spite of their very increased oxidative stability (lower IV, less polyenoic bonds), but because of the higher trace levels of the above-mentioned and other isomeric fatty acids, these oils might yet show an inclination for flavour reversion and become well changed in character: from "green, beany, painty, fishy" to "hydrogenated, waxy".

2. *Low-temperature Hydrogenation*

The minimum formation of *trans* iso-acids during hydrogenation of oils containing 6–10% linolenic acid can be achieved by hydrogenating them at low temperature (105–120°C) and high hydrogen concentration on the catalyst (say 0.07–0.2% active catalyst, vigorous stirring, pressure 3–5 bar gauge). Although this produces somewhat indiscriminate saturation and, therefore, a low linoleic selectivity ($S21$), there is a low rate of isomerization. If the oil in question is not saturated below IV = 105, a reasonable partial saturation of linolenic acid to linoleic is still obtained without the formation of much (15–20%) *trans* acid. With this special hydrogenation (e.g. in the case of soya-bean oil) at an IV reduction of not more than 30 units the oil contains a residual level of linolenic acid of 1–2% and has a "melting point" of 20°C. This means an increased resistance against oxidative deterioration during, for example, frying, notwithstanding the still ample chances for slight off-flavour formation in the cold.

3. "Iso-suppressive" Hydrogenation

In order to avoid as far as is possible the formation of liquid and hardened oil off-flavours, one has to rely on the so-called "iso-suppressive" type of hydrogenation, which works with 0.05–0.15% somewhat used catalyst and up to 5 bar gauge pressure, the working temperature being higher than before (*c.* 120–150°C). The IV drop with soya-bean oil is 40 units (IV = 95), whereas the melting point is raised to 25–30°C. The danger of off-flavour formation is, therefore, fairly well reduced. If the IV drop is even higher and the working temperature is 140°C, then the product with a melting point of 30–40°C has a long plastic range (a flat SFC curve) and may be used for bakery fat purposes.

If the only slightly hydrogenated oils are fractionated hereafter, in order to remove the solid triacylglycerols formed, a liquid (salad) oil with good physical stability characteristics for even refrigerator storage (SFC at 5°C should be low, say 4%) and yet having increased flavour stability is produced. How successfully this operation can be carried out in practice depends not only the legally permitted residual level of linolenic acid, but also on the local taste preferences and the eating habits of a given market.

Carbon monoxide traces in the hydrogen gas used should be excluded when undertaking low-temperature hydrogenations, because of the temporary, but quite effective poisoning (inactivation) of the catalyst by nickel carbonyls which only dissociate above 150–160°C.

Only a two-stage hydrogenation (see below) of very unsaturated fish oils (pilchard, anchovy, menhaden) and a more drastic one-stage hydrogenation of whale and herring oils to lower IVs (below 60–70) can decrease the unpleasant off-flavours originating from the very unsaturated fatty acids of these oils. Although some hydrogenation off-flavours are still formed (not necessarily the above described linolenic acid hardening flavour), they will be considered more acceptable than those of the original liquid oils. This was one of the biggest merits of Normann's discovery.

The old dilemma of producing a partially hardened oil with a new off-flavour which is considered by consumers to be more acceptable than the initial one will of course persist, until even more selective catalysts and better techniques are available.

4. *"Flash" Hydrogenation*

In order to remove sulphur compounds from coconut oil (from so-called smoke-dried cup copra, if dried by sulphur containing fuel gases) which might cause some rubbery off-flavours even after deodorization, and to improve the flavour stability of lard (from pigs fed with fish meal) containing more than normal amounts of linolenic acid, these commodities can be submitted to a low-temperature (150°C) or normal temperature (180°C) hydrogenation of very short duration (1–3 min). Using fresh nickel catalyst (0.05–0.15% nickel on oil), this "flash" ("brush" or "touch") hydrogenation should decrease the IV by only 2 units.

5. *Exclusion of Cyclic Acid and Polymer Formation by Special Two-stage Hydrogenation*

A very important aspect in the chemistry (and technology) of hydrogenation is the avoidance of cyclic and polymeric side products. These can be formed in oils containing high amounts of fatty acids with 3, 4 or more double bonds, e.g. the already mentioned very unsaturated fish oils and, if cheap, some technical oils such as linseed oil. These unwanted side products may be formed due to hydrogen shortage and/or too high processing temperatures.

Low stirring intensity, bad gas dispersion by the sparger, insufficient hydrogen generation and supply all cause a shortage of hydrogen. In the case of these oils there is at higher (although for less unsaturated oils normal) temperatures (160–190°C) an increased chance of conjugation of the polyunsaturated fatty acids. A too abundant hydrogen supply causes very rapid saturation of these very unsaturated oils. Because hydrogenation is exothermic it may lead to an unintentional rise in temperature, and this may run out of control if insuffcient cooling surface is available, e.g. because of poor reactor design. The high temperatures promote conjugation and geometrical isomerization. Furthermore, conjugation can lead secondarily to intra- or inter-molecular cyclization or polymerization of the acids involved.

The incipient, mainly dienoic conjugation can be detected by means of UV spectrophotometry in the wavelength range 230–234 nm. The extent of cyclic or polymeric acid formation can be analysed by means of the urea adducts of the hydrolysed (freed) fatty acids in the sample: the cyclic and polymeric fatty acids remain in the non-adduct residue.

The chance of occurrence of these reactions can be highly reduced if before submitting the oil to normal hydrogenation the polyunsaturates containing more than two double bonds are (not too vigorously) pre-saturated. Therefore, the best remedy is the already mentioned two-stage hydrogenation process. Using 0.1% fresh nickel catalyst in the first stage the temperature is kept below 150°C until a certain IV has been achieved. To compensate for the low reaction temperature (120–140°C) the hydrogen pressure can be increased (3–5 bar gauge) as a possible counter measure. This method limits the amount of cyclic or polymeric acids formed during the hydrogenation of highly unsaturated oils to below 0.1%. The necessary first stage IV drop can also be calculated for individual oils, as

$$\Delta IV = 0.002 \text{ (original IV)}^2$$

If the refractive index of the fat is measured at 60°C (using the sodium D line as the standard light source), then one unit of IV is roughly equivalent to one fourth place unit in the refractive index, which is generally used for hydrogenation control on the factory floor (besides slip melting point or density measurements).

The IV at which the temperature for the second stage is allowed to rise is called the switch point. In the second phase the process can be continued at elevated temperatures (180°C), but never exceeding the 190°C limit. (In this second stage the pressure can be reduced to 1 bar gauge.) If fish oils of high sulphur content are subjected to this type of hydrogenation, the poisoning of the active sites of the catalyst should be compensated for by using 0.15–0.25% fresh nickel catalyst. (In the second stage the sulphur poisoned catalysts will produce very selectively hardened fish oils with steep SFC/temperature curves.)

High-temperature treatments of long duration (e.g. with fully hydrogenated fats) promote free fatty acid formation which then require (unnecessary) post-neutralization of the fats with inevitable extra losses.

6. Hydrobleaching of Palm (and Other) Oils

The hydrobleaching of carotenoids, mainly in palm oils, is a side effect of fat hydrogenation; the saturation of the small amounts of carotenoid pigments in oils by hydrogen causes a definite bleaching effect. Since green chlorophyll and pheophytin pigments are unaffected by hydrogenation, a previously yellowish oil may be greenish in colour after completed hydrogenation. Green pigments should be removed before or after by means of, for example, acid-activated bleaching earth. A deepening of (soya bean) oil colour by, for example, 5,6-chroman quinon (tocored, an oxidation product of tocopherols) or similar oxidized accompanying fatty components of oils, can often be eliminated by the reduction of the oxidized forms to the original one during the hydrogenation process.

The actual hydrobleaching of palm oil can be done on 0.5% pre-poisoned or used (selective) nickel catalysts, preferably below 170°C. In practice it is done at 180°C, with the inevitable formation of tristearin (raising the original slip melting point of this oil from 35–45°C) and with the loss of more than 1% linoleic acid, which are the ultimate limits of acceptability. Hydrobleaching is a sort of high-temperature "flash" hydrogenation technique of somewhat longer duration. The "colour selectivity" might be used as a measure of the efficiency of this operation.

The techniques described so far are aimed particularly at avoiding unfavourable organoleptic and chemical side effects. The techniques described below are designed primarily for producing oils with distinct melting and solidification (crystallization) characteristics.

7. Medium-temperature Hydrogenation

Medium-temperature hydrogenation is a "non-selective" direct one-stage hydrogenation of oils at medium temperature (150°C), in order to produce vegetable all-hydrogenated shortenings with a broad plastic range. The catalyst used should be not less than 0.05–0.1% fresh nickel, which compensates for the lower rates at the medium temperatures used.

8. Trans *Promoting Hydrogenation*

Trans promoting hydrogenation is the classical one-stage selective hydrogenation technique executed at normal temperatures (preferably at 180°C), but often between 160 and 190°C as at these temperatures the *trans* formation begins to attain its equilibrium high level (about 70% of all remaining double bonds in the *trans* form as measured by IR spectroscopy). The melting points are 40–44°C and so the hydrogenation requires 0.25–0.4% poisoned or sulphur containing catalyst in order to attain the higher melting point in a reasonable time. If the melting point is set at 32–38°C, 0.05–0.15% fresh catalyst will also induce isomerization (the inevitable "poisoning" and the temperature making it *trans* promoting). The final products of this hydrogenation may contain disaturated elaidinate or saturated dielaidinate type triacylglycerols, which can cause certain defects in the so-called one-oil margarines.

9. *Standard Two-stage Hydrogenations*

Standard two-stage hydrogenation can be applied to both oils with high unsaturation and to standard oil commodities. The temperature in the first stage is 110–140°C and 0.1% fresh nickel catalyst is used to suppress *trans* acid formation to 20–25%, while reducing the linolenic acid content to below 1–2% (iso-suppression). The second stage is done at 180°C in order to compensate for the lower S21 selectivity achieved in the first stage.

10. *Full or Near Total Hydrogenation*

The production of fats with a final IV of 2 is sometimes required, e.g. the production of totally hydrogenated lauric fats for soft (tub) margarines or vegetable (non-lauric) "stearines" in bakery fats of high plasticity. This low IV can only be achieved, if the catalysts used have not been contaminated by fats hydrogenated to intermediate IVs. This can be achieved by using 0.2% fresh catalyst. Coconut and palm-kernel oils are generally less fully hydrogenated, the final IV being kept at around 3–4. For the full hydrogenation of palm oil, used catalysts can be applied in 0.5–3% amounts, depending on their starting activity and their previous contamination by lower IV fats.

Producing vegetable stearines from liquid oils with high starting IVs necessitates the use of the two-stage hydrogenation technique, with, if necessary, the addition of extra catalysts at the second stage. Very long residence times at higher temperatures may cause (very easily with the lauric oils) the formation of 0.3% free fatty acids, which should be avoided for the obvious reasons. The use of more active catalysts in this process has already been advised. If not fully hydrogenated, but only higher melting fats are required (for example for tropical margarines, needing more "body" or "stand up"), 0.5% used catalyst can be used. The temperature in a one-stage process will be the normal 180°C.

F. Catalysts

The hydrogenation catalysts most commonly used are based on finely dispersed nickel, supported on sillicaceous or other inorganic carriers, and are activated by reducing in a separate dry "step". In the past catalysts based on nickel formate and "wet" reduced (activated), i.e. in the oil in which they were to be used, were quite widely used. The mode of action of these nickel catalysts is quite similar, the differences being in the mode of their production. As has already been discussed the level of nickel used per batch depends on the initial and final IV and the intended use of the fat, the type and activity of the catalyst, the temperature and pressure applied, and the purity of the oil and industrial hydrogen. The amount generally varies between 0.015 and 0.5% (exceptionally 3%), this based on nickel. In repeated use after reduction, 1 kg is capable of hydrogenating about 5–10 ton of fat before being discarded or regenerated.

Besides the nickel based catalysts copper catalysts are also used, being precipitated and reduced in a similar way to the nickel ones (see Section F.1). Copper has a high linolenic (C32) selectivity (8–15) and stops the hydrogenation at the monoenoic acid level. This is a real advantage if the aim is to remove linolenic acid without obtaining too much *trans* and saturated fatty acids. Oils of this composition are more heat stable and can be kept in refrigerators. The general drawback of nickel catalysts is their tendency to form high levels of *trans* acids, which can only be circumvented by applying the iso-suppresive hydrogenation technique.

Copper, or barium or magnesium promoted copper–chromium or copper–magnesium–silica (siliciumdioxide) catalysts make possible the production of soya-bean or rape-seed oil based liquid oils (of IV = 105–115) with less than 1–2% linolenic acid content (this being the precursor of many reversion off-flavours). These catalysts are, as yet, not very often chosen. Copper is a less active hydrogenation catalyst than nickel and so it cannot be reused as often as nickel and the amounts needed per batch are 5–10 times higher. Copper–magnesium silicate, which is reducible at lower temperatures than nickel, is mostly "wet" reduced *in situ* (in the oil charged into the hydrogenation vessel). "Dry" reduction, if applied, needs very careful control of the conditions. The S32 selectivity of the copper catalyst is diminished, even when there are only traces of nickel contamination. The hydrogenation can only be carried out within a reasonably short time if hydrogen is admitted to the system in an autoclave under pressure (2–5 bar gauge). The process temperature is 150–180°C.

Although nickel has some pro-oxidative activity on oils in the presence of air, with copper even more severe measures must be taken to exclude any air contact during the filtration and storage of the filtered oil. If not, the oils tend to develop (reversion) off-flavours. Therefore traces of copper (of course also nickel) must be eliminated, preferably to below 0.02 mg kg^{-1} (for nickel below 0.1 mg kg^{-1}), by means of thorough post-refining, e.g. by the use of citric acid and bleaching earth.

Nickel/silver catalysts have the same claimed selectivity as copper, whereas homogeneous hydrogenation of oils, primarily by means of chromium (and to a lesser extent by cobalt) carbonyl complexes promises both low *trans* acid formation and high selectivity. Until now low activity, high losses and handling problems and costs have severely hampered the industrial application of nickel/silver catalysts. Raney nickel (a nickel aluminium alloy, from which the aluminium is dissolved by aqueous caustic soda to leave a highly active nickel skeleton) and noble metal catalysts (platinum, palladium, used hetero- or homogeneously) have not as yet been used in the fat industry.

1. Catalyst Production

(a) *The wet-reduced catalysts.* A wet-reduced catalyst is produced by the double decomposition (precipitation) of a concentrated aqueous solution of a water soluble nickel salt (sulphate or nitrate, preferably not the chloride) by a sodium formate solution in the cold, nickel formate being relatively insoluble under these conditions. The precipitate is filtered and sparingly washed and dried. The nickel formate salt contains, if crystallized, two molecules of water of crystallization. Starting with freshly precipitated nickel carbonate, its dissolution in concentrated formic acid is also a possible production route. The (fresh) formate salt (after drying it should retain all its crystal water, this being a prerequisite for good activity) is then finely mixed (milled in a paint-grinding mill) with a vegetable fat, preferably composed of C18 fatty acids, already hydrogenated. The practically chosen proportions of nickel salt to oil, are (1:2)–1:4–1:9. The mechanically stirred mixture is then heated in an appropriately vented closed vessel. Oxygen is kept out by introducing hydrogen gas, or by creating a vacuum, these measures also serving to displace the water and the gases formed (1 kg nickel formate produces 366 l of gas). First, the mixture is heated to 160–170°C, where the salt loses its crystallization water, the thermal dissociation of nickel formate starting at 190°C. The salt dissociates to fine nickel particles and three gases:

$$Ni(OOCH)_2 \cdot 2H_2O \xrightarrow{heat} Ni + 2CO_2 + H_2 + 2H_2O$$

The temperature should never exceed 260°C (250°C being safer) as at higher temperatures there is a chance of fat degradation. This is also the reason for using a previously hydrogenated very stable fat, as the "wet" suspending/protecting medium.

The reduced catalyst particles, being metallic, should have a black colour. Sometimes particles of colloid dimensions are formed, and therefore, filtration of this catalyst can present problems (black run). In this case a filter aid (diatomaceous earth, German: Kieselguhr) can be added after cooling to 90°C and before filtering and/or cooling. The filtration may be carried out on, for example, closed filter presses. The filter separates most of the carrier fat and solids, or, without filtration, the whole mass can be directed to cooling drums on which the fatty mass solidifies. The solid fatty mass is scraped off and broken into tiny flakes which then are packed as fatty catalyst into paper or plastic bags. The activity of this catalyst is the highest if used relatively shortly after production. The $S21$ selectivity is very high.

The decomposition step can be carried out in any (hydrogenation) vessel, provided it contains heating elements (devices) by which the necessary temperature of 250°C for *in situ* reduction (dissociation) can be attained. If such a vessel is available then, instead of already reduced fatty catalysts (flakes), fresh green formate catalyst can be directly purchased and decomposed by the fat processor. However, the bad initial filtration properties of the freshly decomposed catalyst has resulted in its use becoming less attractive.

(b) *The dry-reduced catalysts.* The dry reduction of catalysts preferentially produces a nickel carbonate on a porous support, mainly silicaceous in nature (diatomaceous earth, Kieselguhr; acid precipitated silica gel from waterglass), although aluminium oxide can also serve as a carrier.

Usually a hot aqueous solution of a water soluble nickel salt is reacted under vigorous mixing with a water soluble basic precipitating reagent like sodium (hydrogen) carbonate, keeping the pH of the solution at about 9. After the basic nickel hydroxide has precipitated it is mixed with Kieselguhr. The temperature of precipitation and the calcination of the precipitate on the carrier determine the structure (amorph or crystalline), the optimum dry reaction temperature and the final activity of the catalyst. The process is completed by filtering, washing with sodium (hydrogen) carbonate solution and blowing with steam on the filter press. Nowadays a part of the added silica gel or Kieselguhr is replaced by a sodium silicate solution, which acts as both a reactant and as the support, sodium hydroxide solution being the other reactant. All these more-or-less complex basic nickel silicate precipitates are the so-called "green bases".

The wet or dry precipitations are carried out in mechanically stirred, jacketed vessels, preferably of stainless steel or coated iron (iron is an undesirable contamination). Other requirements are pumps and chambered or plate-and-frame type filter presses or any other suitable type of filter. The green (formate) catalyst bases are then dried; e.g. by spreading them out on trays in forced hot-air driers or on belts in drying tunnels.

The amount of support added is determined such that the reduced catalyst has a final metallic nickel content of 20–25% after being mixed with fat. Although apparently inert, the support can be considered as an activity-promoting part of the reduced catalyst. If the correct reduction temperatures are maintained, a slight interaction (fusion) of the silica with nickel oxides which are formed as intermediates, may increase their stability in both the mechanical and chemical sense.

The precipitated supported catalysts are first dried in hot air on a moving wire-mesh continuous screen and reduced in a hydrogen atmosphere to black metallic nickel, e.g. in ovens similar to those used for pyrites roasting. These ovens are composed of more stacked, mixer arms stirred round cast iron trays ("Teller" roasters). The oven is heated from the outside to not higher than 450–500°C. Very high temperatures cause, instead of slight fusion, real sintering of the support resulting in the closing of the pores or a reduction in their useful diameters. The stirred catalyst mass passes in S form through the slits in the bottom of the trays from the top to the bottom tray. In another variant a kiln-like cylindrical horizontal oven is heated from the outside by burners placed inside an insulating mantle. The catalyst mass is shovelled by helical baffles on the wall from the charging to the discharge point. In both cases the reduced black catalyst is first cooled in a chute by hydrogen gas, entering the systems in counter-current flow. No immediate admission of air is permissible even after this cooling stage because of the pyrophoric nature of the black catalyst. The cooled catalyst can be made passive by step-wise (slow) admission of air to a gradually less pyrophoric powder, but for the sake of better protection against air it is generally mixed with molten hydrogenated fat of high melting point (e.g. 50°C), the fatty mixture being further cooled on drums, scraped, broken into flakes and packed, as already described.

The mechanical properties of the carriers after reduction should be scrutinized carefully with an eye to the proper internal structure (pore and channel dimensions), mechanical wear and particle size distribution. Sometimes previous washing, roasting and winnowing of the different guhrs is required. The particle-size distribution of the reduced supported catalysts should be in the range 1–15 μm, as particles below 10 μm and particularly the "fines" cause filtration difficulties. To achieve good mechanical properties during re-use requires a narrow average particle size of around 10 μm, one of the most difficult objectives of catalyst production.

Sodium hydroxide, sodium hydrogen carbonate and many other oxygen or non-oxygen containing precipitating reagents can be used instead of sodium carbonate. Instead of the most commonly used nickel sulphate, nickel nitrate can also be used. The chloride salts are considered to be less favourable, because of residual trace chloride ion contamination. Traces of sulphate may be reduced at high temperatures to sulphide, a permanent deactivator of catalysts. A strictly agreed (normalized) activity testing of the catalyst in a stated type of testing oil is an important part of the quality control. The concentration is always calculated on the basis of the metallic nickel content, which should always be stated, when purchasing and using hydrogenation catalysts.

The activity of these supported catalysts can be deliberately reduced by adding flowers of sulphur (5–7% on nickel) before the dry reduction of the green mass. This selective (because more isomerizing) catalyst has a more constant quality and better activity than does a haphazardly poisoned catalyst (containing 2–3% sulphur/nickel) prepared on the hydrogenation floor by using fresh (active) catalysts for the repeated hydrogenation of poorly refined fish or rape-seed oils containing variable amounts of sulphur compounds.

(c)   *Electrolytic precipitation.* According to different patents (Sieck, 1936; Patterson, 1938), if metallic nickel plates are immersed in 1% sodium chloride solution and are subjected to electrolysis, the plate, connected as anode, is corroded by the chlorine atoms developed there and forms nickel chloride. The plate connected as cathode produces metallic sodium, which first reacts with water, to give sodium hydroxide, which in turn reacts with the nickel chloride already formed. This reaction reforms the sodium chloride electrolyte and produces nickel hydroxide, which is the precursor of the catalyst. The nickel hydroxide, constantly mixed with equal amounts of supporting silica gel or Kieselguhr, is removed from the system by filtration. In order to keep the pH at 9–9.5, carbon dioxide gas is constantly bubbled through the electrolysing system, which counteracts the formation of basic nickel chloride and reduces the chloride ion level of the electrolyte. The system is warmed to 50°C and is constantly mechanically mixed. The electric current density is 5 A cm^{-2}. The catalyst washed, dried and reduced by hydrogen gas is discharged into hydrogenated fat and has a high activity.

The original application of the anodic oxidation of nickel was developed by Lush (1923) for the *in situ* electrolytic corrosion of metallic nickel shavings placed in a metal basket and connected as anode in a sodium carbonate solution. The cathode used was a nickel plate.

2. Commercial Catalysts

Catalysts are produced by many producers of bulk and/or speciality chemicals (amongst others Girdler, Harshaw, Sud Chemie, Unichema International, van den Bergh and Jurgens), most of them as 20–25% reduced nickel on a fat basis, chemically or electrolytically precipitated, passivated or not, with a fatty carrier added. Instead of the hardened fat carrier, new lozenge or drop-like carriers have also been developed, which have a smaller surface/weight ratio. The reduced metal on the carrier is less abrasion sensitive and the product has better flow properties.

Table 5.2 *Some commercial nickel catalysts*

Type	Girdler	Südchemie	Harshaw	Unichema
Highly active	G-50		Nysel HK-12	Pricat 9900
Selective	G-15	KE-NF20	Nysel DM 3	Pricat 9906
Isomerizing	G-111		Nysel SP 7	Pricat 9908

Commercially produced catalysts have a more pronounced high activity, a better selectivity, or better isomerizing properties. Some commercial nickel catalysts are given in Table 5.2.

G. Hydrogen Gas Production

The hydrogen gas needed for fat hydrogenation can be purchased from external producers in cylinders, tank cars under high pressure (even liquefied) or conducted to the production site, like town gas, in pipe lines. The other option is the on-site production of hydrogen by the fat manufacturer.

The gas needed for fat hydrogenation must fulfil high initial quality requirements: at least 99.5% purity; CO <0.05%; and H_2O <0.1% at 7 bar and 17°C. If produced electrolytically, the hydrogen should contain not more than 100 μg mercury per 1 cm^3. It should not contain sulphurous gases, these being irreversible catalyst poisons. Oxygen can be effectively eliminated from the hydrogen by using a palladium catalyst bed to convert it to water. Some other useful hydrogen data are: density 0.069 (against air = 1); good specific heat 3.41; heating value (lower and upper limits) 2573–3053 kcal N m^{-3}. In addition, hydrogen has a heat conductivity 7, (against air = 1) and 1 N m^3 hydrogen weighs 89.4 g at 0°C and 1 bar.

For the fat industry, the most important methods of hydrogen generation are: (i) electrolytic water cleavage, and (ii) steam/natural gas (hydrocarbon) reforming. The steam/iron contact process, not discussed here, is one of the oldest industrial methods, and is cheap and easily accomplished, if modern generators, good quality (rugged) coke of refractory quality with low sulphur content and good quality iron ore (Fe_3O_4, haematite, 70% iron content at least) are available. If electricity prices are high and natural gas resources are scarce, then under the given conditions the steam/iron contact process is a feasible alternative. Approximately 0.4–0.7 kg steam and 0.3–0.6 kg coke are required to produce 1 m^3 of hydrogen gas, whilst for the production of 1 m^3 of hydrogen gas 2.3–2.4 m^3 of water gas and 4 kg of steam are needed.

For small-scale hydrogenation needs, the on-site conversion/cracking of (i) methanol in the presence of water (e.g. Proximol process of Lurgi) to hydrogen and carbondioxide, or (ii) of ammonia directly to hydrogen and nitrogen, depends on local methanol or ammonia prices and the unavailability of other sources of hydrogen.

Hydrogen bought from outside sources might originate from the caustic, the hydrocarbon reforming the defence industries. Liquefied hydrogen is easy to transport and can be obtained from the space-exploration programmes. For liquid hydrogen the transport/storage containers and any connecting pipelines must be double walled and vacuum insulated, because of the very low temperature involved. It can be easily vaporized by conducting it through coils immersed in steam-heated water baths or similar.

Less than 1% of the direct industrial needs are met by the electrolysis of water.

1. *Electrolytic Water Cleavage*

Water can be cleaved electrolytically, according to the law of Faraday, into two parts of hydrogen and one part of oxygen (by volume). The water used in the electrolytic devices of the fat-hydrogenating industry is made conducting by 20–30% solution of potassium hydroxide in distilled or de-ionized water. The direct current needed is

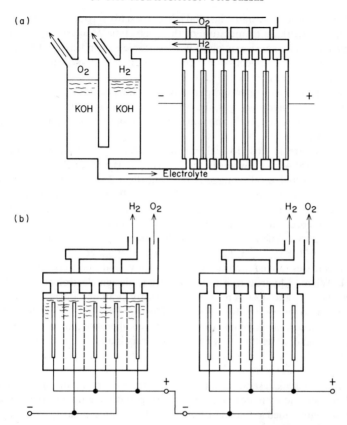

Fig. 5.7 Cells for electrolysis: (a) bipolar; (b) unipolar.

produced by a.c. driven motor/generators (dynamos), by power transformers and huge (mercury or other) rectifiers.

(a) *Bipolar electrolysers*. Bipolar electrolysers are the most often used equipment. They look similar to plate and frame filter presses and, because of the alkaline medium of operation, are made of mild steel. This similarity arises from the fact that they are composed of 40–160 "bipolar" cells attached together. These cells are separated by liquid-permeable synthetic high quality woven asbestos cloth diaphragms, pressed between nickel-wire fabric. The interior of the cells is divided by a steel plate into two parts with the two electrodes placed one on each side of this separation. The negative electrode (cathode) is a perforated nickel plated steel sheet; this is the site of the hydrogen production. The positive electrode (anode) is a perforated, ferric oxide coated steel sheet; this is the site of the (separate) oxygen production. The direct current is connected only with the first and the last electrode of the system, so that all anodes also serve as cathodes in the neighbouring cells. In this way each electrode transmits the total current serially through the battery of cells [Fig. 5.7(a)]. The working temperature of the electrolytic process is about 70–90°C, which means that the two gases evolved must be cooled when leaving the cells with a part of the electrolyte. The electrolyte is constantly circulated and, after giving up the gases evolved in the cells, it is pumped back with freshly added distilled water which both replaces the cleaved water and partially cools the assembly. About 800 g water is needed for 1 m^3 of hydrogen gas, the cooling (of the electrolyser, of the gases and of the small distilled water producing unit) needing a minimum of 30 l water per 1 m^3 of hydrogen gas.

The electrodes are so close to each other that very high current densities can be applied (2000–3000 A m^{-2}), whereas the cell voltage is not much above 1.8–2.2 V. The minimum reversible cell voltage is theoretically 1.23 V and the thermoneutral one is 1.48 V. To produce 1 m^3 of hydrogen (and 0.5 m^3 of oxygen) 2390 A are needed, which means that the energy consumption with 95% electrical yield (based on the 1.23 V theoretical cell voltage) is about 4.5–5.0 kWh m^{-3} hydrogen gas (at 1 bar gauge and 20°C). The Zdanski–Lonza filter press type electrolysers (1951) work at a pressure of 30 bar which, because of lower overvoltage and small gas-bubble diameters, decreases the required cell voltage to 1.8 V, and the electrical power needs to about 4 kWh m^{-3} hydrogen. Future aims are to design electrolysers which work at 1.6 V cell voltage and 4000 A m^{-2} current density (Van den Borre, 1983). Instead of alkalis, solid polymer electolytes (General Electric, Du Pont), which work in acid medium to produce hydroxonium ions for hydrogen liberation, are being developed.

The installation of at least two bipolar electrolysers is advised, as they may have to be disconnected from the mains once every 4 years for maintenance by the suppliers.

(b) *Unipolar electrolysers.* Electrolysers which have common tanks, in which separated anodes and cathodes, fixed in a housing and hanging into the electrolyte, are immersed, are called unipolar. Electrodes of the same polarity are coupled in parallel and the tanks are placed in a row to form a series [Fig. 5.7(b)]. The dimensions of both the cells and the electrodes are larger than those needed to generate equivalent amounts of the two gases by a series of bipolar cells, mainly because of the much lower current density (500 A m^{-2}) employed. Maintenance is easily carried out by factory personnel on one tank at a time, without the need to stop the whole installation.

Unipolar and bipolar electrolysers are very similar in terms of their electrical efficiency of hydrogen production, regarding the current utilization. However, the floor-space requirements are not equivalent: at equal gas production levels these are 10:1:0.3 for the uni-polar : bi-polar : pressurized bi-polar electolysers. Not solely because of this there are still local preferences for one type of electrolyser or other: in Europe mostly bipolar, and in North America and Canada mostly unipolar ones are erected. This is mainly because of the early fears of diaphragm rupture, insulation problems and electrolyte pollution with the bi-polar ones. The double diaphragm bi-polar electrolysers devised by Oronzio De Nora of Milan can provide better security in this respect.

The common big advantage of any of these electrolysers mentioned is the high purity of two gases produced simultaneously (hydrogen 99.9%; oxygen 99.7%; with some carbon monoxide and nitrogen). The gases begin to evolve 1 h after connection to the mains, which also means high flexibility. The disadvantage is the price of electricity and the relatively high installation costs, including the necessary motor generators, transformers, contact or single rotary converters and silicone or mercury vapour rectifiers. They become competitive with reforming at a production rate above 200–300 m^3 h^{-1}.

Both the hydrogen and the oxygen can be immediately compressed, cooled (water removal) for storage and transport under very high pressure (160–200 bar) in cylinders, the gas required for immediate use being compressed at 10–30 bar or more and stored in medium- to high-pressure spherical gas holders or uncompressed in telescopic wet or dry gas holders, called gasometers. The plants should work under constant control of cell voltage, gas and supply (distilled or deionized) water quality, electrolyte concentration and temperature, with alarms for warning of almost full or empty gasometers or medium-pressure gas holders, high-pressure hydrogen tanks, etc.

2. *The Steam/Natural Gas (Hydrocarbon) Reforming Process*

Mainly natural propane or butane, but also naphtha can be easily converted (reformed) to a very high purity hydrogen gas (99.7%) by this process. The natural gas (propane), containing low amounts of nitrogen, is first liberated from organic sulphur compounds (mercaptans) by washing with caustic lye and then conducting above a metal oxide (bauxite ore) at 370°C. The gas is then liquefied by compression. The propane and measured amounts of steam are mixed and conducted under 24 bar compression at 830°C through a nickel catalyst, which is filled into vertical tubes contained in an oven heated externally by burners. The reforming reaction is described by the equation:

$$C_3H_8 + 3H_2O \rightleftharpoons 3CO + 7H_2$$

This so called "reformed" gas mixture evolving from the reactor is cooled by steam to 350°C and conducted to another reactor, filled with an iron–chromium oxide catalyst which converts the carbon monoxide and water to 90–95% hydrogen and carbon dioxide.

$$CO + H_2O \longrightarrow CO_2 + H_2$$

This is the so-called "shift" reaction. Two subsequent "shift" runs over the iron oxide/chromium oxide catalyst convert the gas to the high purity required. The carbon dioxide is washed out by the use monoethanolamine solutions in scrubbers between the runs, the carbon dioxide (300 l per 1 m^3 of hydrogen) being recovered by boiling it out. The carbon monoxide contamination of the gas is $\leqslant 0.001\%$. It is a cheap and flexible process, although the minimum volume of economical hydrogen gas production is, depending on natural gas prices, 50 m^3 h^{-1}. The water needs are about 500 l per 1 m^3 of hydrogen gas.

Instead of using two "shift" conversions, the carbon monoxide can be eliminated by methanation on a nickel catalyst at 300°C. A part of the hydrogen produced is used, but at least methane is only a diluent and not a catalyst poison:

$$CO + 3\,H_2 \longrightarrow CH_4 + H_2O$$

The purity is better than 99.5% hydrogen with 0.45% methane, $<0.01\%$ CO_2 and $<0.001\%$ CO.

3. *Gas Purification by the Pressure-swing Adsorption Process*

Modern installations generally produce hydrogen gas of initially high purity or at least sell them as such. In the fat processing industry the final purification of the (non-electrolytically produced) hydrogen is generally executed by the heatless pressure-swing adsorption (PSA) process, using two parallel-mounted and alternately used pressurized cylinders containing different types of molecular (e.g. Linde) sieves (natural or artificial zeolites such as sodium, calcium and/or barium aluminium silicates), activated alumina and activated carbons. The residual amounts of gaseous impurities of higher molecular diameter than that of hydrogen are very effectively adsorbed. The saturated adsorbents can be reactivated by periodic heating and venting of the adsorbed gases to the air. Molecular sieves can also be applied between the "shift" and methanation reactions.

H. Oil and Hydrogen Quality

Oils which are to be hydrogenated must be free of all substances which, because of their chemical or sorptional affinity, might diminish or alter the catalyst activity and selectivity. These substances are called "catalyst poisons".

Considering their importance, organic sulphur compounds present in mustard, rape-seed, fish and whale oils are the most notorious (but sometimes deliberately used) catalyst poisons. Such sulphur compounds include alkylisothiocyanates and sulphur containing amino acids, like methionine. Either reduced by the nickel/hydrogen system or directly accessible to the catalyst, these organic compounds irreversibly occupy the active sites of the metal thereby causing a decrease in the catalyst activity, but (as we have seen) an increase in selectivity. For example, 5 mg sulphur poisons 13 m^2 of nickel surface. The specific surface of finely dispersed nickel catalyst might be 100 $m^2\,g^{-1}$, in contrast with the specific surface of nickel particles of, say, 5 μm diameter which is only 0.15 $m^2\,g^{-1}$. If we use 200 g of freshly reduced (but porous) nickel catalyst (metal) per ton of oil which contains 5 mg sulphur compounds per kg oil, this amount of sulphur (5000 mg ton^{-1}) will poison 65% of the catalyst added, even though this process has produced a highly isomerizing (thus "selective") catalyst.

The residual "gums" or "mucilages" present in oils, whether they are phosphorus based (phospholipids) or proteinaceous, are also catalyst poisons. It has been found that even 10 mg kg^{-1} residual phosphorus decreases the selectivity of the catalyst quite considerably, 4 mg kg^{-1} being the limit of acceptability (a normal limit even when applying physical refining methods). Alkali-metal soap traces are notorious poisons, irreversibly occupying the active sites on the catalysts. Free fatty acids are not primarily catalyst poisons, although they might deactivate catalysts by direct chemical reaction (dissolution of nickel in the form of nickel soaps/salts), if the conditions are appropriate (high temperature).

The pre-refining (post-desliming, caustic neutralization, but at least activated-earth bleaching) of oils to be hydrogenated low contaminant levels (preferably no sulphur, $<4\,\mathrm{mg\,kg^{-1}}$ phosphorus, $<10\,\mathrm{mg\,kg^{-1}}$ alkaline soaps, $<0.25\%$ of free fatty acids) of oils to be hydrogenated is thus a realistic demand. Any compromise (saving) on pre-refining is penalized by increased catalyst consumption or by reduced activity and the resulting considerable extension of hardening time. Although sulphur poisoning (preferably a deliberate and not an accidental one) might be beneficial to selectivity, flavour defects can be strongly promoted by the significant extension of the cycle time.

The poisons found in hydrogen gas depend on the method of its generation and purification. Hydrogen produced by the steam/iron contact process might contain considerable amounts of hydrogen sulphide, sulphur dioxide and organic sulphur compounds (e.g. mercaptans), if not completely removed by purification. As we have seen most of these compounds attack the crystallites of the catalyst directly, causing irreversible poisoning. Carbon monoxide is a temporary poison, which can be displaced from the active sites by the addition of fresh (pure) hydrogen gas to the system.

Substances like water (both in the gas and in the oil), carbon dioxide, methane and nitrogen are not poisons; their presence only diminishes the accessibility of the hydrogen gas to the catalyst by the dilution effect. These gases accumulate in the reaction vessel and, because they dilute the hydrogen content (sometimes up to 50%!), they must be either "washed out" in the external gas circulation systems or "blown-off" from the headspace of both the external oil circulation and of the "dead-end" systems. The latter can be achieved by a slight, but constant "bleeding-off" (see later) of the accumulated gases.

A very thorough drying of the pre-refined oil immediately before starting the process by heating under vacuum and the use of dry hydrogen ($<0.1\%$ water) gas can eliminate the need for constant or temporary "bleed-off" during processing, thus reducing the hydrogen losses.

I. Hydrogenation Equipment

The execution of heterogeneous fat hydrogenation needs equipment which makes possible the least hindered transport (mass transfer) of hydrogen gas and liquid oil molecules to the surface of the solid catalyst surface by intensive mixing, at suitably selected and controlled temperatures. In the edible-fat industry, these processes are primarily carried out batch-wise.

The most important batch systems use converters: (i) with external gas circulation, with or without mechanical mixing; (ii) with external liquid circulation; or (iii) with internal gas circulation and mechanical mixing in the "dead-end" mode.

1. *External Gas Circulation (Normann) Systems*

In the external gas circulation (Normann) type systems the oil and the nickel catalyst are contacted with the hydrogen gas in a low- to medium-pressure resistant tall mild steel (preferably stainless steel) converter or autoclave. The height/diameter ratio is 2.5–3. The converter has heating and cooling coils, the necessary fat filling and discharging lines with valves, as well as manholes, a thermometer, a pressure gauge and an oil sampling valve. The batch sizes (useful capacity) vary between 6 and 20 ton of oil, the gas head space being about 30–40% in order to reduce oil entrainment with the circulating gas.

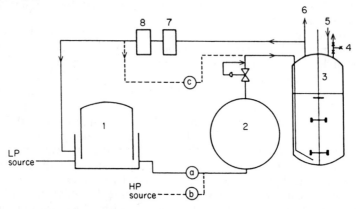

Fig. 5.8 Dead-end hydrogenation with (optional) recirculation. 1, Low-pressure hydrogen holder (gasometer); 2, high-pressure hydrogen holder; 3, converter; 4, vacuum connection including vent cock to atmosphere between two stop valves; 5, balance (hydrogen or inert) gas connection to replace hardened oil drawn from converter; 6, vent to atmosphere or recirculation line with (not shown) control valve to low-pressure holder; 7, fat catcher; 8, cooler/scrubber; a and b, compression and cooling of hydrogen gas; c, option for recirculation. [Reproduced, with permission, from Patterson (1983)]

Near to the bottom of the vessel (Fig. 5.8) there is a hydrogen gas inlet pipe connected to the pressure end of the pumping device. The pipe is attached either to a sparger with perforated holes directed downwards or to a gas lift (mammoth) pump. The hydrogen gas is pumped in via a centrifugal pump, a piston or turbine-driven compressor from the gas holders. These may be either wet or dry telescopic gasometers, medium- to high-pressure (10–30 bar gauge or more) spherical gas holders or high-pressure cylinders, the higher pressures being reduced to the operating one by reduction valves. The fresh and the unabsorbed hydrogen gas is circulated from the very beginning of the process through the oil at a pressure of 0.3–5 bar gauge. The gas also serves as the (pneumatic) stirring medium of the system.

After starting the hydrogenation reaction, the gas not consumed by the oil sucked out is first led via fat catchers (with or without, for example, pipe-bundle coolers) to two closed scrubbers, filled with, for example, Raschig rings or Berl's saddles, the circulated washing fluid in first scrubber being just plain water, and in the second 25% caustic lye solution. The carbon dioxide and fatty acids removed are bound by the washing fluids (to be renewed regularly), the contents of the fat catcher being emptied into, for example, the oil/catalyst mixer from time to time, not forgetting the type of oil to be hardened. The cleaned gas is then mixed, before the circulating pump (compressor), with fresh hydrogen. If nitrogen and/or methane, etc., has accumulated to unacceptable levels (more than 25%, although the reaction may proceed even at an inert-gas level of 50%), the headspace gas from the autoclave is purged by blowing it off to a used-gas reservoir (heating purposes) or, less elegantly, to the atmosphere.

Heating is accomplished by means of steam of intermediate pressure (6–14 bar gauge), in which case the (often) stainless-steel coils can also serve for cooling the oil at the end of the process. Heating is sometimes done by low-pressure mineral oil or by synthetic organic heating fluids (their use is discouraged). When decomposing formate catalysts in oil, but also for thorough drying of the oils to be treated, and when stopping the hydrogenations, a well-dimensioned vacuum line is a useful provision and safety measure. The residual pressure of 50 mbar can be produced by a mechanical pump, e.g. water-ring sealed, or by a steam-ejector driven pump system via a barometric condenser.

There are general measures that apply to all systems. Cooling by amply dimensioned coils (4–5 m^2 per 1 ton oil) is a prerequisite in order to have effective control of the process. A temperature decrease of 5°C min^{-1} is not considered to be too high. The maximum permissible hydrogenation rate is not higher than a decrease of 2.5 IV unit min^{-1}, as this allows the use of cooling water of average temperature (50°C) and an effective ΔT of 130°C (the hydrogenation temperature maximum being 180°C). With fish oils this rate should be decreased to 1 IV unit min^{-1}, with a temperature maximum in the first phase of 150°C. To avoid scaling of the internal cooling surface, soft water or condensate (e.g. from surface condensers and not from the cooling towers) should be used for cooling. Automated valves for closing down heating at set points (e.g. 150°C) and giving a signal when the pumping of the cooling water starts are very good safety measures if the genuinely exothermic nature of the reaction causes unexpected temperature rises.

(a) *Mechanical mixing.* The problem of sedimentation of the suspended catalysts in the gas-circulation systems, and the need for a more intense dissolution of the hydrogen gas in the oil (reduction of the bubble resistance), led to the early application of mechanical mixers in the Normann-type converters. The goals

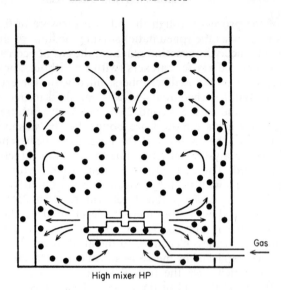

Fig. 5.9 Dispersion of gas bubbles by a radial-flow flat-blade turbine in a baffled vessel. [Reproduced, with permission, from Patterson (1983)]

mentioned above are somewhat controversial: large diameter impellers, such as paddles at low speed, effect a better suspension of the catalyst, whereas the high-speed small blade ones are especially well suited to dispersing the hydrogen gas in the oil. On the other hand (see later), a very effective mechanical mixing in "dead-end" converters without external liquid and gas circulation is imperative to obtain quick but selective saturation of oils. Paddle type mixers of fixed or variable (slow and medium, thus 45–150 rpm) speed and high-speed (400 rpm and more) marine-type propeller mixers were the first to be installed in external gas circulation converters, but these have now been replaced by turbine mixers.

The most commonly used turbine mixers are the radial- or axial-flow inducing flat-blade ones, of either open or disc-mounted design. These mixers force the neighbouring liquid layers to move in a shearing type action, which not only helps the gas bubbles to dissolve in the oil, but produces a uniform flow within the liquid mass. When starting the turbine, the gas may escape to the surface of the oil, but with increasing power input the gas is found near to the wall of the converter where it is directed vertically up and down in a turbulent pattern.

The turbine mixer is fixed at the bottom and should be a vertically bladed one of disc plate-mounted design so that it induces a radial flow. As the goal here is good gas dispersion, the mixer should be placed directly above the sparger. The holes in the sparger should have diameters of between 3 and 6 mm, placed upward; the diameter of the sparger should be roughly equal to that of the turbine (Fig. 5.9).

The turbine mixer fixed at the top should be of a different design. It should have axially downward directed (i.e. inclined rectangular) blades mounted on the shaft, and without a disc. It should be positioned close to the oil surface in order to create a downward vortex, sucking back the headspace gases to the bulk of the oil. The correct positioning is somewhere near to 0.1 of the vessel diameter below the oil surface; trial and error is often necessary to find the optimum arrangement. The top turbine can also be placed below a short suction pipe, directly under the oil surface, where it sucks the collected headspace gases back into the liquid.

The placing of the two different turbines on the same axis is the customary solution for vessels having low height/diameter ratios, these being modern "dead-end" systems without external gas circulation. The placement of a third turbine of the axial type in the middle of the oil phase is advantageous if when using taller vessels half charges are also required. In this case the heating/cooling coils should also be split into two sections. In tall vessels not used for half charges a second turbine of the radial type can be mounted in the middle, to improve mixing at this level.

Slowly rotating modern mixers with specially structured axial/radial blades are also available, e.g. the Multistage Impulse Countercurrent Mixer and the Interference Flow Stirrer of Ekato, both of which are claimed to have good hydrogen and catalyst dispersion.

Mixing should produce peripheral speeds of $2-6\,\mathrm{m\,s^{-1}}$. Medium- (60–150 rpm) or high-speed mixers (400–1200 rpm) preferentially have stageless or three-stage variable gears. Normal power requirements are around 1–2 kW per 1 ton of oil, this securing a gas/liquid interface of around $300\,\mathrm{m^2\,m^{-3}}$.

If using turbine mixers with, for example, three vertically placed baffles, these should be mounted on the wall above the heating and cooling devices and displaced at 120° from each other. The geometry of this vessel should be adapted, preferably having a diameter/height ratio of 1.4 or 1.25, the power requirements for mixing in any case being much higher than those mentioned before.

2. *External Liquid Circulation (Wilbushevits or Loop) Systems*

In these external liquid circulation (Wilbushevits or loop) systems the mixing of the three components is achieved by intensive external circulation of the oil/catalyst mixture. The converter is a tall cylindrical autoclave (height/diameter ratio 2.5–3) with a conical bottom of quite steep slope (Fig. 5.10).

The hydrogen is introduced by a sparger close to the bottom, surpassing the liquid head formed by the oil collecting here. The liquid is sucked out from the deepest point of the cone by a pump and is redirected into the top part of the converter. It is sprayed via nozzles into the headspace which is filled with collected and unused hydrogen gas. The circulating oil/catalyst stream is heated outside the autoclave by heat exchangers, although the vessel itself may have jacketed walls for (additional heating and) cooling.

In its original construction, this system had to withstand 10–15 bar gauge, as 3–5 autoclaves, each with its own pump and sparger, were interconnected

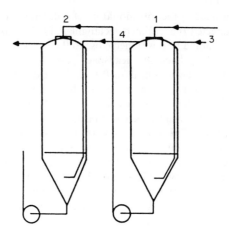

Fig. 5.10 Wilbushevits oil spray. 1 and 2, Oil and catalyst spray; 3, hydrogen feed sparger; 4, hydrogen exit to next stage. [Reproduced, with permission, from Patterson (1983)]

in series, producing a continuous (cross-current) hydrogenation system. The oil/catalyst mixture was moved from the first to the next autoclave by gravity. The autoclaves were positioned in step-wise descending heights, called cascades. Five such units were necessary to approximate a plug-flow. Because of the high system resistance and mechanical breakdown (high pressure, gas leakage through the pumps), the coupling of autoclaves was soon abandoned. The use of single autoclaves, working at 3–7 bar pressure was the preferred solution. The main disadvantage of this and similar systems was and remains the (strong) mechanical abrasion of the oil-pump rotors and housings by the solid catalyst particles and the attrition of the catalyst itself. The appliances on the converters are similar to other systems. With this system it is also necessary to blow-off (remove) the diluting inert gases and/or poisons at the slowing down of the process.

The most interesting modification of this system was made by Wilbushevits himself: the head end of a long and narrow vertical guiding pipe, placed centrally in the converter, served as a sort of venturi jet. The recirculated oil/catalyst mixture collected at the bottom was forced by the recirculation pump at high speed through the narrow upper part of this guiding pipe. The headspace gases were supposed to be sucked into the oil stream, being absorbed by the oil and conducted back to the bottom (Fig 5.11).

A modern variant of this "loop-reactor" is now produced by Buss. The oil/catalyst mixture is circulated by a special centrifugal pump with a liquid sealant system through a modern, venturi-tube based "injection mixing nozzle", where the fresh hydrogen gas and the gas in the headspace are both introduced. The latter is collected by the suction of the venturi tube exerted through side ports (Fig. 5.12). The system retains the great advantage of the

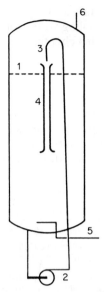

Fig. 5.11 Wilbushevits mixing jet. 1, Normal oil level; 2, oil/catalyst circulating pump, gland sealed with an oil compatible with the charge oil; 3, ejector nozzle; 4, oil/hydrogen mixing tube; 5 bottom gas connection to sparger; 6, top gas (balance gas). [Reproduced, with permission, from Patterson (1983)]

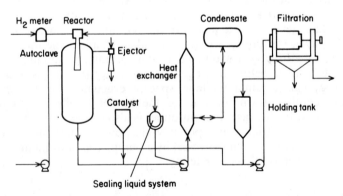

Fig. 5.12 Loop hydrogenation plant. [Reproduced, with permission, from Hamilton and Bhati (1980)]

Wilbushevits system: i.e. the necessary heating and cooling of the oil during the process can be regulated outside the reactor, which not only makes heat exchange possible but also the partial use of the considerable reaction heat evolved. Buss claims that this system is capable of producing 0.07 ton of low-pressure steam per 1 ton of oil. By using extra holding tanks and heat exchangers for hydrogenated fats and liquid oils, it is also possible to save 0.16 ton steam and 2.5 ton cooling water for every 1 ton of hydrogenated product (see Section I.J.4 of this chapter).

3. *Mechanically Mixed "Dead-end" Systems with Internal Gas Circulation*

The innovating idea of the "dead-end" system is the intensive transfer of hydrogen gas to the oil which is achieved by its internal circulation within the kettle by intensive mixing under pressure. This excludes the long(er) pathway which is intrinsic to external oil or gas circulation systems. To achieve these goals, all the factors discussed in Section I.1.a (this chapter) concerning the very effective turbine or other types of mixer are highly relevant.

The "dead-end" converters are mechanically mixed autoclaves (Fig. 5.8), generally built to withstand at least 10 bar gauge pressure. Their internal and external outfit is similar to the already described Normann-type converters. Regarding their geometry, two types are described: (i) one with an oil height/diameter ratio of 2 with a headspace of 15–25% of the vessel's overall height; and (ii) a baffled one with an oil height/diameter ratio of 1.4 (preferably 1.25) with a headspace of about 45% of the oil charge volume. Because of the maximum allowable rate of hydrogen uptake and the prescribed headspace volume, the pressure used is generally in the range 5–7 bar. Instead of the hydrogen circulation outlet, a corresponding vent outlet serves for purging (blowing off) the accumulated diluent gases and poisons leaving good quality oils at the end of the operation. Alternatively, headspace gases can also be bled-off (leaked) during the whole course of hydrogenation via a 0.5–3 mm diameter hole plate placed in the vent line. A vacuum connection is very desirable, but great care must be taken to avoid the mixing of air residues and the hydrogen within them, this meaning a minimum residual pressure of 50 mbar before hydrogen or any other (inert) gas admission. Hydrogen gas can remain in the inner space, if evacuation/heating up of the liquid and cooling down/filtration of the hardened oils is executed in separate vessels. In this case the vent outlet is also connected to a hydrogen balance gas holder (this being possibly the low-pressure tank), the gas replacing the oil drawn from the vessel by a pump. The same method can be used for any other inert-gas blanketing.

J. Hydrogenation Processes

1. *Continuous Processes*

Continuous systems offered by equipment contractors are, with some exception mainly used for the total hydrogenation of fatty acids, representing large volumes of almost invariable quality restraints (low IV). The total hydrogenation of certain edible-grade oils, like the production of vegetable "stearines" from palm and other oils, predominantly the laurics, is also a feasible goal if the production volumes are appropriate. The cocurrent Procter & Gamble (downward flow) and Lurgi (upward flow) columns (stacked reactors), the cross-current Buss (modifying the Wilbuschevits cascade idea on the loop reactors with a common gas headspace), the cross-current Pintsch-Bamag stacked trays and the Dravo- (formerly Blaw-Knox) King tubular reactor with a fixed-bed catalyst are well-known variants. Buss also offer groups of two or four of their single-loop reactors, this representing a semi-continuous variant. The fixed-bed catalyst variants are more sensitive to catalyst poisons than are systems which use a slurried catalyst.

Many batch-converter based (semi-)continuous systems transfer the oil containing the suspended, previously admixed catalyst slurry through cascades of at least 5 (approximation of plug flow) mechanically or hydraulically well-mixed converters (compartments), usually admitting the hydrogen gas in cross-current fashion. If the activity of the catalyst and that of the starting oil (huge volumes of the same sort must be used) are uniform, the continuous system has distinct advantages. There are no filling and discharging periods, and the preheating and post-cooling can be done outside the system, mostly by heat exchangers. The disadvantage is the low flexibility as regards the changeover to another feed stock or a different end product. During this period (three times the mean residence time in the system) the products are non-standard. Selectivity can also vary, the products having less steep SFC/temperature plots, resulting in products which are unsuitable for certain margarines and for confectionery fats in general.

2. Batch Processes

For batch-type hydrogenations (not necessarily in dead-end systems) about 1/3 to 1/5 of the oil charge is first (pre-)mixed with the correct amount of fresh/used catalyst (scraped off from the filters). Knowledge of the activity and selectivity of the catalysts used is a prerequisite for planning the constant final product characteristics. This is not a problem with fresh catalysts, used only once (single use), which is often the case in the U.S.A. practice. With multiple use, higher quantities (not uncommonly 6–10 fold) of used (poisoned) catalysts have to be added in order to obtain the same final refraction/IV in the same batch time, but not with exactly the same melting and solidification characteristics. Experience of raw-material types and catalysts should lead to the use of a balanced mixture of fresh and used catalysts, resulting in products of uniform quality. Replacing about 10–20% of previously used catalyst by fresh catalyst seems to be a good economy from the viewpoint of uniform product quality, provided the same raw materials and uniform process conditions are applied.

First the 2/3 to 4/5 of the oil batch is pumped (sucked) into the evacuated autoclaves and deaerated/dried/heated. (If evacuation and heating up was already done via modern external heat exchangers, the vessel may be kept under a hydrogen or inert gas blanket.) Then the above-prepared (and if possible preheated/dried) oil/catalyst mixture is pumped (sucked) into the well-stirred bulk of oil. The vacuum (50 mbar residual pressure) is broken by hydrogen, after venting some hydrogen to the atmosphere to purify the headspace. Gas addition may start from 120°, and in the case of low-temperature hardening from 105 to 115°C up to the prescribed temperature. During the heating-up process the exothermicity of the process should be taken into account turning off heating valves at, for example, 20°C below the set temperature. The temperature can be regulated by cooling, the cooling water being switched on automatically if, for example, a temperature of $150 \pm 5°C$ is attained (two-stage or medium-temperature hydrogenation). Besides cooling, reducing the hydrogen access (e.g. by switching to a supply line with a smaller reduction orifice plate) or decreasing the stirring intensity (variable or three-speed reduction gear of the mixers/turbines) are means of reducing the danger of overshooting the predetermined IV end-point. Reasonable skill and experience is required on the part of the operator to keep the process within the specification set for a given product.

(a) *Stopping of hydrogen supply*. On reaching the end-point the gas inlet valve is turned off and the mixture is evacuated and cooled to 70°C, at which temperature the vacuum is removed by admitting air (risk). Either the converter or, much better, a separate pressure-resistant drop tank can be used as a buffer tank. Filtration can be done on a closed filter by pumps, hydrogen gas (admitted at the top of the autoclave or the drop tank) or, preferably, by nitrogen. In the last case the oil is cooled only to 90°C. A separate drop tank is also useful as a buffer and heat-recovery tank before external cooling and filtration (designed, among many others, by Chemetron, EMI Corp.). Some other energy-saving provisions are discussed below (see Section I.J.4 of this chapter).

The end-point of hydrogenation is given either in terms of a final refractive index (IV) or slip-melting point, the real goal being a fat with a given solid fat content at a given temperature (see Section I.L in this chapter). Besides the constant control of the refractive index or slip-melting point, a more-or-less exact volumetric end-point control can be achieved in all reactor systems if the oil charge to be hydrogenated has been previously measured (weighed) exactly. The amount of (1 bar) hydrogen required can be calculated according to:

$$\text{Hydrogen required (m}^3 \text{ per 1 ton oil)} = \frac{\text{IV decrease}}{1.14}$$

The amount of hydrogen actually consumed by the process can be mass totalized by, for example, positive-displacement flow meters, in which the gas flow rotates two interlocking elements. Each rotation is equivalent to a set volume, corrected for temperature and pressure. Differential pressure transmitters placed around a single orifice plate are also useful. The admission of hydrogen can be stopped manually or automatically by presetting the required volume of hydrogen gas. In circulation systems, the entering and exiting amounts of gas must be measured, whereas in the other two types of system only the entering amount need be measured.

In dead-end and loop systems the necessary pressure (4–7 bar gauge) must be built up with the batch sizes determined as optimal being rigidly observed.

The entire process or only some parts of it can be controlled either by programmable logic controllers or dedicated microprocessors (see Chapter 8, Section III). Automated processes are offered together with the appropriate equipment by many manufacturers, e.g. De Smet, Chemetron, Wurster and Sanger, EMI Corp., Buss, Franz Kirchfeld (precise gas-metering control and adjustable stirring speed).

(b) *Catalyst separation by filtration.* The transport of the oil to plate and frame or chambered filter presses is by means of centrifugal pumps (or by gasses). The (fully automated) horizontal rotating-disc filters (described previously) are seldom applied for primary filtration, mainly because of the lower amounts of solids to be filtered, the potentially higher amount of fines (small-size particles) and the high initial investment costs. The horizontal or vertical leaf-matting filters with fine metal gauze are used with or without a filter aid, the initial black-run being recirculated to form a primary filter layer. Whilst the addition of 1–2 kg of diatomaceous earth (or a similar inorganic filter aid) per 1 ton of oil can significantly decrease the filtration difficulties caused by some almost colloidal nickel particles, this leads to the (unnecessary) dilution of the catalyst with inert material.

The more old-fashioned filter presses are covered by non-synthetic filter cloths (see Chapter 4, Section IV) which are able to retain after some black-run (returned to the buffer or a separate holder tank) the 1–10 μm particles of the supported catalysts. If for any reason the residual nickel content is higher than 10–20 mg kg^{-1}, the filtered oil should be "polished" on separate filters with filter cartridges of an appropriate texture. Paper should (could) be used as an overlayer on the cloths; this is strongly advised when using formate based catalysts which have no added filter-aid "support". The use of electrostatic fine filters is an expensive investment, although it does lead to very low residual nickel levels.

To reduce the amount of manual labour needed for cleaning/servicing hot filter presses, at least the opening/closing of the frames/elements on these presses is mechanized by hydraulic/electromagnetic devices. Before opening the presses at the end of filtration, they are blown (preferably) with compressed nitrogen gas (also very often by steam) to remove most of the oil. The catalyst from the opened presses is scraped down onto trays with ducts, which direct it back to the catalyst/oil pre-mixing vessel or separate closed storage containers. Good book-keeping of the amount of catalyst recovered from the filters, their previous use (type and sort of oil) and frequency of use is the best way of maintaining the required product quality.

(c) *De-oiling of spent catalyst.* Inactive (spent) catalysts can be extracted by solvents or boiled with hot water to render them oil "free" [see Chapter 4, Section IV A.6(a)]. Incinerating the residual fat and leaching of the ashes by sulphuric acid for nickel regeneration is seldom practised locally. Spent catalyst is generally sold as a fatty mass to the catalyst manufacturers, who buy them back for their nickel content (this being not less than 10%).

3. Hydrogen Economy of the Different Processes

In the gas-circulation process the hydrogen gas actually present in the system is (re-)circulated 3–10 times h^{-1}, this meaning the recirculation of, e.g. 40–120 m^3 of gas per 1 ton of oil per hour. Recirculation of the gas requires very powerful non-leaking (e.g. oil sealed) gas circulation pumps (compressors), an extra energy factor. Because of the extension and complexity of this system, quite considerable gas losses are possible. The losses in an external liquid circulated, or even in an optionally recirculated and thus not optimally used, dead-end system are lower. The hydrogen consumption of any system will be high if the hydrogen gas and oils used have not been well purified and dried.

The deliberate intermittent purging or constant bleeding of hydrogen during processing and the removal of the considerable headspace volume of unused hydrogen in both the circulation and dead-end systems cause inevitable losses, these being 3–6% of the total. The non-deliberate leakage of vessels, pumps and long circulation lines can increase these losses considerably. Therefore, the hydrogen "factors" for circulation systems are 1.06–1.10 and for the dead-end systems are 1.03–1.06, these figures meaning minimum losses of 3–10%. Of course, instead of purging them to the atmosphere, headspace gases can be recovered for other purposes, e.g. in the past for iron-ore reduction in the steam/iron process.

When deciding on a hydrogenation installation for a newly built plant, besides the usual considerations, one should give ample consideration to: the reduction of the length and complexity of the external circuitry; the installation of a vacuum system in order to apply thoroughly dried starting oils, thereby reducing the need for temporary gas bleed-off; the improvement of gas and catalyst dispersion in the oil; and the high purity of the hydrogen gas to be used.

4. Heat Economy of the Process

Hydrogenation is an exothermic reaction and the considerable heat evolved during the process should be utilized for practical purposes. In modern installations the reaction heat is used for generating low-pressure steam in an external heat exchanger during the hydrogenation process and/or by the final heat exchange between the hydrogenated oils and the liquid oils entering the system.

Hydrogenated oils ready for dropping leave the converters at a temperature of, on average, 150–180°C whereas the colder neutralized/bleached oils enter the hydrogenating system at temperatures of c. 60°C. The heat exchange between these two components must be done before filtration the temperature of which is about 90°C in closed systems. The main problem is the presence of catalysts in the finished oils. Only heat exchangers which cannot become clogged by the solids, or if so, can be easily emptied and cleaned, can be used. When producing large volumes of hydrogenated oils, apart from vertical tubular heat exchangers, the welded-channel spiral-blade heat exchangers which have perfect countercurrent flow and minimal clogging tendencies are also very practical. If the mass of oil flow on both sides is equal, the incoming oil flow will show a temperature increase which is somewhat larger than the decrease in the outgoing hydrogenated one. This is because of the higher specific heat of the hydrogenated oil versus the unsaturated liquid oils.

With a buffer tank before the batch converter (connected to a vacuum system for oil drying) and a hot oil-drop (filter) tank after it, (both tanks being double the volume of the converter) the new batch can be heated up by the previous batch already dropped, by leading the latter through a heat exchanger to the filtering devices. This gives ample time for the portion already heated up (first collected in the pre-buffer) to become hydrogenated in the converter. With one buffer tank before the converter, which serves also for liquid-oil drying by vacuum, a drop tank before and one buffer tank after the heat exchanger, the volumes of these tanks can be kept equal to that of one hydrogenated batch (e.g. Simon-Rosedowns Ltd). Furthermore, two batch converters and one heat exchanger can be joined to one processing unit (e.g. De Smet, Buss and Process Systems Inc). These provisions allow the process to be made (semi-)continuous, with pump and heat-exchanger (investment) savings, and as optimal conditions for the heat exchange. However, the energy and time savings realized on a given production volume must be weighed carefully against the investment costs, which all these extra installations incur, in the usual way.

The high heat conductivity of the circulating hydrogen gas also causes heat losses. Bearing in mind the exothermic nature of the reaction this is not always a disadvantage, although in the recirculation system this heat is mostly wasted.

K. Post-treatment of Hardened Fats

The oil leaving the filter press should travel in closed ducts and collecting side troughs in order to avoid air contact both here and in the filtered oil tank, where different charges are collected and controlled for quality. The nickel residues (partly nickel soaps if the free fatty acid level of the original intermediate was or has become higher than normal) should be removed by immediate post-refining treatment. This could comprise a post-neutralization with dilute (0.8 N) lyes (followed by washings, drying) in the worst cases, but at least a post-bleaching, e.g. with 0.5% acid-activated bleaching earth or the combination of an aqueous (10%) citric acid solution (1% of the oil weight added) and the earth, both at 100°C under vacuum. The residual nickel and copper levels in the oil should be <0.1 and <0.02 mg kg^{-1}, respectively.

L. Quality Control

1. *End-point Determination*

The off-line measurement of the refractive index (n, at 65 or 60°C) is the quickest means of obtaining some idea of the progress of saturation. The refractive index is virtually proportional to the IV and decreases linearly down to IV = 70. If batches with a similar final IV are planned, the on-gas hydrogenation times against successive refractive-index readings can predict the correct time to stop the gas supply. More exact is an early-warning alarm system based on preset flow-meter consumptions in closed systems working with or without constant bleeding. Specific-density measurements (to the fifth decimal place) also give the necessary information on the progress of the IV decrease (hydrogenation causes a maximum 1% increase in the oil weight, resulting in the decrease in the specific density with increasing saturation). In daily practice the open capillary softening (slip-)melting-point determinations without stabilization are carried out regularly for routine factory floor control. The differences in the solid fat content of samples taken at, for example, 20°C can be quite considerable due to IV fluctuations and, to an even greater extent, to slight melting-point differences. Hydrogen consumption readings taken from the flow meters are also useful for early warning purposes. The best control on the final quality of the products intended for margarine and shortening purposes would be a quick off-line solid-fat content determination at, for example, 20 and 30°C by pulsed NMR. The melting of the sample at 60°C will give non-equilibrium values, which must be accepted if time is a pressing factor. For margarine and confectionery fat production, large solid-fat content fluctuations at distinct temperatures are unacceptable. For other purposes the batch differences are less critical and batches above and below the optimum values can be mixed before further processing to average the differences out. These products are then considered as if they were hydrogenated less selectively.

2. *Liquid Oils*

The generally neutralized/bleached liquid oil should be controlled for soap, free fatty acid and peroxide contents, these giving a clue as to the pretreatment and storage required. If physically pre-refined oils are involved, the absence or level of phospholipids and sulphur (fish, whale, rape-seed and mustard-seed oils) residues should be controlled.

3. *Catalysts, Hydrogen Gas*

The alkaline reaction of formate catalysts in the ash (maximum 0.5%, as sodium hydroxide) and the absence of water soluble nickel salt residues in the same ash (e.g. by dimethylglyoxime) must be controlled as a measure of quality. The nickel content, activity and selectivity of the catalyst can be tested by using standard laboratory hydrogenation methods. A good measure of hydrogen gas purity is the difference in thermal conductivity between a dried hydrogen production sample against a pure standard. This can be measured using a katharometer, the well-known conductivity detector employed in early gas chromatographs. The amount of oxygen can be determined by burning the gas in standard volumetric tubes and measuring the hydrogen volume decrease by the amount of water formed. Care should be taken to control explosion hazards by applying hydrogen gas leakage detectors in the production halls.

4. Waste Control

Nickel present in waste water originates from the soapstock split by acids, the washings from post-refining and from filter presses which have been deoiled by steam. If biological sewage treatment is applied, about 30–50% of the nickel may accumulate in the biological sludge. A nickel level of 100 mg kg^{-1} (dry weight) precludes the sludge from further agricultural use. If burnt, the sludge ash should not contain more than 5000 mg kg^{-1} nickel. The bleaching earth, if not further de-oiled, contains the remainder of the lost nickel. If de-oiled by boiling water/alkaline lye, the ionic part of the nickel is rendered water soluble, which gives rise to the problem already described. The disposal of heavy-metal contaminated wastes is subject to local regulations, 5 mg kg^{-1} being the limit for waste water in the U.S.A. practice.

II. Fractional Crystallization and Winterization/Dewaxing

Historically the first and economically the second most important fat-modification process is the fractional crystallization (simply fractionation) of (semi-)solid fats. Solid fats contain either already precipitated or still dissolved solid triacylglycerols, which under controlled conditions of cooling can be induced to full or partial crystallization. The crystals formed can be removed by, for example, filtration during which the fat or oil is separated into two or more fractions, at least one having a higher melting point than the starting oil and one remaining liquid at the temperature of intended use, not far above that of the crystallization. The lower the separation temperature chosen, the lower will be the melting point (cloud point) of the two fractions.

As fractionation is basically a physical phenomenon, in the eyes of the food technology purists it is considered to be the only admissible fat-modification process. In more sophisticated variants auxiliary chemicals, such as wetting agents or organic solvents (completely removed in later stages of the processing) can be used to improve the quality, increase the productivity, or create speciality fats (see Chapter 7), like cocoa butter equivalents/extenders and cocoa butter replacers. The need for huge amounts of more sophisticated solid fats makes the omission of hydrogenation and interesterification, both very often in combination with fractionation, impossible.

The fractionation of tallow into hard oleostearine and semi-solid oleomargarine, the classical raw material of margarine, strongly declined with the introduction of fat hydrogenation in 1902. The high volume of palm oil produced in the 1970s gave a new incentive to build modern fractionation plants, which produce huge amounts of liquid palm oleine, the cheap local substitute of the traditional, but relatively expensive, coconut and palm-kernel oils in tropical countries.

For the sake of clarity it should be repeated here that "stearine" is the higher melting solid, and "oleine" is the liquid fraction of fats, both of which are composed of triacylglycerols. In the non-edible fat industry the term "stearin(e)" means a eutectic (see later) mixture of solid saturated fatty acids (mostly C_{16} and C_{18}), whereas "olein(e)" is the liquid unsaturated fatty acids (mainly oleic) obtained by hydrolysis and subsequent fractionation of mostly animal fats.

The relatively high solid fat content of groundnut oil means that it becomes opaque if kept below 15°C (cloud point). Cotton-seed oil, because of its much higher content of "stearines" clouds quickly in the winter season or if kept in a refrigerator. Olive residue oils may also contain some stearines. There is also a market for more flavour stable, but still liquid soyabean and rape-seed salad oils produced by iso-suppressive hydrogenation which increases the solid triacylglycerol content. This demand for low cloud point oils necessitates the removal of the solid triacylglycerols present by some kind of fractionation: this operation is called "winterization". The process of removing dissolved waxes from sunflower-seed, maize and rice-bran oils, which become opaque if kept in a refrigerator, is also carried out by cooling and crystallization, followed by filtration, and is called "dewaxing".

A. Theoretical Background

At a given temperature solid or semi-solid fats can be considered as mixtures of high melting, mainly saturated triacylglycerols and lower melting (or liquid) more unsaturated analogues. Although it is an oversimplification, fats containing C_{16} and C_{18} fatty acids can be subdivided in four triacylglycerol types, representing four melting-point ranges: trisaturated (SSS) ones melting around 60–65 °C, disaturated–monounsaturated (SSU) ones melting around 40–45°C, monosaturated–diunsaturated (SUU) ones melting around 18–25°C, and triunsaturated (UUU) ones melting around 5°C. These ranges are at melting point intervals of approximately 15–20°C. As all these (and any other) triacylglycerols in the molten (liquid) state are completely miscible, they can also be considered as solutions. The degree of solubility of the higher melting components in the liquid ones is different at different temperatures. Mainly for practical reasons, such as the decrease in viscosity (easier crystal growth, quicker transformation into a more stable crystal modification) and the better washability of crystals, the solubility differences can be extended by the use of an organic solvent. However, this does not greatly affect their behaviour with respect to the basic principles to be described.

1. *Melting and Solidification of Fats*

The melting point of a triacylglycerol crystal is the temperature at which it starts to change into a liquid at atmospheric pressure. The freezing point is the temperature at which both the liquid and solid phases of the triacylglycerol can co-exist at atmospheric pressure, if the previously totally melted liquid is cooled. These two temperatures are only the same when equilibrium conditions have been reached. As rapidly cooled (shock chilled) fats easily surpass their usual freezing point without crystallization, the fat becomes undercooled, the solution becomes supersaturated, and the liquid phase now contains more dissolved solids than it would under equilibrium conditions. The supercooled liquid is both kinetically and thermodynamically (see later) in a metastable (labile) condition; as soon as crystallization begins, the temperature rises to the freezing point.

2. Phase Diagrams

The crystallization of triacylglycerols can be better understood, by first considering the phase behaviour of a hypothetical two-component mixture during its melting and solidification. As the possible interaction of the components cannot be denied, at least three different systems should be described for daily practice.

(i) If two compounds, A and B, which are completely miscible in the liquid state are cooled they may solidify as a "continuous series of solid (crystalline) solutions". A hypothetical phase diagram, like that shown in Fig. 5.13(a), of temperature versus composition will show this very often encountered crystallization (phase) behaviour of (fatty) compounds. Above the upper curve the whole mass is liquid. The upper curve starts from the point of 100% pure A or B and is called the liquidus curve; it represents the beginning of freezing (crystallization). The lower curve, starting from the same points A and B, is the solidus curve which represents the beginning of melting or the end of freezing. Below this curve the whole mass is solidified. The area between the two curves represents crystallized solids in equilibrium with the remaining liquid.

The solids formed during cooling are the "solid solutions". These are not simple crystal admixtures, but new compounds having individual crystal properties. Solid solutions are also called mix(ed) crystals. To be completely homogeneous the "compatible" components must possess analogous chemical constitutions, approximately equal molecular volumes and similar crystal structures. Mix(ed) crystal type behaviour is only possible if the melting points of the two components do not differ by more than 20°C (Hannewijk et al., 1964).

(ii) If compounds A and B are completely miscible (if liquid) they may first solidify during cooling to either pure A or pure B in equilibrium with their solutions. Two corresponding liquidus curves may intersect at an in-between "eutectic" (easy melting) freezing point, where both components crystallize simultaneously [Fig. 5.13(b)]. This eutectic mixture is not a real new compound, the components being only mechanically mixed to form a "conglomerate". Adding either pure A or pure B to the system lowers the freezing point of the other to the eutectic one. An example of two fatty compounds which have an eutectic freezing point is the pair of the pure triacylglycerols PPP–StStSt (St = stearate).

Other pairs of pure triacylglycerols such as PPP–OOO, PPP–POP, PPP–LaLaLa (La = laurate) and StStSt–StOSt show the lowest freezing point at that of the lower melting compound, this being called a "monotectic" freezing point [Fig. 5.13(c)]. When starting to cool the mixture, the higher melting component can crystallize in the pure state until the monotectic freezing point, and thus composition, is reached.

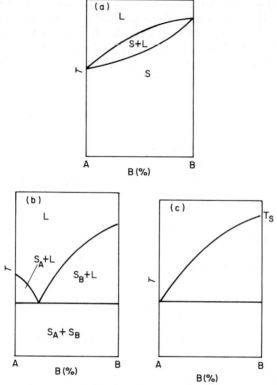

Fig. 5.13 Phase diagrams of binary systems. (a) Continuous series of solid solutions, forming mixed crystals. (b) Eutectic system, no solid solutions (conglomerates). (c) Monotectic system, no solid solutions (conglomerates). S, Mixed crystals; L, liquid; S_A, solid A; S_B, solid B; T, temperature. [Reproduced, with permission, from Boekenoogen (1964)]

It was found again that either the eutectic or the monotectic behaviour is only observed if the difference in the melting points of the components is not more than 20°C.

(iii) The eutectic/monotectic systems are characterized by the more or less "incompatibility" of their two components, allowing the partial isolation of one or other component in the pure state. Some two-component systems having eutectic melting points may solidify in a certain composition as an apparent (pseudo-) compound, the components having a congruent (not necessarily, but sometimes) higher melting point than either of the pure components alone. The compositions of the solid and liquid phases, are, of course, in equilibrium. These components have some "compatibility", although different to that of mix(ed) crystals. For example, blends of liquid palm oleine (rich in POP) with liquid lard oleine or with interesterified palm oleine (both rich in PPO) has been described (Rossell, 1973). After mixing

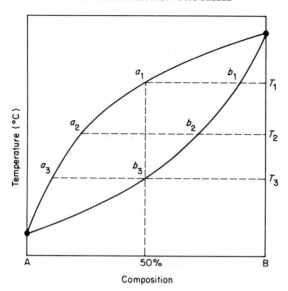

Fig. 5.14 Fractionation of mixed crystals in binary systems. [Reproduced courtesy of Unilever N.V.]

the first liquid with one of the two other ones solid POP/PPO crystals precipitate at ambient temperature.

3. *Natural Fats*

Natural fats which contain only five different component fatty acids might be composed of 75 different types of triacylglycerols. If the fatty acids are of solid or liquid, it is easy to imagine that this fat will contain all four of the simplified main types of triacylglycerols mentioned above. Considering the phenomena of undercooling, slow thermal diffusion and polymorphism (see later) one can imagine that fat crystallization is a very complex process. For the sake of simplicity it is customary to consider solid natural or hardened fats at (moderate climate) room temperature as being physically single compounds and to consider softer types of fats as binary mixtures of solid compounds with liquid oils.

The practical consequence of forming solid solutions is illustrated in Fig. 5.14 for a 50:50 mixture of two types of triacylglycerols, A and B, which is slowly cooled (slow thermal diffusion) to temperature T_1. The first mix(ed) crystals are obtained at point a_1 on the liquidus curve, their composition being given by the corresponding (along a tie line) point on the solidus curve, i.e. point b_1. As the solids are richer in the higher melting component B, the liquid part will become relatively richer in the lower melting one, A. Separation of the liquid and cooling it further leads at the lower temperature, T_2, to a new mixed crystal deposition at a_2, which contains much more A, as given by b_2. Only by further repetitions of this "fractionation" operation can pure A be obtained at point A.

On the other hand, the same fractionation causes the enrichment of one or the other of the main types of the component triacylglycerols in the liquid or solid fractions. It will be seen later that this very common crystallization behaviour of fat blends can be used as an advantage when processing margarines and shortenings by the shock chilling technique. The situation is different if only mechanically mixed crystals behave according to the eutectic or monotectic models or if the melting-point differences between the simplified main solid and liquid types of triacylglycerols are more than 20°C. In such cases the (not complete) separation of certain higher or medium melting triacylglycerol (also wax) fractions from liquid or semi-solid fats during the fractionation/winterization and dewaxing operations becomes much more feasible.

B. The Mechanism of Crystallization

Crystallization depends on two processes: the formation of crystal nuclei and the growth of the crystals on these nuclei (the precipitation of solids on them from the melt). Both processes have the same driving force, i.e. the supersaturation of the melt. For easy visualization of this process the empirical Miers theory defines three regions on a temperature versus concentration diagram: (i) the labile supersaturated region above the super-solubility curve (CD), (ii) the stable undersaturated region below the normal solubility curve (AB), and (iii) the metastable region between the two curves (Fig. 5.15).

Nuclei formation is spontaneous only in the kinetically labile super-saturation region. In the metastable region there is a difference, whether the solution is originally free of solid particles (unseeded) or contains only some of them. If there are no crystal "seeds" in the system, it will remain liquid. Induced nucleation can be considered as a chemical process: a barrier, the "activation energy", must be overcome in order to start it. Rapid cooling to produce local temperature differences is a means not only of creating the concentration differences necessary as the driving force of nucleation, but also of compensating for the high activation energy. On the other hand "seeding", by adding small crystals or even impurities such as dust, rough surfaces or air bubbles, or strong mechanical impact (stirring) can initiate the nucleation process in the metastable region.

Supersaturation is also the driving force for crystal growth on the nuclei. In the metastable region crystal growth occurs at a higher rate than does nucleation, if the supersaturation level is low. Once crystal growth has started, the concentration of the solute in the melt will decrease. Growth will still continue, if the concentration of the solutes is kept between the supersaturation and saturation regions by adjusted cooling and by the creation/addition of new "seeding" material.

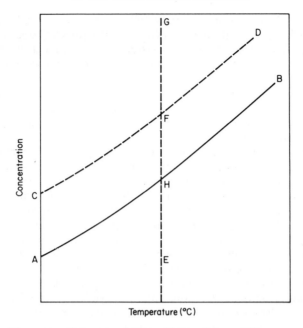

Fig. 5.15 Diagrammatic representation of Miers theory: CD, supersolubility curve; AB, normal solubility curve. [Reproduced, with permission, from Perry (1950).]

Crystal growth is either diffusion or particle-integration controlled. In non-mixed crystallizers the process is controlled by diffusion, while in well-mixed ones particle-integration dominates. The rate of growth in the former systems is crystal-size dependent, in the second it is not.

If the rate of nuclei formation is relatively high as compared with the rate of crystal growth, the solidified mass will consist of small crystals; in the converse case a smaller amount of larger crystals will be obtained. Quick cooling (shock crystallization) leads to unstable, easily agglomerating small crystals, because at the high viscosity achieved the growth is slow. Slow cooling produces larger, more stable crystals. Stability in the thermodynamic sense refers not only to the solid/liquid phase equilibria, but also to the polymorphic transformations of the solid fat crystals formed (see Chapter 6, Section I.C.5). Triacylglycerols tend to change their crystal modification from the lower melting metastable α form (formed during the shock chilling) into the higher melting, more stable β-' or β forms, attained during recrystallization. The α modification, although thermodynamically metastable, cannot be supercooled, whereas the higher melting β-' form can.

By appropriate planning of the supercooling, the time/temperature profile (the temperature of crystallization and the cooling rate) and the mixing rate of the process it is possible to control the texture and properties of the crystallized fat masses over a wide range. "Small" crystals have diameters of

50–100 μm and these sizes are generally difficult to separate by, for example, filtration. The larger crystals, diameters of 500–1000 μm can be easily filtered. The aim of the fractionation process is to obtain a maximum growth rate, i.e. a low nucleation rate. In practice a mixture of large and interlaced small crystals gives the most favourable filtration results. The larger crystals are formed either by transport (precipitation) of further material or by agglomeration of two or more growing crystals. The smaller crystals are either hindered in their growth or are formed by the mechanical abrasion of the larger crystals (attrition).

Adequately placed and dimensioned cooling surfaces are required to maintain the low temperatures needed for the induction of nucleation, particularly as a considerable amount of latent heat of crystallization is released during the process. Appropriate mixing enables a more uniform heat transfer (cooling) to be achieved, based much more on convection than on conduction. Mixing produces more nuclei and keeps the crystals suspended giving them an opportunity to grow. Mixing can also help the formation of crystal aggregates (see Section II.D in this chapter), but, if excessive, may cause the production of fines by attrition.

Minor natural or added components of oils, particularly surface active and thus emulsifying ones like mono- and di-acylglycerols, might affect the rate of the crystallization process, generally by delaying it. However, except with palm oils, where the amount of partial acylglycerols might be 4–8% and with partially hydrolysed hard butters, their role is of secondary importance to that of the triacylglycerols.

As has already been mentioned, in order to achieve effective and reproducible separation there must be a temperature difference of preferably 20°C but at least 10°C between the melting points of the two (or any) fractions produced. The lower the fractionation temperature, the lower is the melting point of both fractions, and *vice versa*. Transitions from the lower melting crystal modifications to the higher melting (more stable) ones can be achieved by keeping the system slightly below the temperature of melting of the higher melting ones.

C. Fractionation Techniques

The most important practical methods of fractionation have the following four process steps in common:

(i) The solubility of the more highly saturated triacylglycerols in the (preferably previously totally molten) fat is decreased by cooling. This may or may not lead to (super)saturation.

(ii) The cooled (super)saturated melt (solution) is precrystallized by means of mixing/seeding, in order to form crystal nuclei and induce crystal growth.

(iii) The crystal growth is sustained by suitable time/temperature profiles and mixing, with the cooling counteracting the heat of crystallization liberated;

(iv) The crystals are separated from the liquid phase by mechanical means.

The main methods of fat fractionation are:

(i) dry fractionation;
(ii) wetting agents mediated or Lanza fractionation;
(iii) winterization/dewaxing;
(iv) crystallization followed by (hydraulic) pressing; and
(v) solvent-aided fractionation;

all of which are based on fractional crystallization.

Table 5.3 *Some typical fractionation temperatures for different oils*

Oil	Temperature (°C)
Tallow	35–41, 25–27
Palm oil	36–40 (SSS), 28–30 (S_2U), 18–20 (SU_2)
Palm kernel oil	8–15
Cotton seed oil	5–7
Iso-suppressive hydrogenated soya bean oil, IV 110	1–3

The temperature at which fractionation is done is a very important parameter of the process; some typical fractionation temperatures for different oils are given in Table 5.3.

1. Dry Fractionation

The original, most simple form of this process is the "topping" of semi-solid fats (such as palm oil in tropical countries). Topping is the slow fractional crystallization of originally molten tank contents which, if left alone, proceeds more or less uncontrolled, but is preferably carried out at 30–32°C for 5–7 days. The stearines settle on the bottom and the liquid oleins can be skimmed off (decanted) with a swinging pipe which is attached close to the bottom at the side of the tank by simply lowering it below the liquid surface. With palm oil, 60% oleine can be removed in this way.

There are two modes of controlled cooling: slow and quick. According to the first, palm oil is slowly cooled from, for example, 70–75°C to 38°C and the oil is then kept at this temperature with slow stirring and seeding during 8 h ("seeding" by SSS type triacylglycerols). The oil is then transferred to the stirred crystallization vessel and kept for 18 h at 28°C, the temperature being strictly controlled by water of appropriate temperature. Filtration should be executed preferably on automated "membrane" or "cushion" filter presses (see below) or on continuous rotating drum or band filters. Filter presses are less feasible because of the manual labour needed, if heated/cooled frames from which the stearine can be melted down are unavailable. Pumping should be carried out with the least possible damage to the crystals already formed. The crystals can also be protected by forcing the crystallized mass out of pressure-resistant crystallization vessels by air pressure.

Among the commercially available slow-rate crystallization systems one might mention the Tirtiaux process, which uses a patented method for avoiding undercooling and the Florentine horizontal continuous-belt vacuum filter. The palm oil is kept in a stirred precrystallizer (seeder) at 50°C, from which pairs of stirred crystallizer tanks are filled in sequence, after emptying the one attached to the filter. The double-walled tanks of 12–40 ton capacity are provided with stirrers and (hot water) heating and cooling coils.

Although many programmes are possible, only one is included here for illustration: the oil is filled and heated to 70°C for 2 h, cooled to 40°C with water for 6 h, and then held at this for 4 h ("seeding"). The cooling is continued to 20°C in 6 h (with semi-refined oils to 17°C), and filtration is done in 6 h. The total cycle time is 20 h. Control units are employed for the exact cooling cycles.

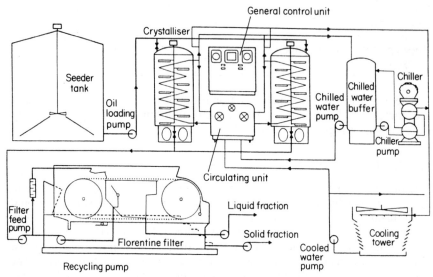

Fig. 5.16 Fractionation plant. [Reproduced, with permission, from Hamilton and Bhati (1980)]

The Florentine filter (Fig. 5.16) is equipped with a coarse-mesh stainless-steel perforated belt as a filtration support. Close to the crystal mass feeding/separating device, at the edge of the vacuum surface, a collector with a recycling pump picks up the solid crystals which were sucked through the belt when the first oleine was removed. The cake is scraped off at the end of the belt. Cool air of controlled temperature is drawn through the belt in order to prevent melting of the stearine crystals. The cleaning of the scraped belt is done by electrical radiant (infrared) heating and subsequent cooling of the lower part of the belt.

Not only palm oil (single or double-fractioned) and edible tallow (in a single fractionation at 36°C, followed by a double one at 20°C and 43°C, respectively), but also iso-suppressive hardened oils and interesterified products (lard) can be processed using this system.

De Smet employ a quick cooling rate controlled crystallization in specially designed and patented crystallizers, using blade stirrers, the speed of which is adjusted to that of the cooling rate. The cooling-water temperature defines the temperature of the wall according to a defined programme. The crystallization takes about 5 h because the cooling surface used is large compared with the volume of the oil treated. The product is separated into stearines and oleines via buffer tanks on membrane filter presses.

Bernardini use modular (three compartment) scraped-wall crystallizers, and a quick cooling rate temperature program (from 45°C via 30°C to 20°C in 4 h) with an additional 1-h filtration from a post-crystallization (slowly stirred buffer) vessel, all positioned in the same column. The filtration occurs on a filter-cloth covered continuous-drum (band) vacuum filter.

2. Wetting Agents Mediated or Lanza Fractionation

The basis of this fractionation was laid down in early patents of the Lanza brothers (1905) and Twitchell (1909) using surface-active agents for the selective wetting of solid tallow fatty acid (stearine) crystals suspended in the liquid phase (oleine). Since 1957 the cooling for solid triacylglycerol fractionation has been done under controlled conditions. In certain routines the oil mass is quickly cooled by means of, for example, the so-called scraped-surface heat exchangers (see Chapter 6, Section I.2(b)) to actual crystallization temperatures. The oil is then kept for some hours under gentle agitation with slow paddle mixers. Just before the separation of the crystal mass formed, the oil is mixed (e.g. by knife mixers) with 0.1–0.5% (on fat basis) sodium lauryl sulphate (the wetting agent, others may be also applied) dissolved in an aqueous magnesium sulphate solution of a given concentration; in the case of syphoning this concentration is about 20%, and in the case of centrifuging about 2% magnesium sulphate, the solution added being 30–45% of the total original fat mass. The wetting solution displaces the interstitial oil (oleine) layer on the surface of the crystals, whereas the magnesium sulphate increases the density of the aqueous phase, which then encloses the solid fat crystals.

There are other possible methods of separation. In a simple method the oleine separating on the surface is simply syphoned off. The remaining crystals, when melted, separate from the wetting solution, which is reused for the next batch.

In a more sophisticated variant the mixture is pumped into hermetically closed centrifugal separators, which separate the selectively wetted aqueous solid fat-crystal suspension from the liquid oil. In the Alfa Laval variant, patented under the trade name Lipofrac, the solid crystalline part is melted by steam which separates the solid fat from the wetting agent solution, if kept in an intermittent tank. The stearine and oleine fractions are washed in order to remove the last traces of the wetting agent:

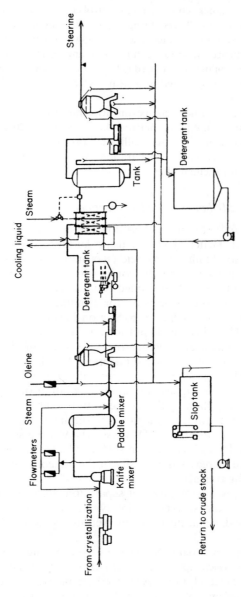

Fig. 5.17 Lipofrac fractionation plant. [Reproduced, with permission, from Hamilton and Bhati (1980)]

15–40 mg kg^{-1} of the wetting solution is used which is later completely removed during the final refining. The fractions are subsequently freed from water by centrifugation and the wetting agent solution is recirculated to the system for reuse. (Fig. 5.17).

The wet fractionation greatly increases the separation efficiency in the filtration stage, which ensures a good yield. It can also be easily automated, which saves on manpower. Heat exchangers can be applied for the regenerative melting down of the stearine which also improves the energy balance.

3. *Winterization/Dewaxing*

When discussing post-cleaning, as an optional step of the refining process, we also mentioned winterization/dewaxing as additional operations, which are often necessary. Both winterization and dewaxing are based on fractional crystallization and the melting point differences between the fractions to be separated are always far above the temperature of 20°C which is essential for sharp separation.

The winterization of (semi-refined or fully refined) oils intended as salad oils means cooling them down gradually, but not always slowly, to the temperature at which the oil is required to stay free of sediments, e.g. 6°C. The cold test of the American Oil Chemists' Society is not always carried out at the prescribed temperature of 0°C, but it is always of 5.5 h duration. For groundnut oil the sediment free temperature can be set at 15°C, the time of incipient opaqueness also being registered. The temperature of incipient opaqueness (the cloud point) is inversely proportional to the time required to become opaque (cloudy) at a low temperature. The chilling and crystallization processes are carried out in huge tanks over a period of some days. The oil remains at the final temperature for 0.5 day before it is filtered on paper-overclad plate and frame filter presses or pressure-leaf filters which have been precoated with a diatomaceous earth. The tanks are at an elevated level and the large surface filter presses are at a lower level, which gives a natural pressure gradient to this slow filtration, and are kept in cooled rooms or are cooled by cold brine circulated in coils in closed tanks. Slow or intermittent mixing is favourable to obtain large crystals, preferably transformed into the β′ modification. The stearines, produced in iso-suppressively hydrogenated soya-bean oil (say to IV = 110) containing fat crystals of mainly C_{18} chain length, can be particularly easily separated after (quick) cooling and fractional crystallization by means of continuous rotary vacuum drum (band), membrane, or cushion filter presses. Among other manufacturers De Smet offers dry fractional crystallizers with extension vessels (maturators), where the oils remain with slow mixing (e.g. 5 h more) at the required temperature (1–3°C). Winterization in solvents gives better yields, but is generally only used for extracted oils which are to be refined in the solvent, e.g. cotton-seed oil.

The presence of oxidized glycerides can retard crystallization, as can the addition of certain polar crystallization retarders such as hydroxystearic acid (from hydrogenated ricinoleic acid) or fatty acids coupled with substances containing many hydroxyl groups.

Dry dewaxing of sunflower-seed, maize and rice-bran oils should be carried out analogously to winterization, by chilling and keeping the oil at low temperatures (say 8°C) for a relatively long period. The filtration is carried out on special pressure-leaf or plate and frame filters, using diatomaceous earth (0.15%) or paper as the principal filter media. The wetting-agent mediated or Lanza wet fractionation process can be easily applied to these oils, the principles being similar for each (Latondress, 1983). The wax is carried away with the aqueous solution, which can be reused.

Interesting modifications are applied to cold crude (but already degummed) oil, out of which the waxes have already been crystallized at 8°C: the addition of a slight excess (0.3–0.5%) of 18 Baume degree alkaline lye at 15°C not only neutralizes the oil, but the soap formed wets the wax crystals and they are removed together with the soap stock. Self-cleaning (nozzle or hydraulically opened) centrifuges (De Laval, Westphalia) are used for this separation, because of the accumulation of waxes and impurities in the centrifuge. The rest of the refining process (washing of the oil soapfree) follows the normal pattern. Wax contents are reduced from 2000 mg kg^{-1} to 100 mg kg^{-1} or even lower. Some other methods apply small amounts of strong lye and a few percent of water for re-refining an already neutralized oil, chilled to 5°C, the alkaline lye also saponifying a very small amount of the oil in 4–5 h during the slow mechanical mixing. The wax crystals formed are also wetted by the soap (-stock) formed, and they are again led on preferably self-cleaning separators, on which the oil is separated from the soap-stock. The waxes accumulate in the centrifuge, being ejected as solids by periodic opening of the bowl (Haraldson, 1983). Careful heating of the oils to 18°C before separation reduces their viscosity during wax removal.

4. *Fractionation Followed by (Hydraulic) Pressing*

Like tallow, coconut and palm-kernel oils can also be crystallized at low temperatures. The oils can be crystallized in rectangular aluminium pans at 2°C (panning) in refrigerated rooms or by air (at 8°C) on continuous belts. The crystallized mass from the pans can be packed in cloths and further pressed with hydraulic presses. The mass from the belts is collected in a receiver and is pumped on cushion or membrane filter presses.

Cushion or membrane presses (e.g. the Hauser or Lenser built types), already often mentioned, are constructed of a central aluminium plate to which a flexible rubber or polypropylene membrane (diaphragm) is attached on each side. This diaphragm is encased in a stainless-steel mesh cover. The cavity formed is gas-tight and forms the cushion. The filtering media are polypropylene cloths. The cushion is kept flat by applying a slight vacuum to it. When the frames are full with the crystal mass pumped onto them and no more oleine is flowing out, the cushion is inflated by air pressure (6–12 bar), which causes further de-oiling of the cake. After opening the frames one by one by means of an automated hydraulic system, the stearine removed, e.g. by vibrators, falls into a chute below the press (Fig. 5.18).

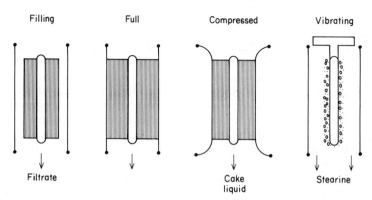

Fig. 5.18 Diaphragm (cushion) filter. [Reproduced with permission, from Latondress (1983).]

The stearines are used as speciality fats for bakery and confectionery purposes, such as butter-oil extenders. The oleines can be used as frying oil or can be hydrogenated.

5. Solvent-aided Fractionation

Either apolar (hexane) or polar (acetone, 2-nitropropane, iso-propylalcohol) solvents may be used for solvent-aided fractionation. The use of solvents reduces the viscosity of the system during filtration in the cold as well as reducing the tendency to mixed-crystal formation during cooling. The melted oil is mixed with the solvent and the resulting miscella (solution) is led via preliminary chillers (e.g. scraped-surface heat exchangers of, for example, the Votator/Vogt or the Chemtec-type) to crystallizing vessels. The mixtures containing the crystallized solids and the dissolved liquids are first separated by filters. The fractions obtained are then freed from the solvent by distillation. In general, multi-step crystallizations are used for preparing speciality fats such as cocoa-butter extenders and equivalents or substitutes/replacers, starting from palm, shea, illipe or hydrogenated oils.

In one of the commercially available installations (Bernardini) hexane is used in a 1:1 proportion for palm-oil fractionation. The miscella is first heated to 45°C and then cooled in a cooling tank to 30–33°C (seeding). The real crystallization begins when the oil is cooled to 20°C. In the second stage the oil is further cooled to 10°C, after which it is passed through a continuous rotating-drum filter in order to separate the stearine from the oleine. The stearine is melted and the hexane separated by distillation. The oleine, which is filtered and collected in a tank is further cooled to 7°C in the first tank of the second cooling line, and then to 4°C in the second tank. Final cooling at even lower temperatures produces a second stearine, from which the solvent, containing the final oleine is separated on continuous rotating-drum filters. The hexane is removed from the fractions by distillation. The second stearine serves as the starting material for cocoa-butter substitutes.

In the Unilever patented acetone process, 1 part of palm oil is crystallized in 3 parts of acetone and the mixture is then cooled. The solution is filtered on continuous band filters. The precipitate obtained is redissolved in acetone and the solution recrystallized at 20°C. The solids are removed on filters, whilst the solution, being rich in SUS type triacylglycerols, contains the fraction needed for producing cocoa-butter equivalents or extenders. The solvent is removed by distillation.

In the HLS process 1 part of isopropylalcohol is added to 1 part of palm oil, before the mixture is chilled in a scraped-surface heat-exchanger tube and passed to the crystallization tank. The process is carried out in two stages: at 20°C a light phase is obtained with 20% stearine in isopropylalcohol and a heavier one with 80% oleine in isopropylalcohol. The heavy phase is recrystallized at 15°C, in order to obtain a second stearine fraction. A special additive (most probably water) makes the necessary phase separations (by decantation) and solubility differences possible.

It has been observed that partial acylglycerols, because of their lower solubility in apolar solvents (hexane), precipitate preferentially in the solids. They may form lower melting eutectics with trisaturates, which co-precipitate with and contaminate the desired SUS fraction. In more polar solvents the mid-fraction is free of eutectics and trisaturates which are mainly transferred to the oleine causing an increase in its cloud point.

D. Regular and Surrounding Equipment

Any crystallization process has its own time/temperature profile. In general, the requirements for a reasonable volume of product with an optimum separation of the fractions are: vessels (sometimes of considerable volume) with amply dimensioned cooling surfaces and effective mixing devices, a strict temperature control for the devised cooling rates and the appropriate filtering devices. A wide variety of crystallization vessels and tanks is used: they are either cylindrical or, less often, of rectangular shape, can be placed vertically or horizontally, be open or closed, and are sometimes pressure resistant if the mass is to be filtered by gas pressure. Depending on the crystallization principle applied, the mixing is provided, with or without wall scraping, by knife-, gate- or peddle-type stirrers rotating either slowly or fast.

Crystal agglomerators or pelletizers have also been proposed as means of crystallization. These consist of a large slowly rotating cylinder of considerable diameter (like a Couette flow device) inside a vertical cylindrical crystallization tank, the cooling being effected through both the cylinder and the jacketed wall. The device agglomerates small crystals into clusters of useful structure and dimension for filtration. At higher speeds eddy streams (Taylor vortices) are produced which lead to fines by attrition.

Heat transfer (cooling and heating) can be provided either by coils (amply dimensioned for efficient and quick cooling and temperature control) submerged in single- or double-walled (jacketed) tanks or by the different (scraped or non-scraped wall) tubular- and plate-type heat exchangers (and precrystallizers) which are of well-known construction. Cooling is not only necessary for the crystallization, as such, but also for keeping cool the separating devices and their surroundings: for example, a cool room is a prerequisite especially for the filter presses. The recovery of heat from the incoming and outgoing oil streams by the use of heat exchangers is a good way of lowering the process costs.

The filtering/separating devices, such as vacuum-operated continuous-drum filters and membrane or cushion filter presses, as well as centrifugal separators, have been described previously for other processes. The pumps used should not destroy the crystal structure already developed.

When solvents are used, they must be recovered by some kind of distillation. Polar solvents form azeotropic mixtures with water, and, therefore, their separation requires the type of sophistication used in ethanol rectification. The general organization and safety measures are similar to those of the solvent extraction installations for seeds. It is evident that many tanks are needed to store the starting, intermediate and final products, as well as the solvents (most of them constituting a considerable fire hazard). The automated temperature/cycle time/filtration controls, partly also microcomputerized, are all part of the more modern installations, whereas in the older types of plant some manual work is needed; hydraulic pressing in open presses is the most labour-intensive operation.

E. Raw Materials and their Quality

Without doubt present-day commercial fractionation is mainly aimed at the fractionation of palm oil into the oleine with the stearine as side product. Speciality fats (see Chapter 7) are of much lower volume. In the simpler processes the split is never perfect and only partial separation is achieved due to the formation of mixed crystals and the mechanical occlusion of the oleine in the stearine. This is also the case with tallow, interesterified lard, hard butters, lauric fats, etc.

There is a lot of discussion on the quality of the oils to be submitted for the fractionation operation, mainly concerning the amount of free fatty acids and partial acylglycerols present in the starting material. It should not be forgotten that partial (mainly diacyl-) glycerols are considered to be crystallization retarders, forming eutectics with trisaturates such as PPP. On the other hand higher melting saturated free fatty acids acting as seeds might positively influence the crystallization process. Mechanical impurities, although sometimes also serving as nucleating (seeding) centres, are best removed in advance. In practice, preferably filtered, pre-deslimed (earth/acid bleached) crude fats should be submitted for fractionation. If for any reason pretreated crudes cannot be used, then alkali-neutralized ones are also acceptable, as this pretreatment has a thorough purifying effect on the fat.

Previous physical deacidification (a fatty acid distillation at high to very high temperatures) has adverse effects, particularly on the crystallization of palm oil and hard butter (shea, illipe, etc.). These adverse effects are: by removing the free fatty acids the role of the higher melting ones as crystallization promotors is cancelled; by removing a part of the monoacylglycerols the predominance of the crystallization retarding diacylglycerols increases; by partially redistributing (interesterifying) the component fatty acids in the triacylglycerols, slowly crystallizing SUS/SSU type pseudo-compounds, amongst others, are produced; by slight re-esterification the oleine yields may also be reduced.

Oils intended for winterization/dewaxing can be optimally treated after or during (alkali) neutralization, but preferably before earth bleaching. The old practice of winterizing oils after deodourization by keeping them in open tanks for days, may cause flavour deterioration by oxidation.

F. Quality of Final Products and Costs of Fractionation

If winterization of liquid or iso-supressive hydrogenated oils is the goal of the production, the main quality requirements are either the AOCS cold test and/or some cloud-point determinations. In France and Austria the use of any oil containing more than 2% linolenic acid for frying is not permitted, and E.E.C., U.S.A. and Japanese food legislations prohibit any residual traces of detergents (15 mg kg^{-1}) or solvents if used in a process. If dewaxing is the goal of the operation, not only the cold and cloud-point tests, but also an analytical measurement of the actual amount of waxes can be used as a quality criterion, 100 mg kg^{-1} wax being an accepted limit.

If the starting oil is palm oil, the main goal (considering the volume) is the production of a good quality oleine which is used as a cooking (frying) and salad oil. The oleine should have a possibly high IV, e.g. of 60. A long-lasting cold stability might be characterized by the oil becoming clear at 20°C within 2 h, if previously kept in a refrigerator at 5°C for 15 h: a behaviour similar to that of groundnut oil. The smoke point of the palm oleine after full refining should be above 200°C, in order that it remains stable when used for frying. Palm oleine is without doubt the most easily marketable product.

The stearine which is obtained as a side product may cause problems due to its properties. In moderate climates it cannot be used in table and kitchen margarines, which represent the largest volume of margarine production. On the other hand, it can be used directly in speciality bakery margarines, liquid shortenings and cooking fats. There are no objections to incorporating stearines into tropical margarines, if the popularity of margarines against cheap locally produced liquid cooking oils (like palm oleine) or imported soya-bean oils is increasing. Palm stearines, if mixed with liquid oils or with (hydrogenated) lauric fats (coconut, palm kernel) can be (e.g. after interesterification) very interesting components of even moderate-climate margarines, if for technical reasons no hydrogenated fats are available. The resulting fat mixtures which are free of *trans* fatty acids feature a more balanced spectrum of triacylglycerols and also contain the SUU and SSU types, so that they have an acceptable melting and crystallization behaviour (less post-hardening).

Because of its limited marketability in the moderate climate countries the costs of stearine production are often included in the cost of the oleine. This means that the oleine yields must be maximized in order to keep the total costs low.

The oleine yields of the processes increase in the sequence: dry < Lanza mediated < solvent-aided fractionation. The IV of the oleines produced may serve as a good quality index for comparative yield/quality/cost calculations.

It will be clear that installing a house-assembled dry or Lanza fractionation plant requires the lowest costs, whereas the knowledge and equipment sold by engineering companies and machine manufacturers (not only the ones cited for process illustration purposes) will lead to higher investment costs. This higher cost may be compensated for by better quality end products and sometimes, but not always, better separation efficiencies.

G. Quality Control

The melting/solidification behaviour of practical fats and fat mixtures can be determined by using simple dynamic methods. The melting of solid fat crystals needs external thermal energy, while their solidification liberates a considerable heat of crystallization. Therefore, both of these processes cause temperature arrests on the cooling curves. Equilibrium melting and freezing curves can be constructed by registering the arrest or break points during slow cooling of fats or mixtures with different starting composition, but of the same total weight versus the time elapsed (cooling rate). The Jensen cooling curves and the cooling curves obtained by isothermal differential thermal analysis (DTA), also called differential scanning calorimetry (DSC), can be used to determine instrumentally the characteristic temperature arrest or break points of any fat, individual fraction or mixture.

The most important and informative data on the quality of the products processed are: the softening, the (slip) melting and solidification points of the starting oils and of the different solid fractions, and the solid fat contents by dilatation or pulsed NMR. The different clear, cloud and cold points are useful in assessing the quality of the liquid fractions. Winterization/dewaxing results are also characterized by the last two tests.

The determination of the refractive index is a quick way of testing the level of unsaturation, a laboratory IV determination being more time consuming. Fatty acid or triacylglycerol (carbon number) determinations by gas chromatography (GC) are the most informative ones for a thorough examination. The carbon number is the total number of carbon atoms present in the fatty acid (acyl-) chains of the triacylglyceride molecules.

The residual amounts of wetting agents and solvents should also be determined, even though they will be completely removed by the subsequent post-refining operations. A smoke-point determination after full refining is a good control for the absence not only of free fatty acids, but also of excessive amounts of monoacylglycerols.

III. Interesterification: Ester Interchange/Randomization

Interesterification is the third most important modification process of fats. The aim of this process is to produce and concentrate triacylglycerols which have physical properties considered to be more desirable than those possessed by the remainder. Fats, being mixtures of triacylglycerols, are esters. Under the influence of a suitable catalyst, the component fatty acid (acyl-) and glycerol (alkoxyl-) ions may leave their original sites, be redistributed and coupled again with new ones. As different esters are involved in this process, the term transesterification is also used. As a consequence of the process the newly formed glycerolesters will not necessarily have the same physical and functional characteristics as those present in the original fat.

If a common ester component (acids or mono-, di-, tri-, poly-hydric alcohols) is either present or added to the fat in excess, a total redistribution of the system may take place, leading to new, but again predominantly esterified products. These processes are then called acidolysis and alcoholysis.

The nature of interesterification with two acids can be visualized as follows (S, saturated; U, unsaturated acids; the sequence in the formulae represents the 1,2,3-acyl substituted sites of the glycerol moiety):

$$SSS \rightleftharpoons (SUS \rightleftharpoons SSU) \rightleftharpoons (SUU \rightleftharpoons USU) \rightleftharpoons UUU$$

The reason for using this quite simple process lies in its ability to alter both the melting characteristics (a reduction or increase in the solid fat content within certain temperature ranges) and the crystallization behaviour (e.g. reduction of polymorphism caused defects such as sandiness and graininess; post-hardening in margarines) of fats and fat mixtures.

A. Theoretical Background

1. *Chemistry and Reaction Mechanism*

The exchange of the ester bound acyl (fatty acid) groups by interesterification requires the use of catalysts in order to make it more practical, although at high tempratures it can take place without them. The most effective catalyst precursors (in fact catalyst activators) are bases, like sodium alcoholates (methoxide, ethoxide), metallic sodium and potassium as such or an alloy of the two. Sodium glycerolate (sodium hydroxide/glycerol) is also effective. The active basic catalyst is the diacylglycerol anion, formed between the above-mentioned basic precursors and the fat itself (which very often, if not always, contains some free diacylglycerols) during an induction (activation) period.

The length of the induction period depends on the temperature, type and length of

Activation step

$$\text{DG}-\text{O}-'-\text{C}(=\text{O})-\text{R} + (\text{Na})'-\text{OCH}_3 \rightarrow \text{DG}-\text{O}-\text{Na} + \text{R}-\text{C}(=\text{O})-\text{OCH}_3$$
    Substrate   Precursor

Exchange steps

$$\text{DG}-\text{O}^- + \text{DG}'-\text{O}-\text{C}^+(-\text{O}^-)-\text{R}' \rightarrow [\text{DG}'-\text{O}-\text{C}(-\text{O}-\text{DG})(-\text{O}^-)-\text{R}'] \rightarrow$$
Catalyst   Polarized substrate   Transition complex

$$\rightarrow \text{DG}'-\text{O}^- + \text{DG}-\text{O}-\text{C}(=\text{O})-\text{O}-\text{R}'$$
    Catalyst   Product

Fig. 5.19 Interesterification reaction: DG and DG', diacylglycerol molecule residues; R and R', substituting alkyl-, (poly-)alkenyl groups of long chain fatty acids in the corresponding diacylglycerol molecules.

mixing and the particle size of the precursor (activator), but leads to a homogeneous system of the substrate and the catalyst (Fig. 5.19). The actual exchange reaction is based on the attack of a nucleophilic diacylglycerol carbanion [DG—O$^-$ (negative), an alkoxide ion] on a positively charged carbon atom (a carbonium ion) of the carbonyl group of one substituting fatty acid (R') of another triacylglycerol molecule DG'. The temporary unification of the two molecules (in a transition complex) makes possible the (now intramolecular) exchange of two or more acyl groups. A subsequent acyl cleavage liberates a triacylglycerol molecule and an anionic catalyst (DG'—O—), both of which carry at least two exchanged acyl groups (R versus R'). This ionic process is called a substitution, nucleophilic, bimolecular, S_N2 reaction. A chain reaction continues until either all the groups are randomly distributed or the reaction is stopped before completion, e.g. by destroying the anionic catalyst by water or acids.

The reaction temperature required for the basic catalyst precursors (except sodium hydroxide, which needs a reaction temperature of 170°C) is relatively low but varies widely between 10 and 150°C. As already mentioned, the reaction can be spontaneous at high temperatures (above 250°C). Acidic catalysts (like toluenesulphonic acid) at 200°C may be used for the (re-)esterification of fatty acids with glycerol. There is a small chance of ester interchange occurring during high-temperature deodorization/deacidification by distillation of fats (palm oil, vegetable hard butters) without any added catalysts. Extracellular microbial lipase enzymes catalyse not only hydrolysis but also interesterifications. Industrial processes based on enzymes have been described in the literature (see Chapter 7).

2. Fatty Acid Distribution in Triacylglycerols

Fatty acids can be distributed in oils and fats in different ways. Formally triacylglycerols can be classified in three types: (i) those in which all three of the substituting fatty acids (acyl substituents) are of the same sort (monoacidic triacylglycerols); (ii) those in which two of the substituents are of the same sort (diacidic triacylglycerols); and (iii) those in which all three substituents are of different types (triacidic triacylglycerols). These three types are present in all fats with only their relative amounts varying. The distribution of the fatty acids in triacylglycerols is subject to certain rules. Natural oils and fats have a fairly typical fatty acid composition and also a fairly constant fatty acid distribution within their triacylglycerol molecules. These characteristics are thus specific to these species. There is some biological variation and also some variation caused by external influences such as climate (the geographic origin). This relative constancy is even more remarkable if the amazing number of permutations and combinations, by which fatty acids can be formed into groups of three (tri-), groups of 2 (di-) or singly (mono-acylglycerols) by even a small number of fatty acids at the three possible glycerol positions, is considered.

If the ester forming reactivities of the two primary alcoholic groups of glycerol (at the 1 and 3 positions) are considered to be equivalent, many different types of triacylglycerols (types of fatty acid distribution) are expected to occur according to the formula:

$$\text{Number of different triacylglycerols} = \frac{n^2 + n^3}{2}$$

where n is the number of different fatty acids. This formula accounts for positional but not stereochemical (optical) isomers. Therefore, for two different acids there are six and with three different acids there are 18 different triacylglycerol types.

3. Fatty Acid Distribution in Natural Fats

In nature fatty acids are *not* completely randomly distributed within the fat molecules. The specific positioning of some of them can be used to differentiate between different triacylglycerol compositions. In most fats and oils the middle (2) position is preferentially occupied by unsaturated fatty acids. The only exception is pork fat (lard), in which the 2 position is preferentially occupied by palmitic acid. Following Van der Wal (1960) and Coleman and Fulton (1961), the preference of (mainly C_{18}) polyunsaturated fatty acids for the 2 position in vegetable oils is expressed by the term 1-,3-random, 2-random fatty acid distribution.

The specific 2 position occupation of fatty acids can be indirectly analysed by a position-specific enzymatic hydrolysis. Pancreatic lipase hydrolyses only the fatty acids occupying the external 1 and 3 positions of the triacylglycerol molecule. (These positions are preferentially occupied by saturated and C_{22} mono-unsaturated fatty acids, at least if present in substantial amounts.) By GLC analysis of the total triacylglycerol and the residual monoacylglycerol fatty acid composition, the types and amounts of fatty acids occupying the 2 position can be calculated. Although this method is very informative in general, it is not quantitatively applicable to fats containing fatty acid chains shorter than C_{14} (lauric fats, butter), longer than C_{18} (marine oils, high erucic oils) and fats having a melting point higher than 45°C. Physical acylglyceride separation methods are based on selective crystallization into fractions by solvents, as well as on adsorption, liquid and high temperature gas chromatographic techniques. X-ray diffraction and NMR spectroscopy can also give a lot of information, although less easily than the pancreatic lipase hydrolysis method.

4. Random Fatty Acid Distribution

Regardless of the method used to achieve it, complete interesterification, and also the re-esterification of fatty acids with glycerol, leads to total randomization of the fat in question; hence this process is also called randomization. This total (statistical) random distribution means that the fatty acids present occupy all the three glycerol positions according to their mass, governed purely by chance, without assigning the unsaturated (C_{18}) fatty acids preferentially to the 2 position.

Table 5.4 *Calculation of the proportion of three randomly distributed fatty acids in all the possible triacylglycerols*

No. of different fatty acids present	Types of triacylglycerols	Proportion (% mol)
1 (monoacidic)	SSS; OOO; LLL	e.g. $S^3/10.000$
2 (diacidic)	SSO,SOS,OSS; SSL,SLS,LSS SOO,OSO,OOS; LOO,OLO,OOL SLL,LSL,LLS; OLL,LOL,LLO	e.g. $3\,S^2O/10.000$
3 (triacidic)	SOL,LOS; SLO,OLS; LSO,OSL	e.g. $2\,SOL/10.000$

The random distribution can easily be calculated by using the formulae given in Table 5.4: for the sake of simplicity a fat composed of only three fatty acids is taken as an example: saturated (e.g. stearic), oleic, linoleic. Considering all the possible mono-acidic, di-acidic and tri-acidic distributions there are 27 possible different triacylglycerols, because in this example all the positional and stereo/optical isomers have been considered, according to the formula:

$$\text{number of all possible triacylglycerols} = n^3$$

where n is the number of fatty acids.

B. Applications

The profound changes which interesterification can bring about to the melting characteristics and crystallization behaviour of different fats are illustrated below for some selected cases. (The ester interchange within one molecule might also be called intra-esterification.)

Table 5.5 *Comparison of the actual and calculated random distribution in a palm oil sample composed of only two main groups of fatty acids*

Type of triacylglycerol	Mole (fraction)	Palm oil (mol %)		
		Initial	Interest. (actual)	Random (calc.)
SSS (S_3)	S_3	8	13	14
SSU ⎱ (S_2U) SUS ⎰	$2 S^2U$	9	13	13
	S^2U	35	25	26
SUU ⎱ (SU_2) USU ⎰	$2 SU^2$	36	23	24
	SU^2	9	14	12
UUU (U_3)	U_3	7	12	11

1. *Interesterification (Randomization) of Single Fats*

(a) *Palm oil.* For simplicity we will divide the component fatty acids of this semi-solid oil into only two groups: one consisting of the 52% saturated fatty acids (S), and the other consisting of the 48% unsaturated fatty acids (U). As was shown in the above, according to the first (simple) combination rule, these two groups of acids can give rise to six possible types of triacylglycerols; the possible types are given in Table 5.5, wherein S and U now also depict concentrations. The initial and final compositions given in the table are those found by the analysis using the lipase technique; the final-product (interesterified) data are in good agreement with the random distribution calculations, as described earlier.

This example shows that the interesterification (randomization) of palm oil causes: an increase in the trisaturated triacylglycerol content of the end product resulting in an increase in its melting point, e.g. in the given case from 39°C to 47°C; a decrease in the absolute amount of the disaturated triacylclycerols; and a change in the proportion of the symmetrical (and therefore more easily crystallizing) SUS type components to the asymmetric, SSU type components from 4:1 to 2:1. All these changes are slightly compensated for by the increase in the amount of triunsaturated components. Due to the increased level of trisaturated solid triacylglycerols, the interesterified palm oil has a more plastic character: the solid fat content temperature curve becoming more flat and extending to a rather high temperature (50°C).

(b) *Lauric and (partially) hydrogenated fats.* When interesterified, the melting point of coconut oil increases, whilst that of palm-kernel oil decreases, due to the increased levels of triacylglycerols with intermediate unsaturation. When interesterifying fully hydrogenated coconut and palm-kernel oils, their melting points decrease, because the saturated triacylglycerols formed are of lower molecular weight than are the initial ones. The melting points of (partially) hydrogenated (non lauric) oils also decrease, mainly because the amounts of higher melting triacylglycerols [mainly the trisaturated triacylglycerols (SSS), formed during hydrogenation] decrease. These higher melting triacylglycerols cause the originally hydrogenated, >80% C_{18} fatty acid chains containing fats to stabilize in the highest melting β form. On interesterification, this tendency is relieved (see Chapter 6, Section I.G.2.a).

(c) *Liquid oils.* Interesterification of soya-bean oil increases its melting point from −7 to +6°C, which is not an asset as the oil loses its transparency when stored in a refrigerator. This is also the case for other liquid oils, except rape-seed oil, which contain only low amounts of saturated fatty acids, their melting points being unacceptably increased. The increase in the melting point of interesterified cotton-seed oil from 10 to 34°C could be favourable for certain purposes.

(d) *Lard.* Interesterification of lard is favourable, as it eliminates a certain storage defect of this fat, called "graininess". Lard contains, among other triacylglycerols, relatively high amounts of 2 palmito-oleostearin and 2-oleo-palmitostearin in the proportion 8 : 1. The former is the higher melting one and, during storage, the mixture of the two produces large, coarse β-type polymorphic crystals, which cause the grainy structure. If interesterified, the proportion of the two triacylglycerols mentioned above is reduced to 2 : 1, and the mixed crystals formed have a reduced tendency to form a grainy texture.

2. Interesterification of Mixtures

In mixtures of unhardened palm oil with unhardened laurics the proportion of the solid fat content which melts at room temperature is strongly increased by interesterification, whilst the proportion of the solid fat content which melts around body temperature is unaffected. Interesterification of hardened laurics with hardened palm-oil causes a strong decrease in the amount of higher melting solids. This is also the case with saturated triacylglycerols (SSS, e.g. palm stearines) mixed with liquid oils (UUU): the saturated fatty acids are incorporated into the unsaturated triacylglycerols as SUU and SSU triacylglycerides. Instead of unsaturated liquid oils, the laurics also cause a decrease in the melting point of palm stearines. These interesterified mixtures are often used as fat bases for margarines, confectionery and speciality fats.

In most practical cases the sequence of the different fat-modification operations is hydrogenation and/or fractionation as the first step with interesterification generally being the last step. It should be remembered that the melting-point raising effects of hydrogenation and fractionation can be totally spoiled by the subsequent interesterification of, for example, fat and oil mixtures; of course this might sometimes be the objective of the interesterification. The optimum sequence of the three modification processes should always be carefully considered.

3. Directed Interesterification

In order to produce either speciality fat products (see Chapter 7) or margarine and shortening fat bases, the combination of interesterification with contemporary fractionation is a very widely practised method. The method was first proposed by Eckey (1956) and is called directed interesterification.

Table 5.6 *The effect of random and directed interesterification on lard*

Type of triacylglycerol	Interesterification (mol %)		
	Before	After random	After directed
SSS	2	5	17
SSU, SUS	26	25	12
SUU, USU	54	44	37
UUU	18	26	34

First, the total or at least partial interesterification of an oil or oil mixture is accomplished with all the components in the liquid state. The mixture is then cooled, by means of a scraped surface heat exchanger or other type of chiller/crystallizer, and the high melting triacylglycerols newly formed by the interesterification crystallize out. The originally random equilibrium of the initially totally liquid system is disturbed, as the solids formed are excluded from the rest of the process. The balance can also be disturbed by removing components of different molecular weight or unsaturation by distillation or selective extraction. The remaining liquid triacylglycerols are now kept at a predetermined low crystallization temperature for a certain time, during which they attain a new random equilibrium, totally different to that of the initial one.

If crystallization of the already formed higher melting triacylglycerols by controlled cooling is the means of producing the equilibrium disturbance, the catalyst precursors must remain sufficiently active to continue reacting at (if necessary, decreasing step-wise) lower temperatures, e.g. varying between 20 and $-40°C$. A sodium/potassium alloy catalyst precursor is often considered to be the most suitable for performing low-temperature directed interesterifications. The reaction rate can be controlled by the amount of precursor added and the rate of cooling during crystallization (i.e. controlling the formation of nuclei and the crystal growth).

As randomization also means the maximum formation of the maximum number of statistically possible triglycerides, then theoretically a total segregation of a starting oil (or mixture) into trisaturates (SSS) and triunsaturates (UUU) can be achieved if the solubility of SSS in the mixture is zero. This can be demonstrated by the random and directed interesterification of lard (Table 5.6).

4. Displacement by Fatty Acids or Esters (Transesterification)

Instead of crystallization, Koslowsky (1974), applying Eckey's original idea, proposed the transesterification of, e.g. deacidified palm oil with a 50% excess of methyl esters of unsaturated fatty acids, using sodium ethoxide, as the catalyst precursor. Ethyl- and isopropyl-esters could also be used. After completion of the reaction at 60°C, the newly formed volatile transesterified methylesters, mainly saturated ones (C16), are distilled off under vacuum, the remainder being mainly palm oleine (the liquid fraction). The saturated fatty acid methylesters are then added to a new batch of palm oil and the mixture is transesterified (interesterified) in the way described previously. The volatile reaction products (C18), now being mainly unsaturated fatty acid methylesters, are distilled off under vacuum leaving the residue of hard palm stearine. The inventors of this method claim the separation of palm oil into an oleine of IV = 72, which after winterization rises to IV = 79–80 (close to that of olive oil), with a yield being 65%. The stearine has an IV of 5 and consists mainly of tripalmitate.

Displacement and/or directed interesterification can be used for the production of a coconut-oil based product, in which the lauric fats have been displaced by the addition of an excess of stearic acid or methylstearate. This method can also be used to produce oils with higher amounts of polyunsaturated triacylglycerols, after removal of the higher melting ones by crystallization.

C. Processing Methods

In practice the batch, continuous and semi-continuous type processes are all commonly employed, for interesterification. The batch process is carried out in large (40–60 ton) holding/mixing vessels which are very similar to the combined neutralizer/bleacher equipment discussed in Chapter 4. The continuous systems generally use tubular reactors. The semi-continuous process also uses the neutralizer/bleacher type vessels combined with the straight flow-through heating/drying, dosing, premixing equipment of the continuous method.

1. Pretreatment of Fats

The alkaline nature of the precursors and of the process prescribes the use of fats which have been neutralized. This means the thorough removal of gums, if necessary by a pre-degumming operation, to a residual phosphorus level of $< 10 \text{ mg kg}^{-1}$. The level of free fatty acids should not be higher than the usual 0.05–0.1%. The residual level of alkaline soaps should be around 100–300 mg kg^{-1}, as their total removal is not strictly required.

The alkaline catalysts are inactivated (killed) by water, so moisture removal to 0.03% in batch reactors, or to 0.01% in continuous ones is a prerequisite. Hydroperoxides also react with catalysts by their reduction to secondary oxidation products such as keto esters, so their absence or removal to low values (1–5 meq kg^{-1}) is a good preliminary measure to save on catalyst use and to improve oil quality.

2. Heating and Drying of Fats

In batch-type processes heating is executed by the steam coils of the closed (combined neutralizer/bleacher) type kettles, provided with slow rotating-gate type or two-step variable blade stirrers, and in the continuous or semicontinuous process by plate or other types of heat exchanger.

Because the precursor is decomposed by water, extremely low residual moisture levels are required. Drying to these low levels can be achieved in one of two ways:

(i) by producing a vacuum via the barometric-condenser based normal vacuum system (mechanical pumps, the residual pressures produced depending on the cooling water temperature), while thoroughly mixing the oil in the above-mentioned combined neutralizer/bleacher type vessel at 120–125°C for some hours; or

(ii) by producing a vacuum via a two or more stage steam-ejector system (producing low residual pressures), while mixing mechanically at the same temperatures given above. Alternatively the drying can be done by spray-drying, if the oil is recirculated by the (filter) pump returning it to the bleacher through a nozzle installed close to the top of the vessel. In this case the time for drying can be reduced to a few hours (Fig. 5.20).

In continuous systems, cylindrical columns with internal stacked perforated trays, on which the oil cascades from the top to the bottom, the top being attached to the high vacuum system, can be used as dryers, the vacuum being, for example, 55 mbar in the first and 13 mbar in the second column, coupled in series. This drying method can also be used in semi-continuous processes.

3. Addition (and Homogenizing) of Catalyst Precursors

After drying has been accomplished, the system is cooled/heated to the actual reaction temperature: for sodium alcoholate, 60–90°C; metallic

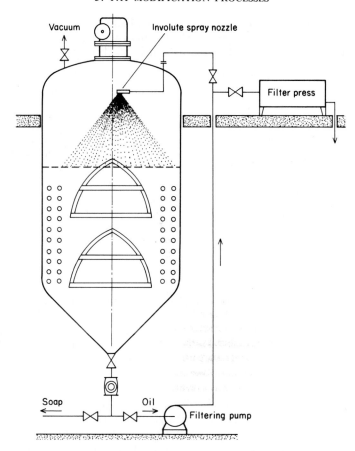

Fig. 5.20 Spray drying of oils in batch kettles. [Reproduced courtesy of Unilever N.V.]

sodium, 130°C; sodium hydroxide/glycerol (spray dried), 150°C; sodium hydroxide/water (spray dried), 170°C; and metal alloy eutectic 0–40°C. Once the reaction temperature has been attained the catalyst precursors, in powder form (e.g. sodium methoxide), are sucked in by the vacuum via a pipe to below the fluid level. Alternatively, the catalyst precursors can be added as a 25% methanol solution, or as fat/precursor slurry. The precursor level is generally 0.06–0.1% of the total. Powdered precursors are originally kept in tight metal drums lined with polyethylene sheets, free of moisture and in the cold, preferably in amounts exactly necessary for one batch. If not purchased, alcoholates can be (although this is not advised) produced by dissolving metallic sodium in methanol or ethanol, not forgetting that complete elimination of the hydrogen gas evolved by boiling and stripping the reaction mixture by an inert gas stream is required. Sodium hydroxide, as precursor, should be added as a concentrated aqueous or glycerol solution before the effective (spray) drying of the oil, in order to avoid the occurrence of saponification instead of the formation of the diacylglycerol–anion catalyst.

NaK_2 alloy, which has a eutectic melting point of $-12°C$, can be added in liquid form or as an oil slurry, whereas metallic sodium can be dosed, like a wire, through hydraulically operated positive-displacement metal extruding and dosing pumps (pre-forming and main dosing presses) directly into the stream of the hot oil on its way to either a batch (semi-continuous system) or a continuous tubular reactor. Before entering the vessel or tube system the slurry is homogenized by a high shear mixer, e.g. Willem's Reactron. In continuous systems the metal or alloy level is 0.03% on oil with a 4-min reaction time or 0.07% on oil with a 1-min reaction time. This process has lost much favour in recent years.

In any case the catalyst level should be low, as each 0.1% excess may result in a loss of 1% triacylglycerol.

4. The Interesterification Step

The actual interesterification reaction takes place in 1–60 min, depending on the type and amount of catalyst precursor and the corresponding optimal reaction temperature. The reaction mechanism and the main reaction products are the same and independent of the precursor, the catalyst always being the same sodium diacylglycerol anion. Sodium methoxide intensively mixed with the oil first forms a white slurry, with a subsequent colour change (a deepening of the original oil colour) which is a visible sign of the formation of the catalyst anion and the start of the reaction. In directed interesterification, shortly after the catalyst has been formed the mixture is cooled to the envisaged temperature of crystallization. The heat of crystallization should be controlled by additional cooling. The liquid part is further reacted and newly crystallized at a lower temperature, the final product having the required characteristics, far remote from those obtained with a single-step random equilibrium. Because of its longer duration (e.g. 1 day), this process is more expensive than the simple non-directed one.

In order to make possible residence times of some minutes, some continuous systems using metallic sodium or sodium hydroxide/glycerol have tubular reactors of considerable length (Fig. 5.21). As mentioned previously, semi-continuous systems may involve the use of, for example, the combined neutralizers/bleachers as holding and reacting vessels.

5. Deactivation of the Catalyst and Finishing Operations

In batch reactors the deactivation (killing) of the catalyst occurs during the washing of the interesterified oil in the kettle. If the reaction temperature is above $95°C$, the system should be cooled to this value before batch-wise or continuous admixing of 10–20% wash water to the oil containing some

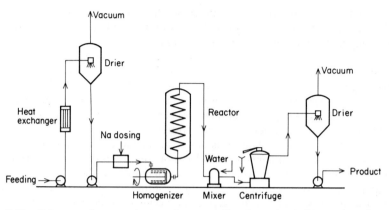

Fig. 5.21 Scheme of a continuous interesterification plant. [Reproduced, with permission, from Ralndaal (1973).]

soap. After a short settling time the water is separated either by decanting or, in continuous systems, by centrifuges. The soap-free oil is then dried under vacuum (55 mbar) before conducting the bleaching operation by earth, which also removes residual traces of the soaps.

In continuous systems with metallic precursors, after completion of the reaction the oil is led to a stripper, where open steam deactivates (kills) the catalyst, liberates the occluded hydrogen gas and transfers it to the open air. The stripper can be of the stacked internal perforated trays type, similar to the driers used in continuous systems. In semi-continuous reactors using metallic precursors, the hydrogen gas is first removed through the vacuum system before deactivating the catalysts by water.

D. Side Reactions

The alkaline character of the interesterification precursors may give rise to side reactions known from the organic chemistry of oxygen containing products, particularly esters. Although in theory there are many possibilities, the "catalytic amounts" of the precursors applied and the thorough pretreatment of the oils results in only extremely small amounts of side products. However, the whole process should be continuously monitored.

Not only the triacylglycerols, but also other fat accompanying substances can be involved in the interesterification reaction. These are primarily the free tocopherols, the natural antioxidants present in most vegetable fats and oils.

Their phenolic hydroxyl groups behave as normal hydroxyl groups and may be also esterified by the migrating alkoxy carbanions. The esterification removes the mobile hydrogen atoms, preventing the stable free radicals which stop the reaction chain. This loss of antioxidant chemical protection should be recognized and counteracted by the addition of (synthetic tocopherols or other phenolic) antioxidants, preferably before the drying/bleaching step, but it can also be done after deodorization. The tocopherols originally present in the oil do not lose any of their physiological vitamin E activity on esterification.

The free sterols present in the oils, most of them also possessing a hydroxylic (phenolic) group, are also substrates of the interesterification reaction. The decrease in the original amount of free sterols can be easily detected by suitable analytical methods. The esterification of free sterols by the acyl anions is a welcome aspect of process control (see below). Non acylglycerol based side reactions, however, are less feasible because of the low level of these substrates.

Because of the use of alcoholates as precursors and alcohols as possible diluents, the formation of small amounts of methyl and/or ethyl esters of the component fatty acids is a quite normal side reaction of interesterification. These esters are easily removed during deodorization, but they do constitute a loss, which can be reduced by the use of, e.g. sodium hydroxide/glycerol precursor. Although the methyl esters are primarily produced from the residual free fatty acids and soaps normally present, in absence of the latter the attack of the precursor on the triacylglycerol molecules gives rise to the formation of diacylglycerols and fatty acid methyl esters. Solid diacylglycerols may function as crystallization inhibitors, whereas the monoacylglycerols produced by the same mechanism strongly lower the smoke point of the final product.

E. Hazards of Handling Precursors/Activators

The reactivity of sodium alcoholates and sodium/potassium metals with water leading to exothermic decompositions and reactions has been amply stressed. The alcohol and hydrogen gas produced can ignite and so low relative humidity (below 55%) and storage in cold (18–28°C) rooms on separate, elevated (not accessible to running water) floors is good practice. Sodium/potassium packages should be stored in a separate (small) building, to which only persons responsible for production should have access. The wearing of safety goggles (face shields), powder masks (the alcoholate powder is highly irritating to the mucous membranes) and gloves when handling these reagents is also necessary. Fires should never be extinguished with water, but with sand and chemical foam. Should it ever be necessary, trained firemen should carry out the fire fighting.

F. Quality and Process Control

The change in the solid fat contents and/or crystallization behaviour during interesterification can be assessed most simply by measuring the melting and solidification points, such as: the softening (slip)-point, the clear point, and the cloud point (turbidity temperature of the melt). These are the quickest methods which can be applied on the factory floor for control purposes. The basic change in the solid fat content at given temperatures or ranges can be determined more exactly by dilatation and pulsed NMR. The Jensen cooling curve and differential thermal analysis (DTA) are dynamic methods for determining the different phase transitions over the whole melting or solidification range. The completeness of the interesterification reaction can be controlled by the disappearance of the free sterol spot, as monitored by thin layer chromatography. High temperature gas chromatography of the triacylglycerols also gives information on the process-caused triacylglycerol changes by measuring the change in the carbon-number distribution. The pancreatic lipase enzyme hydrolysis method, which determines the 2 position occupation, is also used to obtain the necessary information about any changes. All these methods together with the aforementioned ones take more time to carry out than do the first mentioned simple ones. The smoke point of the interesterified product, after full refining, should also be tested for the absence of even low residual amounts of monoacylglycerols, as these strongly lower the smoke point.

6

Production of Edible-fat Products of High Fat Content

In this Chapter the most important emulsified high fat products such as margarines/minimum calory (low fat) margarines and mayonnaise and related products are discussed together with three non-emulsified edible high fat products: shortenings, Vanaspati, and prepared mixes. A general survey of the processes used is given in Fig. 6.1.

I. Margarines/Minimum Calory Margarines

A. Historical Background

The scarcity, the limited keeping properties and the high price of butter around the impending French/Prussian war (1870–71) was for Emperor Napoleon III a substantial reason to induce a search for a butter substitute. He appointed and ordered Hyppolite Mége-Mouries, a previous apothecary assistant and at this time an already famous French inventor, to carry out the research work. As a result of his work Mége-Mouries filed a patent on 2 October 1869. The famous French Patent 86480 claimed the production of a congealed whiteish edible fat product with the gloss of pearls, called margarine (margaritos is the Greek word for pearls). It consisted of the semi-solid part of low-temperature rendered beef and/or mutton tallow fraction, called oleomargarine, obtained by crystallizing, panning and pressing the tallow. The tallow was then mixed with soured skimmed milk. The fat, including finely dispersed milk globules, formed a coarse water-in-oil type of emulsion. This colloid mixture was cooled and subsequently intensively kneaded. Some other additives, like salt and an extract of cow udders were also admixed. The latter was considered at that time to be indispensable for obtaining the butter-likeness. The large-scale production of the new butter substitute started only after the war, in 1873.

One aim of the invention was based on the supposed activity of the cow's udder for extracellular milk-fat formation: Mége-Mouries hoped to perform this direct transformation of tallow to milk-fat by adding pieces of cow udder to the milk/fat

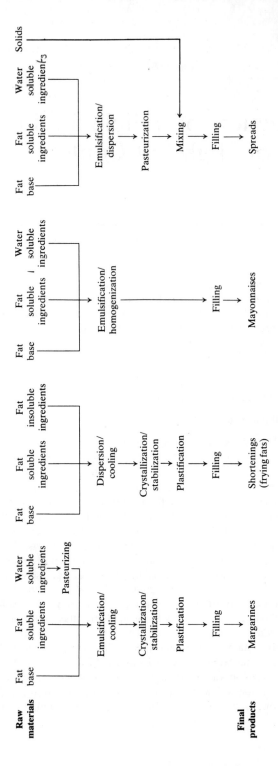

Fig. 6.1 A general survey of the processes aiming at the production of some edible products of high fat content.

mixture; this is of course biochemically impossible. The other aim, to imitate physically the customary appearance of butter, became a big technical success.

Butter is produced from either fresh, or fermented (ripened) and, during the process partially crystallized, cream of 40–50% milk fat content, which in modern processes is increased to 80% butter fat. The cooled cream is mechanically beaten (churned) to cause the separation of most of the serum or buttermilk (the aqueous part of the cream) from the fat.

The milk fat (compositionally and rheologically totally different from the tissue fat of the same animal) is first dispersed in the continuous water phase of the milk in the form of tiny globules. This relation does not change when the fat is concentrated during skimming into cream. Only during churning does the nature of the dispersion change to the opposite form with the inclusion of tiny serum droplets into a continuous phase of butter fat. The butter fat consists of small partly crystallized oil globules and fat crystals. This conversion of the oil-in-water type dispersion into the water-in-oil type is called inversion; this phenomenon was met before when discussing the caustic neutralization of oils. Butter has to be kneaded and then rested for a certain period of time to obtain its typical texture. This results in the formation of a dense network (matrix) of small solid fat crystals, which do not grow beyond a certain size during prolonged storage with temperature fluctuations. This special network provides butter with its plasticity and elasticity.

Although margarine was invented to imitate butter, in the beginning it was a mere substitute for it. We shall soon see which aspects of margarine had to be improved to obtain a product which is closer to butter and whose characteristics made it possible to recognize margarine as a product in its own right.

B. Legal Aspects

Margarine is by definition a food in the form of a plastic, partly solidified or fluid water-in-oil type emulsion/dispersion, produced principally from edible fats and oils, which are not usually derived from milk fat. Under a microscope provided with a light polarizator set (e.g. two Nicol prisms), it can be seen that there are small crystals in the fat phase, mostly larger than 5 µm. These crystals form a stabilizing matrix (frame) held together by different forces, in which the oil and the finely dispersed water (most of the droplets are less than 3 µm, but many of them up to 20 µm in diameter) and some air bubbles are enclosed, like fluids in a sponge. In a so-called non-refrigerated table margarine at room temperature, the oil phase amounts to about 60%, the solid crystals to about 20% and the waterphase to about 20% of the total. If heated excessively the stabilizing fat crystals melt and dissolve in the oil, which is the main reason why the whole emulsion/dispersion destabilizes quickly.

According to E.E.C. legislation, the recommendations of the Codex Alimentarius Committee and World Health Organization (WHO), the fat content of margarines is generally regulated to be not less than 80%, with water (moisture) content being not more than 16%. A part of the water is derived from added milk or whey. The product should be labelled as "Margarine".

If the fat content is around 40% (not less than 39% and not more than 41%) and the water content not less than 50%, the spreadable water-in-oil type emulsion is called in Europe "minimum calory margarine" ("minarine"), "half calory margarine" ("halvarine"), or (unofficially) just a "low calory spread" (LOCAS). This type of margarine is produced principally from edible fats and oils which are not solely derived from milk fat. The labelling of the package should carry "Minarine" (or similar) and/or "Spread", in all cases with the stated fat and ingredients content.

European "health" margarines are produced with high amounts of fat (>80%), with a declared minimum level of polyunsaturated (essential, see later) fatty acids: PUFA (EFA). These margarines contain generally low amounts of salt.

The U.S.A. Federal food standard for vegetable-fat based margarines (FDA, 1973) does not give a maximum for water, but only specifies a not less than 80% minimum fat level. The difference between fat and the actual moisture percentage is constituted by the other compulsory (vitamin A) and optional ingredients which are considered to be additives. There is another standard for animal/vegetable-fat based margarines under the meat inspection regulations of the USDA (1983). Although very close to each other, and the international standard, the name of the product is "margarine" or "oleomargarine". Any claim or advertised aspects regarding PUFA, saturated fatty acid (SAFA) or cholesterol levels are to be labelled on each package.

In the U.S.A. all products containing less than 80% fat have to be considered as "spreads" and are to be labelled as such. Products with at least 53% fat (this means a 33% reduction of the legal fat level) may be called either "reduced calory" or "diet" margarines, if contaiing vitamin A, as prescribed by the law. Of course these "imitation" margarines are legally also spreads. In many countries other low-calory products with 40-75% fat content are on the market, all to be labelled as "spreads". In the U.S.A., bakery fats containing no water but containing vitamin A and resembling butter in consistency might be called "margarines". Margarines, minimum-calory margarines or spreads may, in some countries, contain butterfat. Low(er) fat margarine/butter mixes are called, for example, "dairy spreads" as under Californian law.

All additives, whether fat or water-phase soluble, must be of edible grade and approved by the appropriate health authorities. This means that they are tested toxicologically conforming to the guidelines developed by FAO/WHO Codex Alimentarius Committee (Europe), listed e.g. by the EEC Harmonization Commission in their directives (E number of admission or the proper name of the substance). The emulsifiers in particular should be those described in the monographs of the European Food Emulsifier Manufacturerer's Association (EFEMA) in Europe. The additives for shortenings used in the U.S.A. should be on the list of GRAS (Generally Recognized As Safe) substances of the FDA. Any margarine or shortening ingredients or additives applied in the different countries should always be visibly listed on the packages.

C. Theoretical Background

1. *Margarine/Mayonnaise Emulsions/Dispersions and their Emulsifiers*

In Chapter 4, Section II.A.1 it was stated that an emulsion is a stable and fine dispersion of one immiscible liquid in another. As was observed then, the stablity aspect mentioned in the above-given definition generally requires, besides the solid crystal network as one of the stabilizing elements, the presence of a third substance at the interface of the two fluids, called an emulsifier.

Margarine is traditionally the foremost representative of a water-in-oil type. The fat crystals formed during cooling would stabilize this type of emulsion, but monoacylglycerols must be used as emulsifiers to induce (because of their water-affinity and fat-solubility characteristics) the formation of the above type of dispersion (water-in-oil) before the fat crystallization. Therefore, margarines are produced using these emulsifiers, although other emulsifiers are also in use, serving special purposes in industrial margarines and shortenings.

The main buiding blocks of the margarine-incorporated food emulsifiers are fatty acids (the long alkyl and alkenyl chains constituting the apolar part) and glycerol (the polar part), mostly in the form of a solid or liquid monoacylglycerol (monoglyceride). Modifications are compounds derived from this structure which contain organic acids (acetic, citric, lactic, diacetyltartaric), organic acid salts, and inorganic acid derivatives such as that of phosphoric acid (phosphoacylglycerols, lecithins). Instead of glycerol other alcohols which might be used as bases are polyglycerols, propyleneglycol, sorbitan, sucrose, polyoxyethylenes, etc.

Milk and mayonnaise are the most characteristic representatives of oil-in-water type emulsions. The proteins of milk (caseine, lactalbumine), proteins derived from soya bean and other seed oil meals, and the proteins and lecithin (PC) of egg yolk (the traditional emulsifier of mayonnaise) all stabilize this type of emulsion. Because of the electrolytic dissociation of their hydrophilic groups in aqueous milieu, these stabilizers are more sensitive to increased salt concentration, pH and temperature changes, than are the above-mentioned less water soluble counterparts. The already described inversion of the emulsion type, observed in the traditional preparation (churning, hand beating) of butter is partly explained by (i) the change in the original volume proportion of the two phases by mechanical expulsion of a considerable part of the original continuous phase, and (ii) the removal (inactivation by aeration) of emulsifiers stabilizing the original oil-in-water type dispersion.

This leads us finally to the re-evaluation of the role of different monoacylglycerols in margarine processing. As long as the margarine had to be made by preparing a coarse pre-emulsion, which had to be cooled by ice and mechanically worked (kneaded) in a discontinuous operation, the (pre-) emulsifying action of monoacylglycerols was vital. Mostly solid monoglyceryl palmitate was required to stabilize the pre-emulsion until the solid fat crystals had formed and also embedding hydrophobic fatty acid residues of the emulsifier. The solidified fat thereby took over the emulsion-stabilizing role of palmitate. Present-day margarine machines produce a very fine dispersion/emulsion of water in the fat phase, followed by immediate fat crystallization. In most cases (household margarines) this process greatly reduces the amount of emulsifier required. Because of the different structure and stability of minimum-calorie margarines, the amount of emulsifier used in their production is close to the traditional quantities.

2. Crystallization of Fats

Fats, if left alone without mixing, whether under isothermic or adiabatic conditions, crystallize slowly, with an equilibrium only being reached after several hours. On the other hand, if mixed and cooled quickly (shock chilling), a proportion of the triacylglycerols, because of their limited solubility, will crystallize. The solid phase formed will consist of finely dispersed and generally microscopic sized crystals. The remaining liquid triacylglycerols fill up the interstitial space around the crystals which are interconnected by "bridges", forming a network, also called a (solid/liquid) matrix. Small crystals can form more contact points than can larger ones. The crystallized matrix stabilizes the water added to the system in the form of a water-in-oil fine dispersion.

3. The Mechanism of Crystallization

The driving force of crystallization is supersaturation which governs both the formation of crystal nuclei and the growth of the crystals on these nuclei (the precipitation of solids on them from the melt). The rates of the two processes may differ. If both processes have a high rate at the freezing point, the system cannot be supercooled. Seeding with small crystals and mechanical impact by stirring both induce nuclei formation. If the rate of nuclei formation is higher than crystal growth, the solidified mass will consist of small crystals. In the opposite case a smaller amount of larger crystals will be obtained. By the appropriate selection of the time/temperature profile (the temperature of crystallization and the cooling rate), the supercooling and the mechanical impact, the texture and properties of the crystallized fats can be controlled over wide ranges.

4. Demixing

We have already seen (Chapter 5, Section II) that many mixtures of triacylglycerols, if cooled, form a continuous series of solid (crystalline) solutions, called mixed crystals. We have also seen that at each temperature two different concentrations must be assigned to the components, one for the solid and one for the liquid phase.

The composition of the total mixture in the solidifying crystal mass depends on the thermodynamic equilibrium between the two phases. Equilibrium is attained by molecular diffusion, a slow process which is promoted by slow cooling rates. If the cooling rate is quick (shock chilling), only the surface of the congealing crystals is in equilibrium with the surrounding liquid phase and the composition of the solid mass is locally different, and thus inhomogeneous.

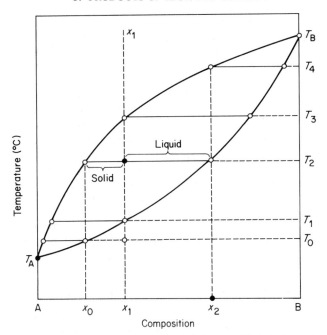

Fig. 6.2 Melting range increase (demixing) of shock-chilled fat bases by isothermal storage. [Reproduced courtesy of Unilever N.V.]

The consequence of rapid cooling is shown by a phase diagram for solid solutions in Fig. 6.2. If the cooling of the mixture of a given composition x_1 is abruptly stopped at a certain temperature T_2 in the region between the two equilibrium curves, the mixture separates into two phases: a liquid one with a composition x_0 given at the liquidus curve and a solid phase with a composition x_2 given at the solidus curve. The relative proportions of the phases are determined by the lever rule. If the phases are separated, remelted and cooled slowly, the fraction of composition x_2 has the melting range T_4–T_2 and that of composition x_0 a range of T_2–T_0. By storing the non-separated mixture at T_2 for some time it obtains the same extended T_0–T_4 melting range of its now equilibrated components. The melting range of the slowly cooled original mixture x_1 would be only T_1–T_3. This phenomenon, called demixing, therefore considerably increases the plastic range of the margarines.

5. Polymorphism

The phenomenon of polymorphism affects many triacylglycerol molecules, and means that they tend to crystallize in different forms (modifications). This affects not only the external (morphological) differences, but also the internal ones, determined most often by X-ray analysis. In general the triacylglycerols may have three polymorphic forms: the α, β' and β forms. In the α modification, the hydrocarbon chains are at right angles to the glycerol plane. In the β one this angle is about 67° to this plane. The β' form has an intermediate position between the α and β forms. The α form is the thermodynamically least stable, the β' form has medium stability and the β form has the highest stability.

On cooling the fats crystallize first in the thermodynamically metastable α form, and later in the β' or β form. More of the metastable α form will be formed if the cooling is done rapidly. One triacylglycerol may block the other sterically and the β' to β transformation (transition) is often hampered. Simply built and symmetric triacylglycerols tend to be transformed further to the highest melting β modification. The transformation goes easily from the lowest to the highest stability form, but not in the reverse direction. Such an irreversible transition is called a monotropic transition (going only in one direction).

Predominantly the melting point, but also the density, the dilatation and the heat of melting increase proportionally with the stability of the crystal modification.

D. Solid Fat Content and its Significance to Product Properties

The solid fat content (SFC) is the percentage of solidified triacylglycerols in an oil at a given temperature; it can be used as a measure of fat crystallization. It can be measured by pulsed nuclear magnetic resonance (NMR) spectroscopy or by the relatively old-fashioned melting dilatation. The dilatation, as measured, delivers a solid fat index (SFI), which has to be converted to a percentage solid fat content by using statistically tested conversion tables (van den Enden *et al.*, 1978). The measurements are normally done at a few selected temperatures. The data, obtained after standardized tempering procedures, are treated as equilibrium results. However, in practice this very often appears not to be the case. Therefore the tempering methods used must be quoted if the data obtained are to be used in practical calculations (e.g. blend properties).

The crystallization behaviour of fats, or fat blends, during processing and storage, can be described by solid fat content versus temperature curves (Fig. 6.3). The lower curve in Fig. 6.3 is for the lower melting (α) the higher curve for the higher melting (β') modification and both are characteristic for a given fat composition. The linearity of the curves in Fig. 6.3 is a simplification, because solubility lines can be bent or composed of segments, due to the mixed-crystal formation tendency of triacylglycerols.

If a given fat sample is cooled quickly the crystallization of the α modification will begin at a certain temperature, this being the α crystallization point (T_α). By lowering the temperature further the amount of α crystals will increase. If the α crystals formed are stored under adiabatic conditions they are (slowly or quickly, see below) transformed to the β' modification. The recrystallization process is exothermic: it gives rise to a measurable temperature increase which causes a part of the α crystals to melt.

The time required for this adiabatic transformation to occur is called the transition time (τ); this is shown by the temperature versus time plot given in Fig. 6.4. The starting temperature of cooling determines the onset and duration of this transition: the lower the temperature, the shorter the induction period (t_{ind}) of α crystallization and the shorter the time needed for the β′ transformation to occur. The steepness of this curve can be

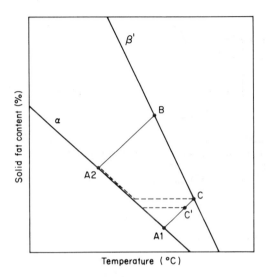

Fig. 6.3 Generalized solid fat content *vs.* temperature (*T*) plot of α and β′ fat crystal modifications formed during margarine and shortening processing: A–C–A–B sequence. [Reproduced, with permission, from Trenka (1987); and courtesy of Unilever N.V.]

considered as a measure of the grade of supercooling of the β′ modification, because the metastable α modification is not able to supercool. The variation of the transition times (τ) with the temperatures at which the α crystals initially formed were "stored" adiabatically, is a characteristic of the given fat blend (Fig. 6.5).

In general, palm-kernel and coconut oils, and hardened fats containing >80% C_{18} fatty acid chains in their S_3, S_2E and similar higher melting triacylglycerol molecules crystallize quickly. Lard, tallows and palm oil crystallize relatively slowly, the latter probably because of the presence of 3–6% of mono- and di-glycerides which act as crystallization retarders.

The solid fat content (weight of α crystals) at the end of a quick cooling step and the increase in the solid fat content (weight of transformed and newly formed β′ crystals) during adiabatic crystallization is roughly proportional to the maximum hardness of the sample. This correlation is of paramount importance when processing modern margarines.

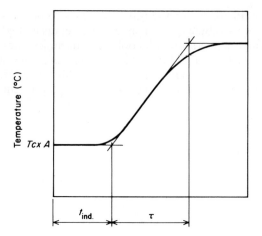

Fig. 6.4 Generalized temperature change diagram depicting the induction time (t_{ind}) needed for α crystallization and the transition time (τ) needed for its β′ transformation under adiabatic conditions. [Reproduced, with permission, from Trenka (1987); and courtesy of Unilever N.V.]

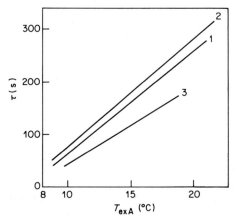

Fig. 6.5 Variation of the transition time (τ) of α crystals with their primary (adiabatic) formation temperature for different fat blends. [Reproduced, with permission, from Trenka (1987); and courtesy of Unilever N.V.]

E. Consistency of Edible-fat Products of High Fat Content

Fats and fatty emulsions have many properties which can be used as quality criteria (see Section IX of this chapter). One of these properties is the consistency of the product, an aspect based on its rheological (flow) behaviour. The processor judges the consistency by the (changes of) hardness during processing, packaging and storage, whereas the consumer judges it in terms of the spreadability/workability and oral melting properties of the product during use. Whatever the criteria for judging, the consistency is connected with a certain change in the shape of the sample in question.

Depending on the conditions used in their processing, fats (in margarines/ shortenings) are generally considered to be plastic/elastic solids and/or viscoelastic liquids (see below). In order to be plastic, fats must have: (i) the co-existence of two phases, a solid one embedded in a liquid one; (ii) a proper proportion of the two, as expressed by the term solid fat content at a given temperature; and (iii) the interconnection/cohesion of the solid fat crystals by "bridges" or "primary bonds". Of course other factors, such as polymorphic crystal modifications, the complexity of crystal composition and the shape of the individual crystals, all affect the consistency of a high-fat product. Large, needle-shaped crystals interlace more easily and these primary bonds result in firm, rigid structures. The same amount of solid fat in a more compact form is bound by less-direct secondary bonds which results in a softer structure. The aggregation of smaller crystals to clusters with substantial amounts of liquid phase also makes the system softer, but more plastic, than these described above.

It is generally not well realized that even the waterphase makes a distinct contribution to the consistency of an emulsion. The contribution of the water phase compared to that of the solid fat (per percent solids) is negligible at room temperature. At higher temperatures (close to the melting point) its relative contribution becomes more pronounced in determining the consistency, with the disappearance of most of the solid fat crystals. This fact makes the consistency of shortenings, margarines and minimum-calory margarines virtually the same at room temperature. However, a minimum-calory margarine close to its final melting point is much thicker in the mouth than are other margarines and these, in turn, are somewhat thicker than shortenings based on the same fat blend.

1. *Rheological Background*

Rheology is the science which deals with the deformation and flow of materials when forces are applied to them. For the sake of simplicity we restrict ourselves in the following discussion to the case where the forces acting on the sample are directed tangentially to its boundary. The force acting is called the shearing force (F), and the force acting per unit area (A) is called the shear stress ($\tau = F/A$).

The deformation under shear stress is called shear ($\gamma = \mathrm{d}s/l$). The rate of deformation or rate of shear (γ') can be written as:

$$\frac{\mathrm{d}\gamma}{\mathrm{d}t} = \frac{\mathrm{d}s/l}{\mathrm{d}t} = \frac{v}{l} = \gamma'$$

where s is the length of deformation, l is the perpendicular distance, v is the velocity, and t is time. The resistance against deformation (hardness) can be characterized by

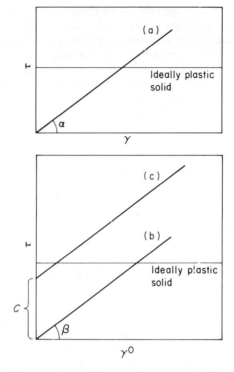

Fig. 6.6 Behaviour of: (a) Hookean elastic solids; (b) Newtonian liquids; (c) plastic solids. τ, shear stress; G, $\tan \alpha$, elastic shear modulus; γ, deformation under stress; $\eta = \tan \beta$, viscosity; γ', rate of deformation or shear; C, yield or stress value.

the relations between the shear stresses applied and the deformation and flow (i.e. the rate of deformation) that occur as a result of these stresses.

In order to understand the rheological complexity of food emulsions such as margarine and spreads, it is appropriate to consider first three simple ideal materials:

(i) *Elastic solids* which behave according to Hooke's law ($F = kx$), which is a linear relationship between stress and relative deformation:

$$\tau = G\gamma$$

The rheological behaviour of this material is characterized by the material constant G ($= \tan \alpha$), [Fig. 6.6(a)]. For incompressible elastic materials, the elastic shear modulus is 1/3 of the Young modulus (E): $G = E3$.

(ii) *Newtonian liquids*, which are characterized by a linear relation between the shear stress and the rate of deformation [Fig. 6.6(b)]:

$$\tau = \eta \gamma'$$

The rheological constant characterizing these fluids is the viscosity, $\eta (= \tan \beta)$.

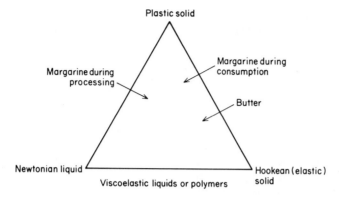

Fig. 6.7 Rheological positioning of margarine(s) and butter in a triangular field [Reproduced courtesy of D. W. de Bruijne of Unilever N.V.]

(iii) *Ideal plastic solids*, which are characterized by the fact that the shear stress levels remain constant during deformation [Fig. 6.6 (c)]

$$\tau = \text{constant}$$

This constant is called the yield (stress) value (C) and is associated here with the hardness or consistency of the material.

In general, the fat emulsions exhibit some features of all of the three types of ideal materials mentioned above, the dominant features being determined by the rate of deformation. This is shown schematically in Fig. 6.7, where the three ideal types of behaviour are located at the corners of the triangle. Materials which exhibit features of all the three ideal materials can be represented by points inside the triangle. In Fig. 6.7 the position of margarines (fats) is at the left-hand side for the high deformation rates characteristic of modern processing. Under these circumstances margarines are adequately characterized as "mixtures" of a plastic solid and a Newtonian liquid. Several models are in use, the simplest being the Bingham liquid model [Fig. 6.6(c)]:

$$\tau = c + \eta\, \gamma'$$

At relatively low rates of deformation, the characteristics of the consumption processes, such as spreading, cutting and chewing in the mouth, are characterized primarily by the plasticity and elasticity of the margarine. Therefore, in Fig. 6.7 margarines are positioned on the right-hand side of the triangle. Butter is usually more elastic than margarine.

The quantification of the rheological behaviour of margarines in terms of material constants, e.g. the yield (stress) value (C), the viscosity (η) and the elastic shearing modulus (G), is difficult in practice. This difficulty is due to the fact that the primary crystal structure responsible for these material "constants" breaks down during deformation because of mechanical working, kneading and spreading. In effect all the possible material constants appear to decrease numerically as the degree of deformation or degree of material working increases.

F. The Hardness of Fats and Fat Blends and its Measurement

The common types of rheological instruments used to measure the hardness of margarines (fats), and thus the deforming shear stress required to cause them to yield and flow, are described below.

(i) In penetrometers a die is forced into the sample either by gravity or by a separate drive. The measure of the hardness is then derived from the force required and the penetration depth attained. In the case of a cone penetrometer with a standardized shape and weight, falling under gravity, Haighton (1959) derived a relation between a consistency (yield or stress) value and the depth of penetration:

$$C = k \times (\text{weight of cone/penetration depth})^{1.6}$$

where k is the cone-angle factor.

Similar to the cone penetrometers is the Stevens instrument which introduces a cylindrical bar of 1–5 mm diameter into the sample with the speed and depth of penetration being preset. The instrument measures the resistance of the material to the penetration. The values obtained are expressed as cone-penetration values.

(ii) A second type of instrument are the rotational viscometers of various configurations. The most practical configurations comprise concentric cylinders for more liquid-like margarines and parallel plates for more solid ones. Viscometers can be used to measure the initial flow characteristics (i.e. the elastic modulus and yield (stress) values) of the unworked structures and the steady flow characteristics (yield (stress) values of the worked structure and the viscosity).

The method of tempering the sample and the temperatures at which the sample was tempered and the measurements made should be always quoted.

1. *Work Softening*

As mentioned previously, vigorous mechanical working, such as mixing, kneading and spreading, destroys a certain part of the crystal network of a solidified fat or margarine. The separated crystals suspended in the oily part begin to float, and the hardness of the sample is decreased considerably. The consequent decrease in the yield value, which characterizes the structural hardness of the sample, can be measured.

The instrument used to measure this softening comprises a thermostated cylinder with a hand-operated kneader containing sieves with holes. The sample is passed through the sieves, the direction of passage being reversed many times. When comparing different samples it is advised to apply samples having similar starting yield values (e.g. 900 g cm^{-2}). When taking samples from different stages of processing, the temperature measured can be the usage temperature or, as a compromise, 10°C. The percentage relative loss of structural hardness is the "work softening" which is an important practical quality characteristic. In general butter cannot be spread as easily with a knife as can margarines. Work softening is thus low (40–60%) for butters [this is in contrast to a controversial statement made by Hoffmann (1986) in his Chapter in Herschdoerfer's monography], low for puff-pastry margarines, and it should be high (70%, but definitely not >80%) for table margarines.

In products showing a low degree of work softening, the deformation in practical situations tends to become severely non-uniform. The deformations actually tend to concentrate in a few slip-planes or cracks; such products are called brittle.

The thixotropic character of fats should also be taken into consideration. If fats are mechanically worked they become more fluid. However, after a certain time they can regain the original firmness if left alone unworked. This effect is based on a temporary, but reversible breakdown of primary bonds. Under mechanical working the loosened fat crystals are oriented in the direction of the stress exerted and the viscosity decreases. With time, the cohesion forces restore to their original, less oriented structures (aggregates).

G. Formulation of Fat Blends

The main building blocks of fat blends are the different triacylglycerols present in the component fats. Without any doubt the most exact way to make products of constant properties would be if all the final-product properties could be correlated with the contribution of individual types of triacylglycerols present in the component fats. The types of triacylglycerols from which conventionally used fats are composed, can often be determined analytically via the 2-position fatty acid occupancy of the component groups. A more detailed description of fats on a molecular basis therefore remains limited. Our knowledge of the influence of different types of triacylglycerols, as a group, on the finished product's properties is empirical and rather scanty.

There are two methods for formulating margarine or shortening fat blends. One method uses the qualitative (empirical) knowledge of how certain individual fats and fat fractions [characterized by their melting point or iodine value (IV)] contribute to the final consistency and future behaviour of the finished products. The other method involves calculating the different quality requirements of the final fat blends (e.g. EFA, PUFA or SAFA contents) as a linear contribution of the individual component fats. Other quality requirements, like the solid fat content and hardness at different temperatures are more complex and depend also on the interaction of the components and the processing, so that they must be determined by the methods which will be described later.

1. *Interchangeability of Component Fats*

In the process of blending, the optimum formulation of a certain type of margarine/shortening fat blend may depend also on the market price of the raw materials being considered. If the complex behaviour of different fats in mixtures can be more or less predicted, and the listing of component fats and oils (except by vegetable, animal or mixed origin) is not required by law, then limited interchangeability of different raw materials will make it possible to produce margarines of fairly uniform quality at competitive prices. Besides the original fat characteristics, fat-modification processes, like hydrogenation, fractional crystallization and interesterification can be used to increase the interchangeability of different fats, and thus produce the required fat blends.

2. *Qualitative Fat Blending*

Triacylglycerols can be roughly classified into four groups which correlate empirically with certain final-product functions or properties. According to this classification the SSS types (Group I, with melting points of 54–65°C) and some of the SSU types (Group II, with melting points of 27–42°C) are the structure builders. The SSU types are also important for the oral properties at close to body temperatures. The SUU types (Group III, with melting points of 1–23°C) are important for oral properties, or as mechanical performance improvers at room temperature. The UUU types (Group IV, with melting points of -14 to $+1°C$) are important for lubrication and nutritional factors (like polyunsaturated fatty acids, PUFA) in the different margarine blends.

It is of course a well-established fact that margarines produced with the correct proportion of higher-IV (soft) oils (Group IV which give no structure), and lower-IV (harder) fats (Groups II and III, the structure builders) are quite satisfactory for most kitchen and table uses. The further addition of some low-IV (hard) fats (Group I) makes the product more plastic over a broad temperature range, and more suitable as a shortening or pastry margarine. A balanced use of oils belonging to the different groups might lead to different products. High amounts of high-IV oils with low amounts of low-IV fats give refrigerator-storable liquid and soft (tub) margarines. Intermediate amounts of high-IV oils with intermediate amounts of intermediate-IV starting fats may lead to soft (tub) or soft-wrapper products. The use of higher amounts of intermediate-IV fats with low amounts of high-IV fats also gives soft (tub) products, whereas a blend of intermediate-IV fats gives typical wrapper products for the kitchen or table use.

Besides these general rules the knowledge of the particular crystallization behaviours of certain fats and fat fractions is also necessary to make feasible blend compositions.

(a) *Polymorphism-based margarine defects.* For a long time it has been known empirically that certain (fully or partially hydrogenated) fats and fat fractions have a strong tendency to (re)crystallize in the highest melting β modification. These substances can be split into two groups: (i) triacylglycerols composed of fatty acids of the same chain length; and (ii) triacylglycerols composed of fatty acids of different chain lengths.

(i) Trisaturated triacylglycerols (SSS), whether formed during hydrogenation or being of natural origin, stabilize preferentially in the β modification. Besides the trisaturates, the disaturated *trans*-monounsaturated (SSE + SES) triacylglycerols, formed during hydrogenation and some (non-hydrogenated) symmetric disaturated *cis*-monounsaturated (SUS) triacylglycerols also show this tendency. Some hardened fats, incorporated in margarines and shortenings, might give rise to the well-known quality defect of sandiness. This defect is caused by the presence of small needle-shaped microcrystals of high melting point in the β modification. These fats contain a high proportion of β tending fractions, because they were processed from oils originally composed of more than 80% fatty acids of C_{18} chain length. This defect occurs very often in the so called one-oil margarines or shortenings, produced from (all-hydrogenated or hydrogenated and liquid oil mixed) sunflower-seed (C_{18} fatty acid content >92%), safflower-seed (>92%), low erucic rape-seed (>91%), sesame-seed (>90%), olive (>90%), soya-bean (>87%), maize (>86%) and groundnut (>80%) oils.

(ii) Unmodified fats containing the genuine solid trisaturated and the symmetric triacylglycerols of the disaturated *cis*-monounsaturated type (e.g. POP, POSt, StOSt) also transform more readily to the β configuration than do their asymmetric counterparts. This applies to formulations containing higher amounts of palm oil or lard. In this case another margarine defect, called graininess, is experienced, where co-crystallizing POSt and PStO triacylglycerols may form higher-melting lumps (aggregates) because of the compatibility of the symmetric/asymmetric components to form (pseudo-)compounds.

On the other hand, fats and the (generally higher melting) fat fractions which tend to stabilize in the β modification, are excellent in speciality products, where the structural stability of the product is a necessity. Examples are fat blends used for puff-pastry, for coating (coherent films) and in liquid margarines and shortenings where they give the necessary body during use. These products should have lower "shortening" power, than the more plastic kitchen, table and confectionery margarines (shortenings), containing fats stabilizing in the β' modification (see later).

There are some general theories which try to predict the β forming tendency of different fats from certain diagnostic clues. For example, Wiederman (1978) proposed the presence of the minimum level of 10% palmitic acid in fats or fat blends as a clue for predicting a reduced proneness to β crystal formation. This theory, based on just one fatty acid, seems to be more vulnerable than the following theory. This alternative theory supposes an increased proneness to β crystal formation when fats or hardened fats with high levels of symmetric solid triacylglycerols content and/or composed of more than 80% of fatty acids having the chain length C_{18} are used. According to this theory the addition of any fat containing substantial amounts of asymmetric triacylglycerols and/or fatty acids with chains shorter or longer than C_{18} reduces the tendency of the starting fat to stabilize in the β modification. Of course fats with a high palmitic acid content (like palm oil, notwithstanding its high symmetric SUS triacylglycerol content) might also relieve the impairment caused by the dominating amounts of C_{18} chains.

Besides blending, the processes of interesterification or fractionation of the starting materials can be used to eliminate the problem. Lard, and even more so, slightly hydrogenated lard, because of the peculiar predominance of its palmitic (C16) acid content in the 2-position of the (mono-)unsaturated PPO + StPO triacylglycerols and the trisaturated ones, tends to crystallize in the β modification. This prohibits its successful use for cake production (no creaming effect, i.e. aeration; see later). Interesterification strongly relieves this tendency, together with the already mentioned consistency defect of graininess. The interesterification of mixtures of obligate β crystallizers with, for example, laurics is also a good remedy. The use of fractional crystallization, to remove the higher melting and β tending fractions can also reduce the problem.

3. Quantitative Fat Blending (the Method of Statistical Solid Fat Content Equivalents)

The quantitative method of fat blending is applied daily by a major international representative of the margarine industry who traditionally produces a uniform final blend at optimum cost from many fats of different quality and price. The optimization of quality requirements and the minimization of raw-material costs can be achieved by using the technique of linear programming on the parameters of production capacities, raw-material buying, stock-holding policies, etc. In this mathematical method the problem is expressed as a group of linear equations and the solution obtained is the optimal one of a range of possible solutions.

The basic assumption when using this method, therefore, is that any property of individual fat-blend components contributes linearly to the final property of the blend, according to the quantity in which it is applied:

$$\sum_{i=1}^{n} A_i X_i = P$$

where A_i is the property of the *i*th component, calculated as a coefficient, X_i is the amount of the *i*th component to be used (the variable to be determined), and P is the resultant common property of the blend aimed at.

6. PRODUCTS OF HIGH FAT CONTENT

The primary task is to formulate a consistent set (a matrix) of linear functions based on the blend components. The so-called objective function will depict, for example, the cost of the raw materials (the price per unit weight multiplied by the as yet unknown amount to be applied, linearly added). Another function will represent the proposed composition of a unit batch. Certain data on quality, like those relating to the maximal use of polyunsaturated (essential) fatty acids (PUFA/EFA), to the minimal use of saturated fatty acids (SAFA) in health margarines, to the minimal or maximal incorporation of certain special fats (high erucic acid containing rapeseed oil or lauric fats) or to any non-complex quality aspect and to production capacity, should all be formulated as "constraints" and expressed as inequalities or with "slack" variables, as equalities. Quality aspects which are influenced in a more complex way by the components, like a minimum and/or maximum hardness (this determining workability/packability, spreadability, oral melt, etc.) at different temperatures, to crystallization speed, etc., must be determined by statistical procedures (see below).

Computer programmes for solving such sets of equations are widely available.

(a) *Statistical solid fat content equivalents.* It was stated previously that under the same processing conditions the solid fat content of a fat or fat blend measured at the same temperature is in close correlation with the (maximal) hardness (C, or yield values), the latter determining many physical and oral-quality characteristics of the final product. The solid fat content values are thus of utmost importance in linear programming determinations of the blending of fats for margarine and shortening production.

The required individual solid fat content (and hardness) data on the available component fats and their blends mixed in different preselected proportions are determined by measurement. From these data the statistical contribution of the individual fat components to the resultant solid fat content is calculated by multiple-regression analysis. The computed coefficients of the regression equation are the statistical solid fat content (SFC) equivalents of the component fats. These coefficients have only a restricted validity, i.e. for the given fats and a block of smaller area, because of the many possible interactions between the triacylglycerols derived from different raw materials.

As an example of how the statistical SFC equivalents (the coefficients of the regression equation) of component fats contribute to the solid fat contents of blends mixed from them at three different temperatures, some data given by Haighton (1976) are cited in Table 6.1. The data were calculated for a block of four components: (i) a hydrogenated oil; (ii) palm oil; (iii) coconut oil, having above its melting point (as a liquid) a negative contribution; and (iv) a liquid oil with the same negative contribution at all temperatures (Table 6.1).

For non-refrigerated margarines and shortenings the blend compositions are changed in accordance with local seasonal temperature variations.

Table 6.1 *Statistical solid fat content equivalents for the calculation of the percentage solids for a given fat blend*[a]

	Contribution (%solids/%fat) to %solids		
Fat	10°C	15°C	20°C
Hydrogenated soya-bean oil	0.71	0.56	0.36
Palm oil	0.42	0.29	0.19
Coconut oil	0.46	0.15	−0.13
Liquid oil	−0.13	−0.15	−0.13

[a] Data reproduced, with permission, from Haighton (1976).

H. Hardness and Solid Fat Content Based Product Properties

The strong correlation between product hardness and the solid fat content of margarines has already been stressed. It is thus informative to look at Fig. 6.8, where this correlation of solid fat content with the acceptable and unacceptable hardness values (expressed by C or yield/stress values) at 10 and 20°C is given, based on margarines processed with modern techniques.

The hardness of the fat (emulsion) in general is responsible for many finished-product properties: general appearance; the ease of wrapping or packaging in general; oral properties (thickness, salt release); workability, spreadability, work softening and creaming properties; heat/form stability (collapse) and oily exudation. Some solid fat content based quality requirements at different temperatures are given below.

The solid fat contents at 0 and 10°C control the short-range spreadability of refrigerated margarines, that at 10–15°C controls the normal-range and that at 15–20°C controls the long-range spreadability of wrapper margarines. The solid fat content at 15 and 20°C is important for product hardness during wrapper-margarine packaging and storage. The solid fat content at 15–25°C is responsible for the creaming properties. The solid fat content at 25°C for the ambient temperatures of moderate climates and that at 35°C for the ambient temperatures of tropical climates is less important for the heat/form stability than is the effective C (yield/stress-) value at the actual distribution temperature. For tub margarine the solid fat content at 30 and 35°C is decisive for the oral-melting properties, and that at 35–38°C for the oiling out.

1. Oral Properties

The melting of the margarine in the mouth simultaneously causes the disruption of the crystal network and the breakdown of the emulsion. For microbiological safety reasons, a uniform water-phase dispersion consisting of water droplets of <3 μm diameter would be ideal. However, this uniformity would give some difficulties as regards the ease of salt and aroma-

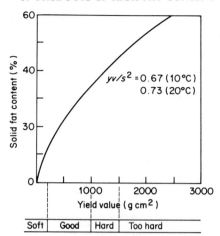

Fig. 6.8 Hardness of margarine as yield value *vs.* solid fat content. [Reproduced, with permission, from Haighton (1976)]

constituents release from this waterphase. The temperature at which the emulsion breaks down (salt release) is also an important quality aspect and can be measured using a conductivity meter. At close to body temperature, margarines should not contain too much unmelted glycerides and associated water phase which cause the feeling of "thickness" on the palate. Viscosity measurement, by dynamic (rotating) viscometers, at the relevant temperatures can objectively establish this quality aspect. Finally, another important oral quality aspect, called coolness, is the coldness felt on the tongue when fat crystals melting at nearly the same temperature absorb heat during their dispersion in the mouth. This phenomenon depends strongly on the difference of the solid fat content levels at 15 and 25°C; the higher this difference the better the cooling effect.

Two other aspects of hardness, namely the packaging hardness and the hardness after storage are discussed in Section L of this chapter.

I. The Different Types of Margarines Produced

Different users have different needs and these differences can be used as a criterion for grouping the many types of margarines produced into three main market segments: (i) household margarines, also used by restaurants and caterers; (ii) margarines produced for the (semi-)industrial users; and (iii) "health" margarines for a special segment of household users. An informative characterization of the different products can be given by ranges of the solid fat content at a few different temperatures. These data, together with the solid fat content of cow's butter for comparison, are given in Table 6.2. In addition to the margarines with high shortening power and/or plasticity mentioned in Table 6.2, there are also real "shortenings" (see

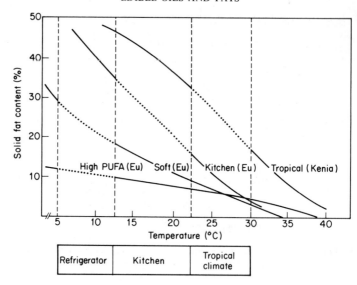

Fig. 6.9 Solid fat content *vs.* temperature for various margarines. [Reproduced, with permission, from Haighton (1976)]

which are used for shortbread, pastry and roll/bread production. By analysing the above solid fat content data given in Table 6.2 and by looking at Fig. 6.9, it can be seen that both the traditional wrapper-packaged block and the tub-packaged soft margarines try to imitate at higher temperature ranges the genuine steep melting (large solid fat content differences) of the butter fat. This type of melting means the total and, if possible, rapid disappearance of the solids in the mouth. Thus the "coolness" of butter and vegetable butters (cocoa butter and cocoa butter equivalents, extenders, and replacer substitutes, which are mainly based on palm oil, hard vegetable butter fractions, and fractions of lauric fats) is a quality to be imitated by premium-quality wrapper margarines.

All household (except tropical) margarines should preferably be kept in the refrigerator. In the distribution chain they should be stored at temperatures preferably below 10°C. The soft and high-PUFA (health) margarines, the minimum-calory margarines, are readily spreadable at this low temperature. On the other hand, the wrapper-packed kitchen and table margarines need to be warmed up to room temperature in order to become reasonably spreadable.

The creaming power is the extent of air retention at the working temperature (20–25°C). Good creaming power requires fat crystals of relatively small surface area which, on whipping, readily enclose air in a fine dispersion. The product must also have a high work softening, when beaten in the mixing device. The main criterion is the yield (specific volume) of the finished cream or cake batter (see Section VI of this chapter).

1. "Health" and Special Margarines

Some margarines are composed of blends with a special aim besides the normal caloric, general nutritional and practical performance aspects. These special margarines may contain:

(i) high levels of the essential linoleic acid. This is not the only, but is the most important, part of the PUFAs. The daily ingestion of substantial amounts of linoleic acid lowers the blood plasma cholesterol level in humans, a statistically corroborated fact. Cholesterol is a risk factor for atherosclerotic heart diseases;

(ii) besides high levels of linoleic, and oleic, (Mattson and Grundy, 1985) acids, definitely stated low levels of saturated fatty acids (SAFA). Saturated fatty acids increase the plasma cholesterol levels more than the essential fatty acids decrease it;

(iii) very high levels ($>95\%$) of medium chain (C_8 and C_{10}) triglycerides. These are obtained by fractional distillation of lauric fat fatty acids and the re-esterification of the relevant fraction with glycerol. This maragarine is produced for small children with some gastric disorder;

(iv) high levels of triacylglycerols with n-6(9,12) fatty acids, such as γ-linolenic acid. This is a precursor of certain prostaglandins. These margarines are not produced, as claims for effects through ingestion are not yet corroborated.

In Europe, margarines with high levels of PUFA and low levels of SAFA are called "diet margarines", this is in contrast with the diet margarines of the U.S.A. which are mostly the reduced (minimum) calory ones (spreads). Further categories of special margarines are:

(v) sometimes advertised to be "free" of cholesterol. If composed of vegetable fats only, the claim is superfluous, as these fats are intrinsically free of cholesterol;

(vi) sometimes also advertised to be "free" of *trans* fatty acids, formed during hydrogenation, or containing only low levels of them. There is no evidence of any unfavourable physiological influence of *trans* fatty acids in humans;

(vii) composed of sucrose polyesters (SPE) on different fatty acid bases. These "synthetic fats" are indigestible for humans, by which they could be considered as a "no calory" fat substitute. They can be used in clinical investigations for slimming diets and on control of plasma cholesterol levels;

(viii) soft margarines composed of high levels of mono- and di-glycerides with liquid oil, but no solid fats added. The α-tending emusifiers form the required gelling matrix, thereby retaining the oil and the emulsified water phase.

Table 6.2 Ranges of solid fat content values for various household and industrial margarines at different temperatures[a]

Type of margarine	Packaging	Intended use characteristics	Temperature (°C)				
			10	20	30	35	40
Household margarines							
Kitchen	Wrapper	Cooking, spreading, cake, shallow frying	30–50	20–30	5–10	0–5	0
Table	Wrapper	Cooking, spreading, cake, shallow frying	20–30	15–25	3–6	0–3	0
Table soft (one-oil)	Tub	Cooking, spreading, shallow frying, not for baking	15–25	10–15	3–5	0–2	0
High PUFA (>50%)		Cooking, spreading, shallow frying, not for baking	10–15	6–10	2–5	0–2	0
Liquid	Bottle	Cooking, shallow frying, not for baking	2–5	2–5	1–3	0–2	0
(Sub-) Tropical	Tub	Cooking, shallow frying, cakes	35–60	20–35	5–15	4–10	3–5
Minimum calory	Tub	Spreading	15–30	8–15	2–4	0–3	0
Cow's butter			45–60	14–25	3–6	0	

Industrial margarines

Product	Properties					
Confectioners' products						
Cakes	Creaming, coolness m.p. <37°C	20–50	15–25	8–15	2–5	0
Icings, cream fillings	Rapid setting, m.p. 38–42°C	20–40	15–25	7–15	2–6	1–2
Bakers' products						
Roll-in/bread	High plasticity, toughness, no oily exudation, low work softening m.p. 42–43°C	25–40	25–38	16–25	12–16	4–8
Short crust pastry		30–40	15–25	5–10	2–5	
Puff pastry		40–80	25–40	10–20	5–10	3–8
Danish pastry	Shortening power, m.p. <37°C	25–50	20–30	10–15	5–10	
Soft dough biscuits						
Short-breads						
Pie pastry		30–55	8–20	5–10	0–3	

[a] Reproduced, Courtesy of Unilever, N.V.

J. Requirements Regarding the Applicability of Different Margarine Fat Bases

The fat bases to be used for margarines/minimum calory margarines and shortenings should be produced from edible-quality fats and oils. This means that the production (fat extraction and modification) and refining processes should comply with the generally accepted high quality standards. In particular, oils and fats of good flavour quality and flavour stability should be used. Freshly deodorized fat blends stored for preferably less than 24 h should be used. Stainless-steel tanks, and blanketing oils with an inert gas or holding them under vacuum all prolong the storage life of blends.

The method of producing high PUFA, low SAFA health margarines is to use appropriate hard stocks, and 85–90% of liquid oils of high linoleic acid content. These hard stocks may be interesterified mixtures of fractions of natural or hydrogenated medium-hard and hard fats (including laurics). Interesterification and fractionation also produce fractions suitable for the structure required at refrigerator and body temperatures.

The low or no *trans* acid containing margarines may be produced by interesterification of totally hardened hard stocks. If a one-oil margarine is required, part of it is fully hardened and part of it is used as liquid oil. Directed interesterification of one oil can separate the randomly formed tri- and di-saturates. The remixing of the separated solids with the liquid residue gives a blend with more body than does simply interesterifying starting oil. The liquid margarines are also based on the judicious mixing of liquid oils with low levels of tri- and di-saturates, in order to obtain fluidity with substantial structure at the temperatures of use. This also applies to liquid shortenings (see later).

K. Margarine Additives

1. *Water Phase Additives*

The water phase contributes certain positive properties during and after melting of the margarine: (i) a part of the flavour release (salty, sweet, sour); (ii) the browning of the sediment on shallow frying; and (iii) the mouth-feel in certain minimum-calory margarines (viscosity). A negative aspect of the water phase when frying with margarines, the spattering, will be discussed with the fat soluble additives, called antispattering agents: the lecithins, (see Section I.K.2.b of this chapter).

The water phase, which comprises 15–17% of the margarine composition, contains the water-soluble ingredients. As will be explained later, because of the microbiological safety of margarines it is favourable if the diameter of the dispersed water droplets is not larger than 3 µm. This fine dispersion is also preferred as it gives a uniform release of water during shallow frying. However, this small dimension and its uniformity have some disadvantages: because of the high packing density of this structure it causes a relatively slow release of water-soluble flavours (mostly the salt and acids) and for the same reason, causes a too fatty impression on the palate. Therefore, a wider distribution of the water droplet diameters (a coarser emulsion) with extremes up to 20 µm is often found in practice. A good release is supported by choosing fat compositions which melt rapidly at body temperature.

The (negative) aspect of the too thin mouth-feel in minimum calory margarines at close to body temperature can be somewhat controlled by the addition of colloidally water soluble proteins, carbohydrates or any other macromolecular edible gelling or thickening agents (gums).

(a) *Water.* The quality of water should be up to local chemical and hygienic standards for drinking water. Water that is too hard (flocculation of lecithins) should be softened to below 8–10 German hardness grades by transmitting it through columns filled with suitable ion exchangers. Any off-odours acquired should be removed by filtering the water through batteries of activated carbon.

(b) *Milk and other protein sources.* The protein composition of the aqueous phase may be from a multitude of sources, such as microbially soured skimmed milk, whey (the fat and casein free serum of milk), milk powder dissolved in water, whey powder, different vegetable (plant) proteins, or none of these (plain water). All the possible variations have been applied at some time.

Cows' milk (the milk of other ruminants is never used) contains about 3–5% butter fat, 2.8% casein, 0.5% albumin, 4.7% lactose and 0.7% mineral matter. Skimmed (defatted) milk has somewhat more of each non-fat ingredient, in proportion to the grade of defatting; its pH is around 6.5.

Milk powder has the advantage of being easily stored for long periods without deterioration and is convenient to use. Generally only the spray-dried milk products are suitable for margarine production. The moisture content ($<4\%$), acidity, microbial condition, water solubility and suitability for souring are important quality aspects of milk powder.

Concentrated milk has the advantage of containing less water, but, being a seasonal product, it is a less reliable source than fresh milk or milk powder.

Whey is obtained by acid and/or rennet caused coagulation of milk, during cheese making. It contains 0.2–1.2% protein, around 5% lactose and 0.2–0.8% mineral matter. The latter may be removed by electrodialysis or ion exchange. Lactose and mineral free whey (albumin) is also available.

The non-concentrated and/or isolated vegetable proteins from groundnuts, soya beans and the serum (milk) of fresh coconuts contain albumins, globulins and glutelins, as substitutes for milk proteins. They also contain carbohydrates.

The animal and vegetable proteins are oil-in-water type emulsifiers and, therefore, in a margarine they tend to destabilize water-in-oil type emulsions. This properly determines their balanced use with monoglycerides and phosphoacylglycerols. These proteins are needed to produce the brown sediment which develops during shallow frying in the pan. The chemical basis of this phenomenon is the so-called Maillard or non-enzymatic browning reaction (this has already been described with the side effects of the toasting of seed meals). The aldehyde groups of the sugars present in milk and the ε-amino groups of certain proteins form condensation products of brownish colour, which may decompose to volatiles of generally agreeable flavour (an advantage). If a margarine is produced for uses other than shallow frying (or not for universal kitchen use), the proteins (a cost factor) can be omitted.

(i) *Bacterial souring*. To reduce the chances of the water phase serving as a medium for infections caused by bacteria, yeasts and moulds, it is generally pasteurized by using one of the available methods. Originally the main purpose of pasteurization was the destruction of the heat-sensitive pathogenic bacteria, especially *Mycobacterium tuberculosis*. The optimum time/temperature relationships causing the thermal death of bacteria determine the method to be used for pasteurizing the milk. Regardless of the method applied, pasteurization should be done as soon as possible after the total solids have been adjusted to the required value and before souring, or addition to the margarine.

Methods currently in use are: the ultrashort pasteurization at 82°C for a few seconds; and the 15–40 s duration high temperature/short time (HTST) pasteurization at 72°C. Plate heat exchangers are used to heat and cool the milk to souring temperature: this is done by pumping the milk through continuous closed systems. Batch long-low pasteurization at 63°C for at least 0.5 h is now a less practised procedure, but gives products without a "cooked milk" flavour, which might develop during a complete pasteurization at 80–100°C. A combination of the continuous-batch processes is also possible in which the milk is flash heated by a heat exchanger to the required batch holding temperature. The heat energy introduced by any method should destroy the pathogenic microflora, and reduce the bacterial, mould and yeast count of non-pathogenics. The heating must not change the chemical/colloid chemical state of the milk or other proteins.

The role of fermentative souring (culturing or ripening) of the milk is to obtain a pH level (4.3–4.5) at which no new bacterial infection can multiply. At the same time a butter-like flavour is obtained by some parallel fermentative processes: this is the natural pH of the so called "sour-cream butters". The souring process hydrolyses a considerable part of the lactose in the milk to glucose and galactose, the latter being converted to lactic acid. Lactic acid has a strong bactericidal action (a property which was discovered by mankind very early: sour milk and sauerkraut production). In the same process some citric acid (0.1% in milk) or even lactose itself is converted to diacetyl, one of the important (but not the only) natural flavour components of butter. The fermentative process also reduces the oxygen content of the soured milk. This is an advantage when the milk is mixed with the fat phase, because some oils are very sensitive to autoxidation. Yeast and moulds can grow only very slowly in a medium of low oxygen content (having reduced redox potential).

In practice souring means the addition of relatively small amounts (1%) of pasteurized milk, which has been preinoculated and ripened for 24 h with cultures of *Streptococcus lactis, Streptococcus cremoris* and *Leuconostoc citrovorum* and *paraci-*

trovorum strains, to the bulk of the pasteurized milk previously cooled to 22°C. The mixture is kept at this temperature for a further 18–24 h. When the necessary acidity has been reached (titrated as 22–35 Soxhlet–Henkel acidity grades, or 0.5–0.8% lactic acid) and the flavour has become substantially strong, the contents of the ripening vessel are cooled to 5°C. The milk is kept at this temperature until incorporated into the margarine at this or a somewhat higher temperature (say 10°C).

(ii) *Chemical souring.* If for any reason fermentative souring cannot be used, chemical souring by adding lactic (to pH 4.5) and citric (to pH 4.3) acids can perform the same bactericidic action. Of course the buttery flavour of diacetyl is not formed and the redox potential of the system is not lowered.

(c) *Other water-soluble additives.* Whether the water phase of the margarine contains any of the above-mentioned protein sources or it is solely based on water depends on the required qualities of the final margarine. However, it should contain some of the following additives.

Kitchen salt at 1–3% of the total margarine is besides being a taste component, a bacteriostatic agent. A salt level of less than 0.2% of the total margarine is considered by some national food legislations as "unsalted"; the salt used should be free of iron (2 mg kg^{-1}; autoxidation), with magnesium (bitterness) and barium (toxicity) traces at reasonably low levels.

A level of 18% kitchen salt (calculated on the basis of 16% water phase) decreases the water activity of the system to <0.90, thereby effectively controlling the microbial safety of the margarine. Salt and preservatives (see below) are mainly needed for reasons of local taste preferences, e.g. when at pH 5.5–6.0 the conditions of a "sweet cream butter" have to be achieved. It has been established that dispersed water droplets with a diameter of <3 μm are the best remedy to repress any microbial growth in modern margarines: this is because there is no space for their multiplication, with the exception of bacterial lipase formers!

Preservatives, such as undissociated free benzoic or (less preferably) free sorbic acids at levels of 0.05–0.2% are bacteriostatic and/or fungistatic (mould inhibiting) agents. The salts of these acids do not act as such if the pH is above 5.7.

The other organic acids (citric or lactic) added as chemical souring agents, also serve as water-soluble flavour enhancers and heavy-metal scavengers. Lactic acid at levels of $>0.2\%$ also acts as a bacteriostatic agent.

The water-dispersible margarine tracers/indicators, such as starch (which on reaction with iodine solution gives a blue colour), are used if prescribed by food legislation.

2. Fat-phase Additives

The main fat-soluble ingredients are: emulsifiers, antispattering agents, vitamins, colouring agents, antioxidants, and margarine tracers/indicators.

(a) *Emulsifiers*. The main (pre-)emulsifiers of margarine are the monoacylglycerols which, either in an enriched form or in mixture with di- and small amounts of triacylglycerols, serve as surface-tension decreasing agents (as discussed previously). The co-existence of mono-, di- and tri-acylglycerols is caused by the equilibrium conditions obtained in producing this emulsifier by adding e.g. 2 mol free glycerol to an alkaline catalyst at high temperatures to 1 mol of triacylglycerol. The component fatty acids are distributed according to the rules of chance. The maximum yield attainable is limited by the immiscibility of glycerol in the fat and the intermediates. There are also molecularly distilled concentrates.

Solid (saturated) monoacylglycerols are used in 0.05–0.15% amounts for modern table margarines, whereas the liquid (unsaturated) ones are used in higher (0.2–0.3%) amounts for minimum-calory margarines. The liquid monoacylglycerols used are mainly monoglyceryl oleates, the ones containing higher levels of unsaturated fatty acids being less preferable because they undergo autoxidation when kept liquid (heated) for longer periods of time. Concentrated stock solutions of the monoacylglycerols in (coconut) oil are prepared for easy handling.

Some monoglyceride based derivatives, such as citric, acetic, lactic, tartaric and diacetyltartaric acid esters and salts, and fatty acid esters of polyglycerol, propyleneglycol, sugar (sucrose), sorbitan, polysorbates and polyoxyethylene, are speciality emulsifiers which are added to margarines (shortenings) produced for the bakery and confectionery industries.

Glycerol-lactylpalmitate/stearate is a well-known dough (gluten) conditioning agent for cake. The same properties are ascribed to the diacetyltartaric acid esters of monoglycerides. Sorbitan or polyglycerol based fatty acid esters and (molecularly) distilled partially saturated monoglycerides, when dissolved in fats, improve creaming properties and the final volume of baked cakes. If 8–12% propyleneglycol monostearate is dissolved in liquid oils, it produces high cake volumes without the presence of the normally required solid fat content at room temperature.

The success of many of these emulsifiers is partly explained by their "lyotropic mesophase" forming properties: i.e. when mixed with water (aqueous milieu) and heated close to, but just below, their melting points they become liquid. A liquid crystalline film (oil-in-water emulsion) is formed which provides the better creaming and dough "tenderizing" qualities (see Section VI of this chapter).

Glycerol-lactylpalmitate/stearate is also an agglomerating and foam stabilizing agent for artificial creams used for toppings. This property is explained by the compound's tendency to stabilize in the lower melting α crystal modification (see Section I.C.5 of this chapter). This is the basis of the β crystallization inhibiting properties of sorbitan fatty acid ester which successfully counteract the "blooming" of chocolates produced with β tending fats (see Chapter 7).

(b) *Antispattering agents*. A uniform release of water from butter is a positive quality of this product when it is used for shallow frying and margarines used for this purpose must also have this quality. Therefore, it is very important that after the melting of the solid fat crystals, the water, which is mainly stabilizing the emulsion, stays in its regular and finely dispersed form and does not coalesce to large droplets or into a continuous water layer. If the water becomes overheated, the large water droplets leave the fat explosively, leading to heavy spattering. Unsalted margarines produced with hard water are particularly prone to releasing their water with heavy spattering.

Keeping the margarines foaming finely at 90°C and above can be achieved by the addition of lecithins to the margarines used for frying, whether "salted" or "unsalted". A very useful component of lecithins is phosphatidyl choline. Although this fraction does not stabilize the original water-in-oil type emulsion, it is insensitive to hard water and salt and can sustain a fine dispersion of the free water droplets until they reach the surface of the oil layer at the top.

There are a number of different lecithin preparations, 0.1–0.3% generally being levels used for good water dispersion and release. Coconut oil may serve as a stable carrier for any type of concentrated lecithin solution needed in the margarine production. The lecithins are dissolved in the coconut oil as colloid aggregates.

(c) *Vitamins.* The vitamin(s) prescribed by food legislation are, for the U.S.A., vitamin A and for Europe both vitamin A and vitamin D_3 (7-dehydrocholesterol). The amounts required are generally around 20–30 International Units (I.U.) of vitamin A and 2.8–3.5 I.U. of vitamin D_3 per gram of margarine. Vitamin A is nowadays fully synthesized and vitamin D_3 at least partially synthesized, making them much easier to handle than the previously used concentrates which were obtained from marine liver oils. The synthesized vitamins are identical to the natural ones. In some countries the water-soluble vitamins, B_1 and B_2, are also added (to the water phase) due to special requirements of local law.

(d) *Colouring agents.* The most often used colouring agent is β-carotene (synthetic microcrystalline), around 3–8 mg of this agent being required for 1 kg of fat. In these amounts β-carotene produces the yellowish colour of summer butter, which also obtains its colour from the carotenoids present in the fresh grass feed of the cows. Other colouring agents which can be used are the colour concentrates produced from very red (Nigerian) palm oils or the alkaline extract of Anatto seeds. The latter contains bixin as a colouring substance which is a dicarboxylic acid of the regular carotenoid chain. All these products are either natural or nature identical. β-Carotene is the provitamin of vitamin A and one molecule of β-carotene produces, by biological oxidation, at least 1 mol of vitamin A. Carrot oil, obtained from carrot seeds by solvent extraction, is less often applied, whereas turmeric can improve the shade of colour obtained with anatto alone.

(e) *Antioxidants.* Antioxidants are generally not used in household types of margarines because the vegetable oils used in them contain, even in the hydrogenated form, substantial amounts of tocopherols (natural antioxidants). The addition of nature-identical synthetic free α-tocopherols or the generally permitted phenolic type of antioxidants is recommended if the fat blend contains overwhelmingly interesterified fats or if the margarine is intended for industrial speciality use. For the biological protection of component linoleic acids around 200 mg of α-tocopherol per 1 kg of linoleic acid is the minimum amount needed. This quantity is amply covered by the linoleic acids naturally present in most of the common fat blends used. The industrial types of margarine may be supported with up to 400 mg of the permitted phenolic types per 1 kg of fat.

When used in the water phase, the generally used food-grade organic acids or their salts may serve as effective metal-scavenging agents and, in this way, synergistically support the antioxidant's protective action. Ascorbyl palmitate and stearate, and (calcium disodium) ethylenediamine tetraacetate (EDTA) are typical metal scavengers.

(f) *Margarine tracers/indicators.* The application in the fat blend of e.g. 5% very lightly refined sesame oil (the carrier of sesamol, the substrate of the Villavecchia–Baudouin colour test) may be prescribed by some national legislations for the easy detection of margarines. This serves to detect the adulteration of butter with margarine.

3. *The Flavouring of Margarines/Minimum-calory Margarines/Shortenings*

The flavouring of margarines is probably one of the most interesting parts of modern margarine production. The primary aim of flavouring is to imitate plainly the "typical" flavour of butter. By using fermented skimmed milk, the early Continental European margarine products were strongly dominated by a sour taste of lactic acid and the flavour of diacetyl. Thus by producing a "water margarine" and adding to it lactic acid and synthetic diacetyl (since 1922) it was thought that the sour-cream flavour of fermented, unsalted butters was imitated. The non-fermented, heavily salted sweet-cream butters of the Anglo-American, Australian–New Zealand and Danish dairies were thought to be imitated only by adding around 2% of salt (on the weight of margarine). This was sometimes enforced by the addition of some butyric acid, to give the rancid character of a somewhat aged sweet butter.

With the discovery of gas chromatography in 1952 a thorough investigation into the components of flavours isolated from genuine butters of different origins was begun. More than 300 partly flavour-carrying volatile components have been identified and many of them are used in producing modern butter flavours. These are different short(er)-chain acids, alcohols, esters, aldehydes, ketones, diketones, mercaptans, etc. The different γ- and δ-lactones, the cyclized hydrolytic decomposition products of different hydroxy fatty acids of shorter chain length, are very important components in modern butter flavours (Boldingh and Taylor, 1962). The γ- and δ-lactones are also important components of butter fat. Modern margarine flavours (generally added to the fat phase as a concentrate together with the colouring substances and other oil-soluble ingredients) are composed of 5–20 or more different components. These components are selected from the range of at least 100 commercially available components. Most flavours are synthetic, but very often they are nature-identical products. Some flavours are isolated from natural sources and some are produced by fermentation as natural components.

It should be remembered that any butter flavour should only be considered as typical within those geographical locations where the butter is produced or where butter of specific origin is imported. Buffalo, caribou, sheep and goat milk, if churned to butter will give quite differently flavoured butters which are well accepted by the local consumers, but disliked by consumers of different regions who are accustomed to cow's butter.

In countries where consumers are not familiar with butter because there is no dairy industry, the newly introduced margarine is accepted as a convenient and stable cooking and baking fat with new types of flavours, e.g. pineapple flavour is well accepted, if pineapples are a locally consumed commodity.

Fantasy flavoured margarines, such as those with sugar, chocolate, hazelnut, peanut, garlic, onion, bacon, cheese, meat, etc., are real possibilities for margarines which could be introduced to new, different segments of the market.

The flavours of shortenings and shallow- and deep-frying fats can be obtained by using the same flavouring agents. It is interesting that some shallow-frying fats may contain water-soluble flavour precursors, or complex (more component composed) reaction flavours. If dissolved in 1–2% water and dispersed in the fat, the fats develop very characteristic (meaty, cheesy, etc.) flavours during heating in the pan.

L. The Technology of Margarine/Minimum-calory Margarine Processing

As was described previously, the fats in margarine become supersaturated and crystallize first in the α modification when cooled. More of the metastable α form will develop if the crystallization is done rapidly. Both the presence of these metastable crystals and the solidification of some triacylglycerols in solid solutions or mixed crystals will give rise to different transformations. One transformation is the polymorphic transition from the α to the β′ form which occurs (except with "slow" oils, like palm oil) quite quickly. Another transformation is the demixing of solid solutions, and a third one is the recrystallization of small crystals to larger ones. The fourth transformation (if any) is that of β′ modifications to the most stable β ones. A considerable heat of crystallization is released during these transformations which requires careful temperature regulation during the partial crystallization process. The different transformations take time and depend on the mechanical working applied during the processing as well as on blend composition and the temperature profile applied.

At least a part of the original α crystal structure must be retained, despite the inevitable and/or necessary structural changes caused by the different mechanical handlings. This so-called working not only breaks down some interconnecting crystal bridges and thus the primary-bond determined structures of margarines, but also influences the flow behaviour of the (emulsified) system under the shearing stress of the rotating machine parts (e.g. thixotropy). The heat evolved by the friction of these parts also causes irreversible changes in the system.

In conclusion it might be said that margarine production is the case history of programmed partial, or as far as possible retarded, crystallization of different fat blends to obtain products with appropriate final properties, such as substantial hardness, but no post-hardening at packaging, storage and use stages, and other quality aspects already mentioned. Margarine production is executed in a sequence of different "unit operations" (mixing, shock chilling, kneading, partial recooling, etc.) with the relevant machines coupled together in suitable "configurations"

Before describing in detail the most important continuous way of margarine processing, it is worth while to look first at the discontinuous method of margarine processing.

1. *Discontinuous Margarine Processing*

In discontinuous margarine processing steps are clearly separated into emulsification, cooling (for nucleation and pre-crystallization of the fat), mechanical working and resting (post- and re-crystallization of the fat).

The process is done in open-air mixed batch/semi-continuous systems. The processing starts in (pairs of) mixers/emulsifiers called churns, where the weighed batches of the fat blend with the added fat-soluble additives, and the weighed batches of water phase blend with the water-soluble additives. The fat phase is previously heated to some degrees above the temperature of its melting point and the milk is at 5°C, the temperature at which it is stored after pasteurization.

The churns (Fig. 6.10) have two gate-type mixing arms, driven by a planetary gear, with a mixing speed of around 200 rpm. If pure milk margarine is to be produced, the milk is first introduced into the water-cooled churn, followed by the first half of the fat blend. The coarse emulsion is formed when the originally yellow colour of the mass turns whiteish. After 15 min the second half of the blend is added to the churn and after a further 20–30 min this emulsion was ready for further cooling to a final temperature about 10°C below the melting point. The churns work continuously with time shifts and the subsequent processing during working and packaging is partly continuous.

The droplet size of the water phase (the fineness of dispersion/emulsion) could be further reduced by homogenizers. These are either rotating colloid-mill discs or spring-operated variable-pressure heads with a small orifice, called texturizing valves,

6. PRODUCTS OF HIGH FAT CONTENT 313

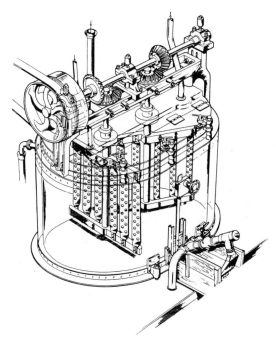

Fig. 6.10 Margarine churn from 1910. [Reproduced, with permission, from Stuyenberg (1969)]

the latter being coupled with a high-pressure positive-displacement pump. Both devices produce large shearing forces, which achieve further size reduction of the droplets.

After reduction of the droplet size the emulsion is transferred to horizontally rotating direct expanding ammonia-cooled drums (of 1–3 m diameter, 2–4 m length, 6–9 rpm), where the emulsion (slightly pre-crystallized in the churn) is fed by a rotating feeder roller to form a thin film. After one revolution the film temperature is about 0°C and the fat is in the supercooled crystallized α form. Metal-knife scrapers remove the fat into aluminium or stainless-steel carriages (trailers) which are then kept in separate rooms at 8–10°C for 20–30 min (ripening, crystallization). Because of the heat of crystallization liberated, the temperature of the emulsion rises to 5°C.

The mechanical working of the "ripened" mass is done using different types of kneading machines, e.g. the ones with three pairs of horizontal (smooth or corrugated) double rollers (the best known was the type "Multiplex") connected to a plodder-extruder, into which the mass is scraped by knives. Here the mass is kneaded between double meat-mincer-like helices, extruded and collected at one floor below in carriages. These carriages are transferred by means of moving-bridge cranes above double Z-armed vacuum mixers. The side door of the tilted carriage is opened and

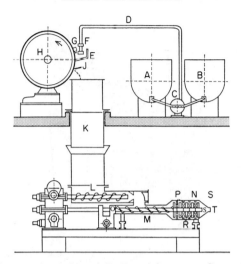

Fig. 6.11 Complector plant. A and B, Premixing vats; C, emulsifying pump; H, chilling drum; J, scraper blade; K, hopper for resting; L and M, feeding screws; P, diaphragms; R, pins; T, extruder. [Reproduced, with permission, from Stuyenberg (1969)]

the contents poured into the machine. Besides the further working, the air enclosed in the mass during processing is partially removed. After tilting the opened mixer, the contents are collected in small carriages, which are kept for some hours at 17°C for ample post-crystallization into the β' form. This part of the process is called tempering. From here the tempered margarine is transported to the printing and packaging machines, which are fed by shovelling the mass manually into the hopper.

Post-crystallization could also be done continuously during the transport of the very thin scraped fat flakes on a continuous belt from the cooling drum to the single or double kneading/mixing helices of different extruders. These extruders, like the Complector (Fig. 6.11) built by Gerstenberg and Agger, the one by Silkeborg, or any other similar types, have sieve plates or different orifices for mechanical working of the mass. Some resting in the hoppers of the machines before kneading for inducing crystallization and in the transport carriages before packaging for inducing post-crystallization is also necessary at this stage.

As the main characteristics of the churn/drum process we should mention the intense supercooling of the fat on the drum to α crystals, after which it re-crystallizes slowly at low temperature. The solid contents are high, and the mixed crystals formed contain a large proportion of lower melting triacylglycerols. These triacylglycerols are separated (demix) later in the course of the processing from the higher melting ones, but less so than in the continuous process.

In summary, the very good pre-crystallization, the strong cooling and the intense working are the advantages of this process. The main negative aspects are the excessive manual labour required, the large space requirements, the impossibility of liquid filling and the higher microbiological risk.

2. Continuous Margarine Processing

Present-day margarines were first produced on machines originally developed for the ice-cream industry. These machines were first adapted by the Girdler Corporation (now Chemetron Corporation, Votator Division) for the continuous production of shortenings and margarines in 1935. A similar ice-cream chilling machine was the Cherry Burrel freezer invented by Vogt. Machines are produced under the names Thermutator [Cherry Burrel, Anco (Albright-Nell Co.)], Zenator, Kombinator (Schroeder), Perfector (Gerstenberg and Agger), Merksator and Unitator (Unilever) or the European Votator, produced under licence of Chemetron (Girdler) by Johnson. These are variations and improved versions of the basic continuous ice-cream freezers.

(a) *Blending, storage and dosing of components.* Blending of the many different fat bases, as calculated by linear programming or by the qualitative method based on manufacturing experience, can be executed either in the refinery department or in the margarine factory.

In the refinery department, the freshly bleached components are blended by weighing or metering and are immediately deodorized for direct delivery into the margarine department. Co-deodorization results in the need for only modest storage facilities in the margarine department. If the storage tanks are always emptied completely before introducing a new batch of freshly deodorized oils into them, i.e. if no topping-up is allowed, the fat basis can be kept at an optimum quality. The closed stainless-steel, aluminium or glass-lined tanks should be large enough to accept one batch. If varying amounts are stored, then facilities for inert-gas blanketing or for vacuum storage are needed. The tanks should be provided with inlet pipes reaching to the bottom (no splashing), heating coils, thermometers and the necessary appliances for internal cleaning.

If individually deodorized oils are compounded in the margarine factory, extensive storage space is required again with facilities to exclude air-access caused damage to the incompletely used batches. The topping-up of the tanks is prohibited in this case also. Individual deodorization is considered to be favourable, because in this way the specific time/temperature related process conditions required by different types of oils (e.g. coconut versus soya-bean oil) can be maintained better. Despite this, co-deodorization is the process most frequently applied.

The aspects to be considered when blending components of the fat and water phases accurately are of widely different in nature. These aspects are: (i) legal (maximum amount of moisture, vitamins); (ii) economic (the mean amount of water, vitamins, etc.); (iii) quality (oil blend, consistency, colour, flavour, milk, etc.); and (iv) hygienic (no microbial spoilage).

The blending of fats and of soured milk and water can be done volumetrically in calibrated tanks, the fat-soluble and water-soluble ingredients being added to the batches separately. It is customary to weigh the additives gravimetrically on (automatic) scales. For automation the weighing tank may be mounted on a load cell with, for example, strain gauges, or on the lever bridge, e.g. the dial may be foreseen by some electronic contact. These parts are coupled via, for example, programmable logic controller (PLC) with the pumps emptying the storage tanks.

In modern factories, instead of weighing, temperature-corrected volumetric blending is carried out either by coupling (the speed of) pumps with flow meters, pressure or level detectors in the tanks via PLCs or by automatic piston or diaphragm metering (dosing and proportioning) pumps. A metering pump is distinguished from an ordinary controlled-volume pump in that its rate of delivery can be pre-adjusted within fine limits and the accuracy at constant temperature can be maintained between 0.05–0.1% of the stroke volume. If more than one pump is used, each of them can pump different liquids into a common delivery line or vessel at preset but adjustable rates of flow and at a constant, but adjustable proportion to each other. The pumps can be used as a blending and mixing system.

The completely molten, weighed and/or blended fats are transferred in the refinery to the deodorizer, in the margarine factory to intermediate buffer tanks (e.g. of 20 ton capacity). Buffer tanks (e.g. of 4 ton capacity) house the soured milk, if any, and the drinking-quality water with a starting temperature of 5–10°C. In some installations, the components of the side streams must first be admixed to these two main-stream components of the final emulsion. These side-stream components consist of water-soluble ingredients dissolved in water and the fat-soluble ones dissolved in (coconut) oil. The two streams are then mixed in the appropriate amounts in the buffer tanks of the processing machines to form low volume pre-mixes.

The dosing of the individual components of the side stream into their two carrier phases can also be carried out by precision metering pumps. Here the two components of the main streams (fat blend and milk/water) and the individual solutions of the ingredients constituting the side streams are all proportioned together into one stream by means of a battery of individual metering pumps. An example of a well-known dosing machine is that developed by Bran & Luebbe (patented), where all the individual dosing pumps in the units are driven by the same motor (Fig. 6.12). The claimed accuracy is 0.05%.

If low-pressure proportioning pumps are used, they are provided with pulsation dampeners (air balance) on the suction side and a small (300–600 l) pressure-resistant balance tank on the pressure side, the pressure inside being regulated by a control loop to the low-pressure feed-pump motor. If high-pressure proportioning pumps are used, then no balance tank is required. The final mixing of the two phases starts in the pump after the dosing unit. This transfers the emulsion to the A-unit, where the mixing blades also do the necessary post-emulsification work. Some manufacturers (e.g. Dover Corporation's Groen Division) advise the use of static mixers for extra mixing and pre-emulsifying purposes.

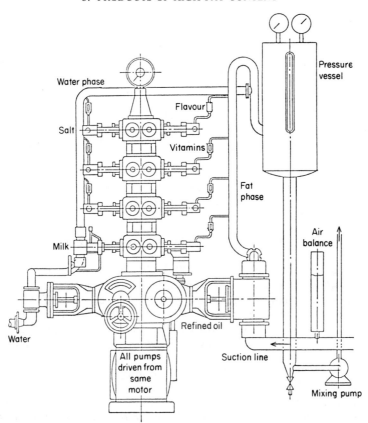

Fig. 6.12 One motor-driven precision multiple-dosing machine. [Reproduced courtesy of Bran and Lubbe A. G. and Unilever N.V.]

(b) *Scraped (swept) surface heat exchangers or A-units.* The basis of the process in scraped-surface heat exchangers is a very intensive heat transfer (refrigeration) to the fat (emulsion) achieved by the direct expansion of ammonia gas. (The U.S.A. terminology is swept-surface heat exchanger.) These "A-units" (Fig. 6.13) are cylindrical, jacketed, externally cooled vessels (tubes), through which the fat and the water phase are continuously fed by high-pressure pumps. The system pressure is around 20–50 bar, but may be higher if necessary. The internal surface is intensively scraped by fast revolving (sweeping) plastic blades (knives), mounted on a thick rotating shaft. The annular space between the concentrically placed rotor and the cylinder wall is 2–25 mm. Lower widths increase the internal pressure if the viscosity is high (high degree of crystallization). If needed, the shafts can be heated to reduce crystal deposition on them.

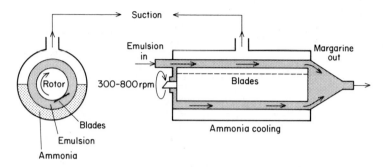

Fig. 6.13 Scraped surface heat exchanger, A-unit. [Reproduced, with permission, from Haighton (1976)]

The input temperature is an important process variable, although for microbial safety reasons this must be kept above 47°C (complete melting). When mixed/scraped intensively (1000–1200 scraping min^{-1}) in the A-unit the mixture is cooled by the evaporating ammonia to a certain lower temperature. Ample cooling surface areas are a prerequisite for the necessary process variability. The shaft speed is 300–800 rpm, depending on the size of the unit and on the number of rows of scraper blades being used. Lower speeds result in laminar flow with supercooling, and higher speeds in turbulent flow with rapid crystallization. The residence time of the mass is 5–10 s. Some designs place the rotating shaft excentrically and a reduced mechanical working action on the mass is claimed. Vertically placed scraped surface heat exchangers are also known solutions (e.g. the DR(C) series units of the Dover Corporation's Groen Division). Refrigeration and power savings are claimed by the manufacturers.

The A-units, coupled in different sequences, are generally mounted on a common frame, together with C-unit crystallizers (see later). The A-units and the crystallizers are belt driven by electric motors. The ammonia pumps, the low-pressure dosing/proportioning, the high-pressure emulsion and recirculating (rework) pumps are driven by variable speed electromotors. Inspection glasses, electrical temperature gauges, etc., are mounted, for example, on the frontside of the machines.

The primary task of the A-units is the rapid cooling of the fat emulsion to the crystallization temperature of the α modification, thereby producing substantial body for the further recrystallization processes. The molten fat/water phase (crude emulsion) is thus directly introduced into a suitably cooled A-unit, where the first crystal nuclei and crystals forming on the wall are in the α modification. In the scrape-surface heat exchanger tube some of the crystals might be first remelted in the bulk of the liquid fat. When the average temperature of the whole mass is at the congealing temperature of the α crystals, the solid fat content, as measured at the exit of the unit, represents this modification (see point A_1 at the α line of the S/T diagram in

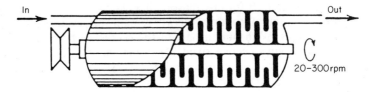

Fig. 6.14 Crystallizer or margarine C-unit (shortening B-unit). [Reproduced, with permission, from Haighton (1976)]

Fig. 6.3). The lower the temperature of the mass arriving at the A-unit, the shorter the induction time (t_{ind}) of α crystallization and the shorter the transformation time (τ) needed for its future re-crystallization into the β' modification. The higher melting β' modifications are absent (i) mainly because they can be easily undercooled, and (ii) because of the short residence time of the mass in just one (the first) cooling unit which prevents their formation. The exit temperature can be calculated from the energy balance of the solid fat phase present, this determining the heat of crystallization, the energy dissipation by shearing (flow) and friction (mechanical mixing), and the heat removed by the cooling medium. The emulsified, nucleated mass generally leaves the A-unit(s) at between 15–28°C.

Depending on the needs of the process two-stage reciprocating, or for the larger factories screw-type, ammonia compressors are available. Common suction mains are normal, and suction separators protect the compressors against liquid ammonia. The energy need is up to 0.2 MJ ton^{-1} margarine or shortening for peak production. Considering also good ventilation in case of any leaking ammonia is a very important safety measure. Cold store room energy needs are not included in this figure.

(c) *The crystallizer or C-unit (shortening B-unit)*. The nucleated, partially α crystallized mass leaving the A-unit(s) is further treated in the crystallizer or C-unit (Fig. 6.14). In its original development the C-unit was called the shortening B-unit. The C-unit is a (generally) non-insulated cylindrical vessel, with a rotor fitted with rows of pins and rows of pins are also fixed on the inner wall of the cylinder. These pins, besides promoting re-crystallization, also accomplish the mechanical working of the mass, i.e. breaking the primary bonds of the crystal structure already developed. The rotor speed is typically 20–300 rpm.

Most fats crystallize completely, if cooled quickly to 0°C in quiescent conditions, within a few hours. When the fat is mixed, the crystallization time can be reduced, at the same low temperature, to 5–7 min. Considering the volume of a C-unit (50–100 l) the residence time is generally 120–150 s. Therefore, under these near adiabatic conditions, a certain portion of the α crystals can be re-melted by the evolving heat of crystallization and, thereafter be transformed into the more stable β' modification (see point C in Fig. 6.3). The β' crystals can also be formed (crystallized) directly from the supersaturated solution. If for any reason all α crystals are transformed into the β' modification, then the margarine will remain soft, even if stored under adiabatic conditions. With only a part of the α crystals being transformed (the β' line not having been arrived at), under adiabatic conditions the further β' crystals form the later newly formed α crystals (in a second A-unit) will develop a certain texture.

Considerable heats of crystallization are liberated in the C-unit in the nearly adiabatic conditions. The remelting caused by this heating should be (over-) compensated in the subsequent (second and third) A-units, where the α line shown in Fig. 6.3 is again arrived at (point A_2). Fast-crystallizing blends containing high-melting triacylglycerols might develop at the relatively low temperatures in these units of high flow resistances, which can lead to high working pressures in the viscous mass.

(d) *Pre-crystallization.* A more controlled formation of higher melting point crystal fractions can be achieved by pre-crystallization. This process variation aims at the selective formation (fractionation) of fat crystals, which melt above the pre-crystallization temperature. To create these crystals either:

(i) the incoming warm pre-emulsion is mixed in a crystallizer C-unit with a recirculated part of the chilled mass leaving the first or second A-unit (recirculation system with pre-crystallization); or

(ii) one or two A-units are solely used as a nucleator/α crystallizer before a crystallizer C-unit, followed by a further A-unit(s) (straight-through system with pre-crystallization).

Favoured processing configurations are thus:

Fat + water phase —C—A—A— (recirculation)
　　　　　　　　↑　　　↓
　　　　　　recirculation

Fat + water phase — A—A—C—A—A— (straight through with pre-crystallization)

Fat + water phase — A—A—C—(T.V.)— (straight through without pre-crystallization)

With longer residence times some of the solid solutions, including the α and β' crystals can demix. The new mixture will contain a series of higher melting (and larger) crystals, than the originally congealed mix. This not only extends the plastic range, but also stabilizes the final product against softening at storage temperatures above 20°C. These new crystals together with those smaller ones already formed in the pre-crystallizer, render the final products generally softer and more plastic, although with some less favourable oral-melting properties.

6. PRODUCTS OF HIGH FAT CONTENT

Not only increased recirculation but also strong pre-cooling combined in the later A-units with light post-cooling will produce softer but more plastic and textured products by demixing of the crystals. The heat stability of such a margarine will be less satisfactory, because of the lower initial crystallization temperatures. In addition, the work-softening of the margarine will be low. Besides decreased recirculation, light pre-cooling (and strong post-cooling) will also produce harder products with a narrow plastic range, due to the "demixing" of the crystals now formed at a higher initial crystallization temperature. The heat stability will be better, but the work-softening higher.

In any case, care should be taken with high crystallization rates, which lead, because of the inevitable destruction of the primary crystal structure in the mechanical-working phases to permanently soft products without any structure. Low crystallization rates lead to primarily soft and unpackagable margarines, which post-harden to non-plastic (brittle) products with high structural hardness (low work-softening).

(e) *Finishing operations.* The margarines produced by the churn/drum process were all predestined for being packaged as blocks (prints) in wrappers. The increasing market share of refrigerator-stored margarines in the large industrialized countries made it imperative to fill margarines in the semi-liquid state into different containers, like tubs, cans, etc., using the technology originally developed for the shortening manufacture. In addition, the introduction of margarines into hot-climate countries forced the use of tubs and/or cans.

(i) These margarines are derived mainly from fat blends containing substantial amounts of higher melting (quick) triacylglycerols which form large amounts of easily transforming α crystals in the A-unit(s). If the product is transferred from here to a crystallizer C-unit, then the mechanical working/kneading action of the pins will decrease the viscosity of the mass by breaking crystal bridges and thus making the mass too soft for wrapper packaging. The temperature of the mass can rise to around 25°C. However, in this straight-through system without pre-crystallization, this soft product can be (nearly adiabatically) semi-fluid filled immediately from the crystallizer via a texturizing valve (TV) into tubs and other containers. If necessary an extra booster pump can be added before the crystallizer (see Section III.A of this chapter).

Table, cake and puff-pastry margarines produced from suitable blends can also be directly liquid filled using this method into preformed cavities containing an inner wrap. The fluid mass stiffens in the package in seconds to give a quite hard product with a dense texture and a relatively narrow plastic range. The table and puff-pastry margarines produced by liquid filling are not tempered after packaging, whereas the cake margarines are well tempered (see below). If cooled to below 16°C, and worked at the latest possible moment, the product is more uniform, more spreadable and quicker melting. The late overworking in the A-units reduces the structural hardness in the product.

In addition, precrystallization of the fat blend with or without recirculation in A-units can be applied, this improves the oral-melting aspects of the margarine. The post-crystallization of the supercooled mass is executed in the C-units, this followed by adiabatic liquid filling of the (semi-)fluid mass, if necessary, through a presure releasing texturizing valve, into the different cups/tubs and other containers.

(ii) *Brick- or cube-shaped wrapper-type margarine block production.* Here the following processing method can be applied. After crystallization the mixture is efficiently post-cooled in further A-units. This supercooling induces the additional formation of α crystals. Because of the short residence time, the already formed β' crystals will not grow further in these A-unit(s). The object of this second deep cooling is to obtain the necessary packaging hardness and to arrive at a final packaging temperature of 14–17°C in the resting tube or static B-unit.

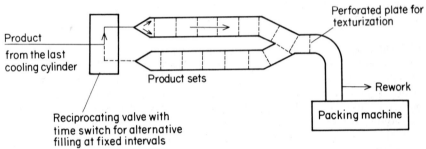

Fig. 6.15 Double-leg resting tube or static B-unit. [Reproduced courtesy of Unilever N.V.]

(f) *The resting tube or static B-unit.* From the A-units the supercooled mass is transferred for post-crystallization to the resting tubes, also called static B-units (Fig. 6.15). The static-B-units are cylindrical tubes, with inserted sieve plates at the end for some additional kneading action; this is the place where most of the final structure is formed. A part of the newly formed α crystals is transformed (recrystallized) in the static B-units to β' crystals, again with some additional β' growth from the supercooled solution (see point B in Fig. 6.3). The α to β' transformation is promoted by the sieve plates, which also break down the crystal–crystal interactions and help to relieve residual supersaturation. As the amounts of the α crystals formed in the last A-unit and the amounts of the β' modification formed during crystallization in the B-unit determine the hardness, so the fat blends and the conditions of processing have to be selected appropriately. The residence time in the static B-units for different types of margarine blends is at least 120–150 s. Longer residence times can be achieved by varying the length of the resting tubes using their flanged sectional construction. Sometimes two alternately working resting tubes are mounted in parallel.

Excess amounts of margarine not taken by the packaging machine are redirected from the resting tube for reworking, e.g. in case of breakdown of the packaging machines. The reworked product (e.g. 15°C) is remelted at 47°C, and if necessary pasteurized in a heat exchanger at 72°C (HTST) and cooled to 47°C and is then added back into the process proper by means of the rework centrifugal pump to the pre-emulsion feeding line. On the other hand the mechanical feeding and the extrusion of the mass into the moulds of the packaging machines can be considered as the logical extension or partial replacement of the resting tube.

3. *Packaging Hardness*

One important objective of wrapper margarine processing is to obtain products which can be handled easily by the packaging machines. There are two practical problems very often encountered in the resting tube.

(i) Because of insufficient cooling in the first A-unit, there are no substantial amounts of α crystals formed in slow crystallizing blends. This will result in very stiff products in (and after) the B-units, with subsequent channel formation, which ultimately leads to products which cannot be packaged.

(ii) Because of too deep cooling in the A-units too many α crystals are formed which, after transformation to the β' form in the C-units, will be totally destroyed by the pins. As only small amounts of additional α crystals are formed in a second A-unit, the final product becomes soft and is called overworked.

These phenomena have thus to do with the pre-crystallization of the blend. Different measures can be taken to counteract the defects. The temperature profile realized in the different units determines the amounts of α and β' crystals produced in the A-, C- and B-units, all this related to a given margarine blend. So the temperatures at which the mass leaves the first A-units as compared with the temperature decreases in the second (and third) A-units and the speed of mixing (scrapings) should be balanced in order to reduce the supercooling. The applicable temperature variations to the lower ranges are strongly limited because of the increasing viscosity of the α crystallizing blend. The total residence time in the C- and B-units should be at least equivalent to the adiabatic re-crystallization time measured on the fat blend (5–7 min). If this time is shorter, the margarine begins to crystallize in the package, causing post-hardening, a well-known margarine defect. In this case the throughput must be decreased.

The useful parameters for controlling the packaging hardness are: (i) the fat blend; (ii) the type of the machines used and their configuration: (iii) the inlet temperature; (iv) the cooling rate in the first and second group of A-units; (v) the individual and total residence time in the different units (throughput); and (vi) the recirculation rate.

4. *Hardness After Storage (Maturing or Tempering of Margarines)*

The hardness after storage is one of the main quality characteristics of margarines. Increases in the solid fat content and other changes may occur after packaging, i.e. during storage. The two phenomena most often observed are post-crystallization and re-crystallization.

Post-crystallization occurs if the storage temperature of the product is lower than that of production. The solid fat content and the hardness of the margarine will become like those of a product originally produced at the storage temperature. Post-crystallization depends on the fat blend and the degree of supersaturation remaining after the resting tube and packaging machine. Post-crystallization during storage is typical of margarines produced with slowly crystallizing fat, such as palm oil.

Re-crystallization will occur at nearly constant solid fat content if the product is stored at higher temperatures than that of production. This special storage of packaged confectionery and industrial margarines is called maturing. The process untilizes the re-crystallization tendency of the specific fat blends produced under predescribed conditions. When maturing a product, the fat blend is held above the melting point of its lowest melting polymorphic modifications for a longer time. The crystals will melt; this results in demixing, a new crystal distribution (more larger crystals instead of many small crystals) and polymorphic changes. Maturing improves, among other things, the creaming behaviour and causes a permanent softening of the structure.

Cake and bakery margarines ought to be shock crystallized, intensively worked and, after packaging, tempered for 48 h at 27°C. This treatment improves their plasticity and creaming properties. These margarines should contain their solids mainly in the more stable β' modification. Puff pastry and coating (icing) margarines should contain fat blends producing high levels of pressure resistant, coherently solidifying β modifications. To this end the Votator-produced margarines are rapidly chilled (supercooled), intensively worked in crystallizers and not matured after packaging. If produced by the churn/drum process or on a Gerstenberg and Agger type Complector, the margarine should be well matured for, e.g. 6 days at 20°C.

5. *Minimum-calory (Low-fat) Margarine Production*

The production of minimum-calory (low fat) margarines, which look like soft margarines, is analogous to those of soft margarines produced with liquid filling. Before chilling the initially oil-in-water type emulsion is pasteurized (HTST) and during filling the temperature profiles are kept somewhat higher than those of soft margarines. Care is taken to avoid the incorporation of air and to minimize line-pressure variations. These precautions are taken because of the lower stability of the final water-in-oil type, phase inverted emulsion (see next section), formed with or without the addition of protein, despite the higher levels (0.2–0.3%) of glycerol monooleate used as emulsifier. A possible processing configuration is

$$\text{Fat} + \text{water phase} \text{ —A—C—A— } \text{Liquid filling}$$

6. Phase Inversion and Double-emulsion Margarines

It is possible to produce margarines by a complete imitation of butter production, which is based on the inversion of the original oil-in-water emulsion by churning into a water-in-oil type emulsion. A 40–50% (fat) oil-in-water emulsion (an artificial cream) is first produced by suitable emulsifiers and this is then churned in mixers or in one of the continuous butter-manufacturing machines (e.g. Fritz) into an inverted-phase product. The excess of the water phase (artificial buttermilk) is removed and the residual mass washed and kneaded as usual. The product, although having properties close to butter, has, like butter, a much shorter storage life and is not produced industrially. Margarines based on double (oil-in-water and water-in-oil) emulsions are also possible. The advantages claimed are better flavour retention and release. Higher flavour levels are incorporated for industrial baking purposes, with higher levels of oil phase incorporated as the inner phase (oil-in-water) in the more solid outer phase (oil-in-water/oil).

II. The Production of Mayonnaise and Related Products

A. Mayonnaise

Mayonnaise is a quite different product from margarine. Traditionally it is home-made in France but nowadays it is also industrially produced. Mayonnaise is an oil-in-water type emulsion used as the most prominent savoury type of "dressings" of high oil content. It is used both in the household and in the restaurant/catering business for preparing sandwiches, salads and similar cold dishes.

According to an (incomplete) FDA definition of 1950, manyonnaise is "the semi-solid emulsion of edible vegetable oil, egg yolk or whole egg, a vinegar, lemon juice and/or lime juice, with one or more of the following: salt, a sweetener, mustard, paprika or other spice, monosodium glutamate, and other suitable seasonings".

Mayonnaise contains, by any food legislation, at least 80% vegetable oil and if less the product is classified as a salad sauce or low-calorie dressing/spread. The lower the oil content, the more difficult it is to achieve the necessary stiffness without the use of additional gelling and thickening agents, binding the higher proportion of moisture incorporated. The spices determine the flavour, the carotenoids of egg the colour and the vinegar the sour taste.

The oil should be dispersed in the continuous water phase in small globules, with the diameters of the droplets not being larger than 2–6 µm. The stability of the emulsion, mainly provided by the phosphoacylglycerols (phosphatidylcholine; lecithin), and the protein complex of the 8–10% egg-yolk used is generally sufficient to withstand any mechanical shocks suffered during transport or storage. Some enzymatically hydrolysed egg-yolk lecithins (lyso-lecithins) are eminently suitable for stabilizing the emulsions in mayonnaise and in low(er) calory spreads.

Mayonnaise is generally produced in batch kettles, fitted with slow stirring devices. Continuous production is also possible. The water-soluble ingredients are first mixed with the eggs and a portion of the vinegar at 20°C. The oil is then gradually added to the system and the rest of the vinegar added only when the oil has been "bound" to give a stiff gel. The remaining vinegar will bring about the final consistency. In order to obtain a better dispersion, the system can be additionally dispersed by the shearing force developed in a colloid-mill type disc homogenizer.

Produced in open kettles, mayonnaise contains about 10–12% air by volume because of the stirrer-induced vortex. When made in the modern closed, pressure (vacuum) resistant kettles, the complete composition undergoes colloid-milling and is then deaerated by recirculating it from the kettle back into its headspace, kept under vacuum.

The oils to be used should be (preferably) winterized (free of solid wax or fat crystals) and freshly deodorized liquid vegetable oils of premium quality. This prerequisite usually includes the presence of substantial amounts of natural antioxidants. The oxidative stability of the system can be improved by the use of metal-scavenging (chelating) agents, such as citric acid and calcium sodium EDTA. The microbial control of the eggs (yolk; fresh, dried or hydrolysed eggs), the gelling and thickening agents and the spices (the latter also chemically on heavy-metal traces) is a prominent task of the laboratory.

B. Sweet and Savoury Spreads

Sweet and savoury spreads are relatively low (10–20%), medium (40–50%) or even high (65%) level fat containing products, emulsified with eggs (yolk, hydrolysates), lecithin preparations, other food-grade emulsifiers (mono- or di-acylglycerols), and filled with gelling (gelatine, starches, hydrolysed starches) and thickening (guar or locust bean gums) agents, preservatives (sorbic acid). There are many types of differently flavoured products on the market with basic tastes, e.g. salty, sweet, salty/sour and sweet/sour.

The type of emulsion produced is generally oil-in-water for spreads with a low oil level. Both oil-in-water and water-in-oil (e.g. minimum-calory margarines) emulsions are used, when producing medium to high oil level spreads. The pH is generally kept low, but if this is not possible preservatives have to be used. Pasteurization (HTST) of emulsions is also customary. The water activity of the sweet systems with different sugars is kept low (<86%, the practical growth limit for heat sensitive and resistant food spoilage and poisoning bacteria). As already mentioned, according to many food legislations, minimum (low) calory margarines and minimum (low) calory mayonnaises should also be classified as (savoury) spreads.

The flavouring of the spreads is achieved using natural ingredients, such as chocolate, ground nuts (hazelnuts, groundnuts), sugar(s), salt, etc. (mostly in non-emulsified sweet or savoury pastes), or using approved food flavours, protein hydrolysates, spices and chunks of added fruits, vegetables, onions, pickled or not, etc.

The production of the water-in-oil type emulsions without solids might be done on production lines as described for continuous margarine processing. The other types of emulsion are produced as described for mayonnaise. The chunky ingredients are mixed into the product after the basic type of emulsion has been formed.

A different kind of oil-containing products are the non-emulsified dressings consisting of differently flavoured vinegars and an oil phase which separate in the bottle into two phases (vinaigrettes). Before use these dressings need to be shaken to a coarse dispersion in order to be poured immediately onto salads. Garlic, blue-cheese, exotic (e.g. curry), dill, estragon, etc., flavoured dressings of this type are very popular.

C. Filled Creams

Filled creams are meant to replace dairy cream by vegetable fats, either due to the unavailability of real cream and the abundance of good quality fat (coconut oil, e.g. Philippines), or because of health (high PUFA, no cholesterol) requirements. The need to sterilize these creams and their decreasing whippability with age make the products sensitive to the correct choice of emulsifiers. The creams comprise concentrated (20–40%) oil-in-water type emulsions, produced in a similar way as for mayonnaises. The choice of whey, skimmed milk or vegetable proteins as the proteins, and of lactose, sucrose or glucose as the sugars is made according to local needs or possibilities.

D. Pan-releasing Agents

Pan-releasing agents contain about 30–50% fat in water-in-oil type emulsions and are used in large commercial bakeries for coating the baking trays, plates and tins before placing the bread or cake doughs or pies on or into them. These agents allow the easy release of the baked goods from the cooking vessels. The emulsifiers most often used are lecithin, mono- and diacylglycerols, and ricinoleic acid esters of polyglycerols. The production is similar to that of spreads and mayonnaises. For the sake of easy spreadability the viscosity of the emulsions must be kept low.

III. Shortenings

In a general sense we might say that modern shortenings are usually (but not always) the water-free counterparts of confectionery and bakers' margarines. Historically, lard can be considered as the most early representative of shortenings. The blending of hard tallow fractions with liquid oils has made it possible to substitute lard by "lard compounds". However, only the introduction of hydrogenation provided the basis for producing shortenings fully independently from the meat packing industries and on a "tailor-made" basis. The word "shortenings" quite often also refers to any type of fat used for bakery, confectionery or frying use. In general, four types of shortenings are produced and marketed: the liquid, the fluid, the plastic, and the solid ones.

The liquid shortenings, if this denomination is valid at all, are in fact liquid oils containing low solid fat content at ambient temperature. They are merely lubricants and/or produce a certain gloss on the surface of products, although they are often used as heat-transfer media in certain food procesing (dewatering) operations, e.g. deep frying.

The fluid shortenings contain a low level of suspended solid fat crystals in the β modification which makes it possible to transfer them by pumps. The oil phase, if used for repeated heat transfer (deep frying) must be compositionally similar to that of the liquid shortenings. Both liquid and fluid shortenings have some strict compositional limitations (see Section VI.A of this chapter). The necesary solid fat content is produced by adding highly hydrogenated fats containing more than 80% C_{18} fatty acid chains or β forming fat fractions (stearines) to the oil. Crystallization is done at gradually decreasing temperatures in order to form the β modification, together with the homogenization of the mass. Quick chilling and slow crystallization is another variant before filling the shortening in cans. Their main uses are as bread and roll lubricants. If enhanced by the addition of suitable emulsifiers (e.g. 10–12% propyleneglycol stearate), as already described, fluid shortenings are claimed to be applicable in cake manufacture without the addition of further solid fats.

The plastic shortenings are similar to the industrial margarines used for bakery and confectionery purposes and so their fat blends are composed on the same basis. In general, plastic shortenings are not coloured because the product should have some resemblence to lard, the traditional shortening.

Solid shortenings are close in their nature to prepared mixes (see Section V of this chapter). They contain high levels of plastic fat which is often spray dried if they contain different originally water soluble non-fat components.

A. Shortening Processing

The opaqueness of packed lard, despite its natural grainy structure, is the result of the incorporation of air (aeration) during its processing. Futhermore, after the quick drum (roll) chilling, air inclusion occurs during the kneading in picker boxes by the paddles of rotating shafts and/or by the helical (meat mincer) type mixers of, for example, the Gerstenberg and Agger Complector. In both cases the mass is pressed through orifices for further working.

In modern installations, 10–15% by volume of air, or an inert gas such as nitrogen, is incorporated into the shortenings. The rapid expansion of this gas initiates the necessary and also fine crystallization that supports opaqueness. Whipping results in a 50% over-run (volume increase) and opaqueness

6. PRODUCTS OF HIGH FAT CONTENT

if 33% by volume of nitrogen gas is incorporated under high pressure. The gas inclusion can be achieved by using a high-pressure booster pump and a texturizing valve. These latter two elements can be used for aerating the shortening before the final crystallizer or resting tube. In the case of whipping the nitrogen gas is added under high pressure before the booster pump and the A-units. In all cases the valve releases the pressure of the chilled, crystallized mass from 20–50 bars to 1 bar (atmospheric) at packaging. Instead of the texturizing valve, small-volume C-type crystallizer units (Schroeder) with high-speed (900 rpm) pinned rotors in pinned cylinders can also be used to relieve the pressure and redisperse the nitrogen in the semisolid fat mixture.

The manufacture of shortenings is very similar to that of modern margarines. The configurations used are very often the straight-through system with either C-unit (shortening-B-unit) crystallizers or the resting tube (static B) units. The choice of the latter determines the method of packaging for plastic shortenings as a liquid fill or as a moulded block in a wrapper. Recirculation is also an option, mainly for the moulded blocks. Favoured basic configurations are (BP, booster pump; TV, texturizing valve):

Fat + gas —A—C—A—A—A—BP—TV—static B—Moulded blocks

Fat —A—A—C—gas—BP—TV—C— Liquid filling

Fat + gas —C—A—A—BP—TV—static B— Moulded blocks
↑   ↓
recirculation

Many plastic shortenings used for short pastry and cake making are matured after packaging meaning that wide temperature variations before use may be detrimental to their performance.

IV. Vanaspati

In the broader sense of its definition we should mention a type of "shortening" called vanaspati. Vanaspati was first produced in the early period of fat hydrogenation mainly for India, and subsequently for other countries with Hindu communities or Moslem populations (Arabia, Pakistan). It was conceived as a water-free substitute for buffalo, cow, goat and sheep butter produced in historic India and in Moslem countries. This butter was originally dewatered in open kettles, often directly heated by burning, e.g. dried cow-dung. Dewatering considerably increased the shelf life of the butter fat under the tropical climatic conditions. The flavour of the product is determined by the occluded smoke flavour of the open fire and/or by added spices (e.g. cardammon seeds); the latter probably also act as antioxidants. The dewatered fat is called ghee. The flavour of ghees produced by the modern (closed vacuum kettles) techniques is different, with rancidity playing a role when stored for a longer time. Groundnut oil, sesame-seed oil, but nowadays mainly soya-bean and other cheaper vegetable oils are first hydrogenated by using the iso-suppressive techniques to melting points between 35 and 41°C. If necessary the oils are mixed with stearines and other oils and completely refined. The fat is then filled, in the liquid state, into tins which are then sealed. The contents are allowed to crystallize slowly in order to develop large (coarser) fat crystals. If smaller crystals are acceptable, the normal processing of shortenings can be applied, which makes their production quicker and cheaper. The product can be flavoured to give products with smokey, cardammon, cheese, rancid, etc., notes.

V. Prepared Mixes

Prepared mixes are typical convenience products used by bakers, caterers and housewives. They contain homogenized (or in some cases spray dried) fat mixtures and all the ingredients, except flour, necessary for, e.g. a madeira or sponge cake (sugar, powdered egg yolk and baking powder). They save manual work and time. Shortenings mixed with finely ground sugar only (up to 95% sugar can be homogenized with fat by roller mills) are well known as "bakery creams". If the fats used contain about 50% of cocoa butter or one of its equivalents, or replacers substitutes, together with 40% sugar and about 10% non-fat cocoa matter, then the resultant product is a couverture, or imitation couverture. These latter products are used by bakers and caterers. Almond and hazelnut pastes are easily imitated by mixing sugar, starch and vegetable proteins with refined oils and nut-like flavours.

VI. Fundamentals of Baking

Moistened flour worked into a dough develops a curious protein complex, called gluten. Gluten is a close tangled network throughout the worked dough giving it the properties of toughness and resilience which when baked sets hard. Without the use of yeast (baking powder), as in bread, or without sugar, as in pastry, the baked product remains less palatable or appetizing. Gluten formation can be controlled by the addition of fat before moistening the flour, as in the case of short (crust) pastry. Fat (the presence of fat crystals) prevents the cohesion of gluten strands during the working of the dough. By this "coating" the gluten is made more "tender". The presence of fat crystals in the baked goods also prevents the gluten (and starch) from "cementing" into a hard mass and makes the product disintegrate into discrete, but not too small, particles, when a shearing force is applied. Cookies, biscuits and wafers are products which are heavily rolled by machines into thin sheets. This operation spreads out the fat into layers, thereby preventing the cementation of the baked flour components. The "anti-cementing" effect of the fat is called shortening power. Fermentation by yeast and sugar also makes the dough more tender.

6. PRODUCTS OF HIGH FAT CONTENT

A high final volume with uniform cellular structure is an important quality requisite of many commercially baked goods such as bread and cake. This is a result of the skillful inclusion of a gas during the preparation of the dough which can be achieved either by fermentation (in bread) or by baking powder (in other products). The foam structure of a cake can be produced by using suitable fats which have a high "aeration" or "creaming" potential. This is probably the only shortcoming of the traditionally most ideal shortening, lard. Due to its β crystallizing tendencies it has very little or no creaming power. The carbon dioxide or the air expands during baking and induces the bread or cake to yield the satisfactory final volume with the required fine cellular structure. Another way of thinning out the structure is lamination. This process arranges the dough in paper-thin layers which are kept separate by suitable (puff-pastry) fats of near-waxy consistency over a wide temperature range. This fat, in the form of sheets or pencil-like thin cylinders, survives the severe turning and folding treatments during the preparation of the dough without melting or collapse. The presence of water in puff-pastry shortenings (80–90% fat content) and of course in margarines (>80% fat content) is very important. The extensive release of water vapour on baking enhances the "puff" and the survival of a laminar structure.

Danish pastry is similar to puff pastry, but the dough is yeast leavened and sweet, and the fat is absorbed by it during baking. Cream fillings for sandwiched cookies and icings are mixed with sugar. In the icings high levels of emulsifiers are incorporated as the product must contain water. Shortbreads, pies and soft dough biscuits are traditionally the pastries which should be made with plastic shortenings. No emulsifiers are incorporated in these traditional types of shortening.

Some additives, such as emulsifiers, can increase the volume of certain baked goods prepared with sugar and liquids: these are the so-called high-ratio products. The additives aid the easy dispersion of fats around the gluten strains in fine globules. Fats (margarines) containing these emulsifiers are the high-ratio shortenings: they were previously called "superglycerinated", because of the monoacylglycerols or their derivatives added as the primary emulsifiers. Some modern fluid shortenings might contain, rather than liquid oils, only higher amounts of added emulsifiers of, e.g. the propylene glycol type.

In summary, bakers and confectioners require special plastic fats (margarines) as aerators, lubricators, laminators and, last but not least, as real shorteners. As we have seen, they also need some other special fats (margarines) for icings, cream fillings, ice creams and chocolate couvertures for ice creams, the latter being part of their trade, but, because of its low fat content (<10%), not within the scope of this book.

A. Oxidative and Flavour Stability of Baked/Fried Goods

An important quality aspect of all fat products used by bakeries is their intrinsic oxidative stability at high temperature, which is necessary for the fat to withstand oxidation at the production temperature of the oven, and the long storage periods at ambient temperatures. Despite the fact that the temperature at the centre of bread and cakes hardly reaches 95°C, the parts close to the crust reach the oven plate and air temperature (180–220°C). These temperatures require the use of fat blends with low levels of di- and, better, no tri-unsaturated fatty acids and only medium levels of mono-unsaturated fatty acids. The addition of synthetic (phenolic) antioxidants is one way of increasing the stability of the fats. Care should be taken by the user not to degrade the good quality of the correctly chosen fats by other dough ingredients which may be heavily contaminated with oxidation promotors, such as heavy-metal traces (e.g. those present in flour or those from the attrition of the metal equipment used).

There are well-known forced-oxidation-based fat-stability tests executed at 98°C or even higher temperatures, such as the active oxygen method (AOM test). However, these tests cannot predict the stability of the finished baked product if the real promotors of oxidation are present in the other ingredients added to the fat. The off-flavours developed during oxidation are even less well predicted by the AOM test. The latter depend mostly on the interactions between the different types of unsaturated fatty acids (the precursors) and the pro-oxidants which accelerate the autoxidation during storage. Metal scavengers can somewhat reduce this adverse influence.

Certain liquid and plastic shortenings are also used for the household, catering and industrial deep frying of different fat commodities such as chips, french-fried potatoes, fish-sticks, doughnuts and a wide range of sweet and savoury snacks. Deep frying is actually the heat transfer from the heating source to the product in order to dewater it. The nature of the process (the turnover rate of the fat, the amount of fat absorbed and the required shelf life of the product) is important in the selection of the liquid oils to be used. Only oils with low levels of linoleic acid should be used. The presence of more than 2% linolenic or higher unsaturated fatty acids should be avoided (France and Austria) if the oil is to be used repeatedly for deep frying. The AOM stability test (often camouflaged by legally added antioxidants) again gives no clue as to the flavour stability of the finished product. Iso-suppressive hydrogenation (and subsequent fractionation) of liquid oils with high initial dienoic and trienoic unsaturation is one remedy. More intense hydrogenation of the oils produces a risk of the oil not being accepted because of the possible development of hydrogenation off-flavours. Good natural frying oils are the recently developed high oleic sunflower and the well-established high oleic safflower oils, besides the traditional groundnut, cotton-seed and other oils or fats of similar more saturated character.

VII. Packaging and Packaging Materials

The correct packaging of margaines, shortenings and other high fat food products is not only a natural extension of their processing, but also makes possible the distribution of the products in appropriately sized units, and helps to maintain the required shelf life of the product.

A. Packaging Materials

The finished product should be filled, shaped and formed by means of suitable machinery into individual packages. Some packages are closed and/or sealed, and some are merely wrapped in special paper-based materials. Products are further packaged into paperboard cases for storage and delivery. The primary packaging material used is either of non-rigid or rigid consistency.

1. Non-rigid Materials

Traditionally the most frequently used packaging material was the parchment paper wrapper produced by immersion of the paper sheet in sulphuric acid baths. The paper may have a low or high gloss (calendered) surface, which can be multicolour printed. The paper should be resistant to the oiling out of the margarine wrapped in it. Because of its consistency it is only suitable for wrapping kitchen margarines at moderate climates, and the harder consistency industrial margarines and shortenings. The modern wrappers are aluminium-foil laminated parchment papers glued with, e.g. (solvent-borne) synthetic resins or wax. They may have a glossy polyethylene coating or similar on the outside. They are used in the packaging machines either precut-to-size or in reel form. Thin-gauge polyethylene sheets are increasingly used as a cheap material, especially for large packs and in countries without an indigenous paper industry.

Flexible sachets (polyethylene laminates with nylon) are used for oils and certain fats especially where the desired sales units are relatively small, e.g. less than 250 g.

2. Rigid Materials

Rigid packaging materials are mostly in the form of cups, tubs, jars, cases, containers, etc. There are four basic types of materials used for rigid packaging; these are described below.

(i) Fibreboard or corrugated fibreboard is used for outer cases. Paperboard is used for outer-carton packages. Polymeric plastics layered paperboard with internal waterproof coatings is also used.
(ii) Polymeric plastics, such as: monomer-free polyvinylchloride (PVC); low-density polyethylene (LDPE); polypropylene (PP); acrylonitrile–butadiene–(poly)styrene (ABS) copolymers; and polyethylene terephthalate (PET)/(poly)styrene (PS) copolymerized laminates.
(iii) Metals such as tinplate and aluminium tubes.
(iv) Glass is used for bottles or jars.

The thermoplastic (reversibly melting when heated, re-solidifying when cold) polymeric materials are produced from powders or from preformed sheets by two main processes: (i) injection moulding (for LDPE, PP and ABS); and (ii) thermo-forming (for PVC, PP and ABS). When applying injection moulding the heated powder is pressed by a hydraulic plunger through a nozzle into the mould cavities with the shape of the cup/tub or lid. Sheets of PP, PVC or ABS are made either by extruding the molten material through slit-dies, or by calendering it through a series of nip rollers under high pressure. These sheets are then used for thermo-forming.

The wall thickness of a tub is a compromise between its rigidity (mechanical stability), its light, water and oxygen permeability and the price of the finally processed raw material. The physical properties of the product, its outer packaging and the method of distribution are also important cost parameters. As regards autoxidation prevention, PVC is less permeable to air than the other materials, whereas ABS is more water permeable than the others. Light of the wavelengths < 520 nm, because of its autoxidation promoting action, should be excluded by using deep yellow or, better, brown or black coloured polymer when products are to be kept for longer times without overwrap cartons or outer cases.

Bottles for packaging salad oils are made of PVC, e.g. blow moulding. The molten material is preformed by extrusion into the form of tubes. Portions of the tube are trapped by a split mould around a blowing pin, all placed inside the cavity of a cooled mould. Air is blown through the blowing pin into the sealed tube which then expands into the shape of the mould. A more recent development of this technique is the so-called stretch-blow moulding for PVC and PET which gives the bottles an improved gloss and strength at reduced weight. As more unsaturated oils may be affected by light-catalysed oxidation causing flavour and colour damage, the bottles should be preferably produced of polymers excluding light of < 520 nm wavelength, or the bottles should be kept in outer cases.

B. Filling and Packaging Machines

Filling and packaging machines are produced by many specialized firms (for example, in Europe: Benz/Hilgers; Breckel, Dolman and Rogers; Forgrove Ltd.; Hamba Hassia; Hansa; Kuestner Freres; Mather and Platt; SIG; Trepko). The machines can be divided into one of three types: (i) automatic ones which mould the margarines and wrap them in regular (50–1000 g) rectangular brick or cube shapes (packages), (ii) automatic ones that fill semi-liquid margarines/shortenings by straight-line or rotary-head fillers into tubs or other containers; and (iii) semi-automatic or manually operated ones that fill margarines/shortenings into large packages for industrial use.

An automatic packaging line is actually the extension of the production line and consists of four groups of machines. The first group is the margarine moulding/wrapping or filling machine. The moulding produces a block of solid/plastic margarine in a mould and wraps it. The filling process fills the semi-liquid mass into a moulded plastic tub or into a preformed wrapper (a cavity prelined with an inner wrapper). This group also closes the wrappers or puts the lids on top of the tubs/cups and does the necessary labelling and coding. The second group of machines pushes the units into paperboard cartons, if any, and performs the actual case-packaging. The third group consists of semi- or fully-automatic case-sealing or case-sealing/ taping/labelling machines. Palletizing machines comprise the fourth and last group.

The speed of the block moulding/wrapping machines varies between 90 and 250 packages min^{-1}. In the slower (older) machines, a piston forces a preset quantity of margarine into a single moulding chamber, the machine being fed from a hopper by feed worms. When the moulding chamber is completely filled the piston movement reverses and the block is injected into

6. PRODUCTS OF HIGH FAT CONTENT

the wrapping head for the final operation. The high-speed machines are fitted with drums with four to six moulding chambers. During the filling operation a series of pistons automatically retract, the rest of the operation being similar to that of the single-chamber machine.

Precut wrappers are placed into the magazine stacks in the slower machines. In the high-speed ones the wrappers are in reel form. The moulded block, for example, can be moved over the wrapper to make the first fold in the form of a sleeve, and the final fold sequences are carried out whilst the wrapping head turret is revolving. Reel-packaging needs special registration marks on the wrapper so that it is cut out of the reel accurately. The machines stop automatically with any disturbance of the action.

Instead of moulding, speciality margarines of substantial packaging hardness can be cut into pieces of uniform weight. This can be done by automatic cutting machines positioned immediately after the resting tubes (static B-units). The pieces are then packed, for example, by hand. Tins (cans) can be filled with rotary fillers and seamed with a close-coupled seaming device. Salad and cooking oils can be packaged by automatic fillers of the volumetric (filler chamber) or vacuum types, the height of liquid level in the bottle stopping the filling action. Bulk quantities can be filled directly on a scale into drums, etc. The filling of oils and margarines under a nitrogen blanket or vacuum into tins or bottles can be carried out by special machines developed for this purpose.

The multiple straight-line or rotary-head liquid margarine/shortening fillers are of either the timed flow or volumetric action (piston type) type which can fill up to 300 tubs min^{-1}. These fillers require preformed cups/tubs and lids which can be moulded by specialized machines as follows. The printed plastic sheet material is unwound from a reel and led under infra-red electric panel heaters. A clamp frame places the warm sheet on the top of the mould(s) having the shape of the cups/tubs and the lids. By means of vacuum or gas pressure (or both) the material is "thermoformed" by these mould(s). The filling machine accepts these preformed cups/tubs and lids in separate stacks. After filling the lids are mechanically pressed onto the cups/tubs, by which the latter are closed. Closing of tapered tubs can be done by heat sealing: after filling a (pre-printed) plastic web is placed on top of the tubs/cups and the filled containers are separated by cutting them out of the web, the scrap being reused. Machines that perform the (thermo-) formation of the container, product filling and sealing operations as one unit are called form/fill/seal type machines.

The volumes (weights) of the individual portions dosed and filled into the different types of packages are carefully controlled on-site by constant weighing. Any statistically inadmissible variations, i.e. weights outside the control or warning limits, are immediately adjusted by resetting the geometry or the volume of the dosing units. The same control method applies to the visual inspection of appearance, gas content (shortenings), sealing/closure and physical state of the packages.

The storage, make-up and feeding of packaging material should be done on a different (higher) level to the production area. In particular, in order to reduce the risk of microbial infection and moulds the packaging material (parchment) should be stored under controlled humidity conditions.

VIII. Warehousing, Distribution, Storage Life

Finished packed products are stored on pallet racks of the drive-in-type, leaving sufficient aisle width for truck manoeuvering. Soft and high PUFA margarines and minimum-calory margarines should be stored in the factory, in the wholesale warehouses and in the retail shops preferably at temperatures below 10°C. This is also a recommended maximum temperature for the cool cabinets and/or household refrigerators. Kitchen, table and industrial margarines may have at room temperature a storage life of 4–6 months. Margarines older than 3 months should not be sold if they have been stored at ambient temperature. To reach an optimum flavour quality the "first-in, first-out" principle should be maintained in the warehouses, stores, etc.

Matured (structured) products should never be handled by bulking installations because temperature changes are detrimental to their quality.

IX. Product-quality Assessment

Margarines and shortenings are multipurpose fat products in their own right and their quality is determined by many aspects. The quality of (high fat) food products is composed of many aspects. First there are some intrinsic aspects such as: the caloric aspects (energy, satiation); the physiological aspects (the fatty product is the carrier of essential fatty acids, vitamins, proteins); and the hygiene aspects (the product must be produced from good raw materials by the application of good processing, storage and distribution techniques, all of which give good keeping properties).

A. General Quality Control

Extrinsic quality aspects are those which can be monitored by physical, chemical or organoleptical techniques. First of all it is very important to assess the "mouthfeel", composed of flavour, odour and taste, and the so-called oral-melting properties. The former are assessed organoleptically on fresh production and at 2, 6 and 12 weeks of storage, the samples being kept at 15°C in the dark. The different oral-melting characteristics, such as coolness, thickness, salt (and flavour) release can be also tested instrumentally. The physical appearance (colour, gloss, transparency), the hardness (in many stages and also after standardized working, i.e. worksoftening) and solid fat content can be measured as necessary. Some important derived aspects, such as the stand-up/collapse, the oiling out (during storage, under the pressure caused by the weight of cases stapled above), the surface darkening (the drying out, i.e. moisture loss, of the surface emulsion layer below the lid of the tub or under the inner wrapper layer), and the spreading, creaming, spattering, and emulsion stability are all parts of the product quality that can be tested by existing methods. The gas content of shortenings must be controlled constantly and, if necessary, on-site. Colour and flavour defects of chemical (mainly oxidative) and/or microbial origin are discussed separately (see below) from those caused by the physical effects mentioned above.

A very important aspect of consumer protection is the constant inspection of the moisture level and the weight of the individual packages. This quality assurance is done regularly at the production site by the members of the processing team from the very beginnings of machine packaging. As a result of these instrumental measurements any moisture-level changes are constantly corrected by the exact setting of the volumes delivered by the proportioning/dosing machines (pumps). The weight is constantly controlled by the correct setting of the geometric or volumetric (timed flow or piston-type) elements of the block-moulding/tub-filling machines. The execution of all these quality or quantity related controls and analyses must be based on the general rules of statistical sampling and data analysis.

The quality control of the raw materials involves, for the fat phase, the determination of taste, refraction, colour, the presence of emulsifiers, lecithin and vitamins, and for the water phase, pH, salt content, presence of milk, flavouring agents and acidic additives. This and the control of the packaging materials, and the semi- and fully-finished products is the daily routine task of the main laboratory and/or of a separate quality-control department which is authorized to carry out the (non-microbiological) control of the whole production and storage area.

In order to attain good feedback as to the quality of the final product, besides the quality-control data, market research and research guidance information (consumer panels representing distinct segments of the market and structured on a firm statistical basis) can also be used, in particular to help the product research and development activity of the margarine producer. Besides the consumer tests, regular application tests on the products by, e.g. an experimental bakery and by the users themselves, with appropriate feedback, are also very informative for development work on the quality of a product.

B. Microbiological Quality Control

A specialized section of the main laboratory is responsible for the daily routine microbial controls executed on the fresh milk delivered, on the basic cultures, on the soured milk and on the final products (fresh, 2- and 12-week-old samples). The control of the hygienic conditions of all the receiving and storage vessels, equipment, valves and the connecting pipelines according to the prescribed methods and frequency is the minimum requirement for good housekeeping and factory hygiene. The regular inspection of the personal hygiene of the production operators and the very rigid control of general hygiene (cleaning) of the whole factory is a shared responsibility of the main laboratory and the quality-control department, who both report their findings to the production management.

Some possible aspects to be controlled microbiologically are: (i) the standard (total) plate count and the inspection of the type of colonies; (ii) *Coliform* or *Enterobacteriaceae* counts; (iii) lipolytic bacteria counts; and (iv) yeast and mould counts. All the necessary sampling and analytical methods are well described in the relevant standards.

X. Cleaning Methods

To reduce risks of microbial infections the production lines must remain filled with fat-blend/emulsion during short production pauses. Before the weekly (twice) executed cleaning operation the fats/emulsions are first removed by hot water displacement into the rework-fat containers, the fats being returned to the refinery department.

The most efficient method for cleaning the tanks used for the storage of fat and water-phases is the manual one. The B-units and packaging machines must also be cleaned manually. On the other hand the cleaning of the pipelines, dosing/proportioning and transportation/rework pumps, and the different continuous units is done "in-place" and by a programmed automation of the different stages. Previously prepared hot (80°C or more) aqueous solutions containing either sodium carbonate, or hydroxide, silicates, phosphates and sometimes EDTA, and disinfectants of the active chlorine or quaternary amine types together with hot water of 50°C are pumped in sequence at high speed through the production lines. For this type of cleaning the installations can be divided into two sections: e.g. those situated before and after the proportioning/dosing pumps.

7

Speciality Fats

When describing the different technologies applied to edible-fat and oil products, we encountered various fat products used for general or specialized food-processing operations or distinct types of foods. The only area not yet covered is that of "speciality fats", used for chocolate- (genuine or imitation) or butter-based confectionery products.

Chocolate is a mixture of roasted, fat-free cocoa powder, sugars, milk or other proteins, emulsifiers, (viscosity decreasing) flow regulators, like lecithin, added flavours and fat, which, due to legal requirements, is generally genuine cocoa butter. Some legislation permits the admixture of 5% (on fat), 15% (on chocolate) so-called cocoa butter equivalents or extenders. As we have seen (Chapter 5, Section I), cocoa butter is composed of large amounts of the symmetric triacylglycerols of the POP and POSt types, resulting in very steep and uniform melting at its melting point, which is close to, but below, body temperature. It has a strong tendency to recrystallize after solidification into its highest melting β-modification, which causes some brittleness and, after some repeated and non-deliberate temperature changes during storage, the well-known "bloom" of chocolate products. As we have seen, this can be somewhat reduced by the use of crystallization retarders. The brittleness which causes the "snapping" sound, when breaking a chocolate bar, is a welcome quality aspect of the good quality product.

Although cocoa butter is produced in substantial amounts, because of its price (6 to 12 times that of common fats and oils) it cannot be incorporated in very popular, but generally much cheaper cocoa-powder based confectionery products, like candy bars, etc. For reasons of economy the cocoa butter component is partially or totally replaced by new fats and fat fractions which are produced by a specialized branch of the fat industry as "speciality fats".

Speciality fats which are used to replace part (or all) of the cocoa butter fraction in butter are:

   (i) compatible for blending in any proportion with cocoa butter—the cocoa butter equivalents (CBE);

(ii) compatible for blending in nearly all proportions with cocoa butter—the cocoa butter extenders (CBE); and
(iii) similar, but not compatible for blending with, cocoa butter, but well replace it in the above-mentioned cheaper products—cocoa butter replacers (CBR).

The most commonly used raw materials are: the natural vegetable butters (shea, illipe, acetune, kokum, sal fats and palm oil) which contain high levels of SUS fractions; the two lauric fats, coconut and palm-kernel oils; high palmitic level cotton-seed and ground-nut oils; and any other suitable (vegetable) oil. These materials can be processed by hydrogenation, interesterification, fractionation, or a combination of some or all these techniques. The processing can be done on single oils or on blends. All the above choices depend on the final quality characteristics required, the availability and price of the raw materials and the technological possibilities.

Partial hydrogenation of palm oleine with no trisaturates by selective (two-stage, or high *trans* promoting) hydrogenation techniques and/or catalysts, can lead to useful margarine bases or, after fractionation, to non lauric based replacers.

The nearly full or full hydrogenation of lauric fats (mostly palm kernel, and to a lesser extent coconut oil) will lead to products with drastic solid fat content changes over a narrow temperature range, the steepness of the solid fat content/temperature (melting) plot being a necessary minimum quality requirement for a good cocoa butter replacer. The already discussed aspects (Chapter 5, Section III) of this technique as applied to fat mixtures, gave evidence that the combination of hydrogenation and interesterification leads to useful replacers.

Other possibilities are the partial/multi-step fractionation of hard vegetable butters, high palmitic fats, the laurics and partially hydrogenated fats, which is carried out be mechanical pressing, and dry, wet or solvent-aided fractionation techniques. In the section on fractional crystallization and winterization/dewaxing some possibilities were discussed. The step-wise execution of solvent-aided processes on palmoil or mixtures of it with the hard vegetable butters, leads to fractions which are equivalent to cocoa butter. The equivalence is valid not only in terms of their physical behaviour, but also their chemical constitution, as regards the high levels of symmetric SUS type triacylglycerols. These fractions are thus real cocoa butter and can be blended, without any (except legal) restrictions, with the much more expensive natural product.

The fractionation of selectively hardened vegetable oils by any of the above means can also deliver fat fractions which can be used as replacers in confectionery products.

Cocoa butter equivalents may of course also be produced chemically by reesterification of suitable monoglycerides with the relevant acids, protecting the free hydroxyl groups with suitable (hydrolysable) substituents. An alternative method, which might be quite promising in the future, is the enzymatic interesterification of separated POP fractions from palm oil with stearic acid or tristearine by 1,3 position specific microbial lipases (e.g. *Rhizopus* and *Candida* strains). The lipase enzymes are in the presence of only a scarce amount of water typical esterases and, therefore, under equilibrium conditions give the ester interchange on fats instead of just hydrolysing them. In one variant (Macrea, 1983) the reaction is done continuously at 40°C with, for example, 0.05–5% immobilized enzyme on the carrier Celite (Kieselguhr) impregnated with 8% water. Other methods (Novo Industri A/S) involve immobilizing the enzyme on macroporous ion-exchange resins with particle sizes of 200 μm, facilitating its use in solvent-free continuous systems. A two-step process (Asahi Denko Kogyo, 1984) involves 90% interesterification in aqueous medium and the other 10% interesterification in dry medium. After fractionation the interesterified product is significantly enriched in POS, the genuine symmetric triacylglycerol of cocoa butter.

The re-esterification of the medium length chain (C_8 to C_{10}) fatty acid fraction from (hydrogenated) lauric fats with glycerol gives the necessary fat basis in a special "diet margarine" for children with gastric disorders.

8

Miscellaneous Subjects

I. Storage and Transport of Oils and Fats

A. Inventory Control (Weighing)

The mass or volume quantities of oils entering or leaving a factory tank or storage installation should be measured. The measurements are often done by one of two means: by scales or by volumetric type measurements. The conventional method still widely used is rail- or tank-car weighing, which is done by measuring the brutto weight and tare of the individual vehicles received or dispatched. Oils (fats have to be melted, preferably by hot water because of the necessary pumping) obtained by ship can be weighed semi-continuously by a cascade of three tanks, positioned at different levels, the middle one being placed on or connected to the balance. While the oil is in the middle tank its weight is measured and recorded; meanwhile the top tank is newly filled and the bottom one is emptied.

The middle tank can be installed as a strain-gauge load cell. The strain gauge is a grid of very fine resistance wire attached to a piece of thin impregnated paper. Two pairs of such gauges firmly attached to steel billets or rings are placed below the container. During the deformation caused by the applied weight the gauges are deformed. By compression or elongation the electric resistance of the grids changes, this being transformed into current signals, if, for example, the gauges are members of a Wheatstone bridge. The signals are transferred to a weight recording or indicating meter. Pneumatic or hydraulic capsules developed for measuring pressure changes are also in use and give an accuracy of ± 0.05–0.1%.

Continuous weighing can be automated on a preset time of filling or a preset weight basis. The control actuates the closing/opening of the relevant valves on the load cell and the other tanks and/or starting/stopping pump motors. The system can be easily adapted for the mixing of ingredients or different-quality oils.

Transferring oils from or to different parts of the tank installation and to or from a central distribution point makes it necessary to apply volumetric (positive-displacement type) flow-meters (pumps or interlocking elements, each stroke or rotation being equivalent to a set volume) with due care being taken to exclude any solids (a basket filter) and the necessary deaeration being undertaken. Continuous density or temperature measurements are a necessity; an accuracy of $\pm 0.2\%$ is possible. Flow measurements by area meters (Rotameters) are free from the above-mentioned restrictions, having no moving parts and being insensitive to the necessary purging of lines with low-pressure gases (air) and steam.

Tank levels can be centrally monitored by the use of units which measure the hydrostatic head or liquid pressure, e.g. open dip-pipe/bubbler tube approach or by a diaphragm-type pressure transmitter, compensating for temperature and specific-gravity influences. However, the accuracy is not better than ± 0.25–0.5%. Float gauges, also mechanized, and manual dipping are further possible methods for obtaining the necessary data for an inventory control, a very important part of factory management. The use of well-calibrated tanks with a minimum of three fixed measurement points and an accurate temperature measurement (to convert volumes to mass weights) is still used as a reliable inventory control. The possible bulging of tanks should be carefully recognized and compensated for.

B. Internal Transport

The main internal transport is executed by a well centralized and marked pipeline system connecting the loading points, tanks and process equipment. Solid and semi-solid stocks, as well as different sorts of oils, are often stored at ambient temperature and so intermixing of liquid with (semi-)solid fats must be prevented. This is most efficiently done by a separate pipeline system. Liquid fish oils should not be pumped in the lines used for other oils. The lines should have a slope, with low points from where the residual oil can be collected, e.g. under vertical connecting pieces leading to lines serving tank domes or roof tanks.

Lines should be cleaned after transport of one type of oil, by blasts of compressed inert gases, air (if cold and/or not very unsaturated oils are pumped) or by steam. Mechanical scourers, called "mice", "pigs" or "go devils" (metallic brushes applied on a torpedo-shaped pressure-propelled moving body) are quite useful, if the pipe system has been designed with the necessary large radius bends and its internal diameter is above 50 mm. If steam is used the oil droplets must be removed upwards by small cyclones at the end of the lines where the steam emerges. Air should never be blown through the oil; inside the tank a vent hole at the top of the filling pipe should be situated.

The material of construction is usually mild steel, but care must be taken not to use valves or appliances of any type which contain copper (brass, bronze, aluminium bronze), which is one of the most potent pro-oxidants leading to rapid quality damage. In cases where high-quality demands are to be met, mostly with semi or fully-finished/refined oils, the lines should be made of stainless steel or aluminium. Lines for deodorized oils should preferably be made of stainless steel, in order to avoid the adverse influence of dissolved traces of iron on the keeping quality of many finished (liquid) oils. The pumps used are mainly the centrifugal or gear-type rotary ones.

C. Storage Tanks

The number, size and construction material of the storage tanks are amongst the most crucial points to be considered when planning the operation of an existing or future oil mill or oil-processing factory. The choice of these factors of course depends on the quality of the oil to be stored and what type (designs) of tanks are applied, crude oil having different requirements from intermediate or semi- and fully-finished products.

As regards the crude oils, the one-oil producing or processing mills need tanks only for pressed- and extracted oil storage, pre-deslimed oil storage being another option. Where many oils are to be stocked a series of larger and smaller tanks is required in order to decrease the danger of contaminating/mixing different oils which is a real hazard of tank yards and other storage facilities.

The normal crude tanks have some very common design features:

(i) the feed pipe should run from the top to the bottom, with a vent hole close to the top bend;
(ii) the tank floor should have a slope to a sump, with cleaning facilities (a drain line);
(iii) the take-off of oil should be 0.5 m above the bottom and at the side (sludge settling);
(iv) a heating coil (the line entering from the top of the tank), on the bottom with a vertical inverse-U shaped spin-off at one side of the tank (this also providing space for already melted material, when expanding), the heating being done by hot water;
(v) outside insulation, where needed;
(vi) side-entry mixing propellers for counteracting "layering" in large (500 ton) tanks;
(vii) a plinth protection of the outside foot of the tank against corroding water;
(viii) a bund wall (dike) to prevent oil spillage;
(ix) a welded construction;
(x) protection of mild steel constructions against fatty acids by plastic coatings, if necessary;
(xi) a high height to diameter ratio in order to protect the contents against oxidation by air; and
(xii) rectangular tanks are generally more expensive and more difficult to clean than are cylindrical ones.

Semi- or fully-finished oils are stored in tanks inside or outside the processing building. Important points for the protection of sensitive products against oxidative damage are keeping them under an inert-gas cover or under vacuum, both of which require pressure-resistant tank constructions. The inert-gas protection system, a costly measure, should have a central reservoir or container to which the gas from a filled container can be displaced. This saves costs on the very pure nitrogen (< 50 mg kg^{-1} oxygen) needed for this task. The material may differ depending on the sensitivity of the product to heavy-metal contamination from mild steel via aluminium to stainless steel or internal plastic coatings. Heating, cooling (by stainless steel coils), mixing, cleaning, filling, emptying, venting and metering devices are used according to the needs of the process undertaken. Very careful numbering (identification) of the tanks and marking (colour coding) of the pipelines is an absolute necessity.

No copper (brass, bronze, aluminium bronze) should be used in any of the appliances, either for tanks for the oil, or for pipelines used for steam, vacuum or water, for the aforementioned reasons.

For high-quality demands the tank material used for semi- of fully-finished/refined products is stainless steel or aluminium, whereas for most liquid deodorized oils the use of mild steel is discouraged.

D. Storage Conditions

The real problem of any oil storage operation is the quality deterioration caused by the many chemical and physical influences to which an oil is prone. The dangers are different in different processing stages and so the measures to be taken will also vary. If deposition of "gums" or separation of liquid and solid fractions during storage are the physical problems, it must be ensured that the oil is timely pre-degummed, or that moisture is excluded and the oil is heated/mixed. If the problem is the absence of certain minor constituents, e.g. of free tocopherols in certain crude or in interesterified oils, then there is a chemical problem which may be solved by adding extra antioxidants or by excluding air in the headspace of the tank. Therefore, all decisions must be adapted to the given situation. For this reason some general aspects are discussed first and then some special cases, including external transport are given.

1. *Moisture*

Depending on the amounts of free fatty acids and accompanying constituents and the temperature, oils can dissolve different amounts of moisture. The equilibrium relative humidity of oil determines how far the oil is saturated with water at a given temperature. The solubility of water in an oil varies between 0.1 and 0.5%, but some fine water droplets may possibly also be dispersed in addition to the physically dissolved water. Water is the reaction partner or agent and free fatty acids are the catalysts of the hydrolysis of acyl- and phosphoacyl-glycerols. Hydrolysis is considered to be a kinetically zero-order reaction.

$$\frac{da}{dt} = ka$$

where a is the free fatty acid content (%), t is the number of 10-day periods, and k is the hydrolysis constant which is dependent on the temperature ($k =$ 0.025 at 37°C and doubles for approximately every 10°C of temperature increase; the oil being saturated with water at a given temperature). Only the dissolved water is counted in this equation. The lower the equilibrium relative humidity, the lower is the value of k.

In general crude oils with low (0.1%) moisture and no high hydratable phosphoacylglycerol levels stored at low temperatures keep better.

2. Light

Light catalyses autoxidation of the oils forming, in the presence of "sensitizing" agents such as chlorophyll, singlet state oxygen. This type of oxygen has a lower activation energy when reacting with unsaturated fatty acids than does the normal triplet state of oxygen of the surrounding atmosphere. Even without the presence of chlorophyll, light acts on the α-methylenic groups of the unsaturated fatty acids to produce via the initiation step of autoxidation, the alkyl free radicals which are the initiators of the autoxidation chain reaction.

The presence of radical quenchers, such as carotenoids, counteracts the formation of light-induced free radicals: red palm oil is excellently protected by nature against the radiation of tropical sunshine by the large amounts of carotenoids which give it its orange-red colour.

The protection of oils against light, mainly of wavelengths below 520 nm, is a part of good housekeeping. The total exclusion of light is the best measure; this applies mainly to finished products the storage tanks for which generally have no access to light.

3. Heat

An important practical question is at what maximum temperature can an oil (fat) be kept for relatively long periods without the deteriorating action of heat-induced autoxidation, polymerization and heat-accelerated hydrolysis? The chemical reaction rate doubling action for each 10°C increase in temperature should be kept in mind, this being relevant not only for hydrolysis, but also for autoxidation. More-solid fats, partially hardened oils and certain liquid fats have a tendency for phase separation (like stearines in groundnut oil and waxes in sunflower-seed, maize and rice-bran oils) below a certain storage temperature. A partly or totally solidified (crystallized) fat phase may include more oxygen (air) diffused between the interstices of the crystals, than the normal solubility of oxygen would allow.

If such a fat is (quickly) heated up this air will then react preferentially with the more unsaturated liquid phase loosely wetting the solid crystals. Therefore an oil which is kept liquid in its totality for not excessively long periods of time at temperatures not higher than 10°C above its melting point might suffer less oxidative damage than would an inhomogeneously solidified oil. This behaviour prescribes some insulation of tanks. As a general remark, frequent and extreme heating up and cooling down is not only uneconomical, but is often deleterious to the quality. Unsaturated (crude, pre-deslimed) liquid oils should always be kept as cool as possible when being stored for a relatively long time, this being the best measure for minimizing spoilage of any kind. Condensation of moisture out of the surrounding atmosphere should be reduced by minimizing the (moist) head space. It may be recalled that there is the additional danger of sludge formation by spontaneous hydratation of the phosphoacylglycerols.

4. *Air*

The flavour quality of oils in general, and that of some unsaturated and hardened oils in particular, depends on the oxidative history of the raw material, the storage history being as important as the processing one. Exclusion, or at least reduction, of any undue access of air to oils in nearly all phases of its life is one of the most enervating problems of the quality-conscious processor. This is not only a huge cost factor but also a difficult one when trying to trace back its real effect on the selling figures. As we have seen previously, local taste patterns will primarily determine how much protection is really needed in order to suppress or reduce the rate of off-flavour formation. This will depend highly on the types of fat or composition of the production to be packed and stored.

The problem of oils which are prone to flavour reversion can be quite easily illustrated by a simple calculation. The actual solubility of oxygen in oil is about $22 \, \text{ml} \, \text{kg}^{-1}$ at ambient temperature, being the equivalent of about $1 \, \text{mmol} \, \text{kg}^{-1}$. If this amount of oxygen reacts with, for example some linolenic acid of the oil in question, we obtain theoretically $2 \, \text{meq} \, \text{kg}^{-1}$ of hydroperoxides or a peroxide value of 2. The hydroperoxides are flavourless, as we have seen before. If only 0.1% of these $1 \, \text{mmol} \, \text{kg}^{-1}$ hydroperoxides decompose for any reason to, for example, the aldehyde 3-*cis*-hexenal (MW = 98) with a flavour threshold value of $0.1 \, \text{mg} \, \text{kg}^{-1}$, we obtain an (off-)flavour called "green" (Hoffmann, 1961). This oil may in terms of flavour be considered off-quality if the local preferences do not consider this type of flavour acceptable.

This example shows that the exclusion of oxygen by a vacuum or inert-gas blanketing is only effective if an oxygen level much below this threshold level can be effected, not only during processing and storage, but much more significantly during the shelf-life of the finished product. Obviously, it is impossible to keep products out of air contact after the package has been opened. The quick consumption of a once opened and appropriately sized unit is one of the best remedies, a very welcome solution also from the producers' point of view.

Fortunately most oils that do not contain linolenic, certain isolinoleic and more than triunsaturated fatty acids, do not produce the decomposition aldehydes which are sometimes considered as off-flavours, if the above-mentioned low, or sometimes even lower, threshold oxygen levels are maintained. Therefore, the blanketing of oils with pure nitrogen (or under high vacuum) is most profitable if it is applied to oils which are the least protected by antioxidants (interesterified intermediates), i.e. the ones most sensitive to pro-oxidants (bleached oils) and finished to be sold or incorporated in other products (deodorized oils). The problem of less unsaturated oils and fats is much more a question of protection by added antioxidants (to animal fats and oils), or by the measures already mentioned as good housekeeping during storage and transport.

Oxygen exclusion during and after processing is thus a very problematic and costly operation. The real advantages attained by it for any taste-sensitive product has to be explored by consumer studies based on careful taste panelling, covering the sectors where the product should be successfully marketed. The decision taken will be, as always, based on the balance of costs and the real benefits achieved.

5. *Contaminations*

(a) *Metals*. The pro-oxidant action of heavy- or transition-metal traces, such as copper, iron and nickel, introduced mainly during the processing and storage of oil products from outside (nickel is introduced on purpose as a hardening catalyst) is deleterious above certain levels. The practical limits are: iron <0.1 mg kg^{-1} copper <0.02 mg kg^{-1} in liquid refined oils; nickel <0.1 mg kg^{-1} in hardened oils; and iron, exceptionally, accepted in less unsaturated fats if <0.5 mg kg^{-1}. The contaminations are not always exogenous, as plant and animal tissues contain trace amounts of iron and very low levels of other metals, e.g. as parts of enzymes or other vital constituents. The exclusion of copper, which is the most injurious metal for oil and fat shelf-life, has been stressed amply before. Any measure to exclude and/or reduce the levels of these heavy and other metals to the accepted values diminishes the dangers of quick autoxidation, even in the presence of oxygen. These measures are: the use of suitable materials of construction, the use of refining reagents and additives free of heavy metals [such as bleaching earth which is low in iron, and even in this case the iron is effectively removed (filtered out) after use], and the use of further metal sequesterants (citric acid or similar agents). A very effective final measure is the exclusion of any additives or non-fat components (e.g. flour) in processed fatty products which increase the heavy metal levels to above the limits which were given previously for the refined oils and fats.

(b) *Intermixing with other oils and fats*. The dangers of intermixing liquid oils and solid fats, oils of different variety or quality, (liquid) fish or sea mammal oils and, at least as important, with mineral oils of any kind, should be stressed. The good technical and administrative maintenance of the tank park and all the storage facilities is of utmost importance. The different synthetic heat-transfer media (Dowtherm or similar) often used in the past for high-temperature treatments (deodorization, distillation, etc.) were very often the cause of contaminations, generally caused by cracks of the heating coils inside the kettles; their use is generally discouraged.

(c) *Foreign matter (cleaning)*. The presence of any solid foreign matter should be excluded by preliminary filtration of the oil to be stored, where relevant. If filtering is not possible for any reason, then the deposition of solid sediments in a smaller tank can be temporarily avoided by a mild, but constant stirring of the contents. The temporary descaling of tanks by wire brushes, the removal of sludges (foots) or any other sediment from the tanks by specialist measures such as wetting agents, emulsion breakers and hot water, and the final cleaning of refined-oil tanks by filling them with, and circulating vegetable cleaning oil, help prevent mechanical contamination of the stored oil. A coating of (polymerized) oil allowed to build up on mild-steel tanks is also a good protection for the semi-finished (refined) oils kept in them. The film is only removed if it grows too thick or cracks.

The cleaning of tanks is a dangerous operation, and no person should remain inside any tank without close contact to persons outside. All measures for good ventilation and for excluding fire hazards should be taken before any entry to the tanks.

E. Special Storage Cases

1. *Crude Oils*

Crude oils should preferably be stored pre-degummed, as this lowers the chances of hydrolysis and sedimentation. Autoxidative damage is less prevalent with liquid vegetable oils, which are well protected by natural antioxidants. Palm-kernel and coconut oils, although not rich in antioxidants, can be kept without addition of synthetic (nature-identical or other) phenolic antioxidants, because of their low level of unsaturation. For palm-kernel and coconut oils the exclusion of moisture is more important, as their hydrolysis leads to the preferential liberation of lauric and myristic acids, thereby increasing the refining losses. The diacylglycerols formed can be further hydrolysed in the finished state to monoacylglycerols. The acids set free during this process cause, above a certain level, soapiness, the well-known flavour defect of lauric oils. Crude animal fats and fish and sea mammal oils do not contain any natural antioxidants and, therefore, if they are to be kept unprocessed for any reason (transport, etc.) they should be protected by added (and legally admitted) phenolic type antioxidants and kept in closed tanks with no or only small headspace.

2. Neutralized Oils

Although alkali neutralization removes mainly the free fatty acids, an appreciable amount of lipid substances which contain polar groups e.g. hydroxylic and carbonylic hydroperoxide decomposition products, dimers and polymers, steroids, tocopherols, waxes, etc) and the majority of the residual (hydratable or non-hydratable) phospholipids are also carried away with the "soap stock". Therefore, neutralized oils, if not washed, will contain, besides the normal accompanying components, emulsified aqueous soap remainders, and even if washed and dried, some traces of soap (batch neutralized around 300–500 mg kg^{-1}, centrifuge neutralized about 50–100 mg kg^{-1}). Experience teaches that the presence of traces of soap is advantageous in protecting oils against autoxidation, although it may promote their hydrolysis.

3. Interesterified Oils

Because of the reduction of the active natural antioxidants during the alkali-catalysed interesterification of neutralized oils (their phenolic hydroxyl groups are also esterified up to 90%), these oils are completely unprotected against autoxidation after the elimination of the catalyst, washing and drying. Addition of a phenolic antioxidant (e.g. synthetic tocopherol or something similar) before drying the intermediates is a necessity, if liquid oils were also included in the process. Exclusion of air contact also seems to be a good measure at this stage. Quick reprocessing is, in any case, the second-best measure after antioxidant addition.

4. Fractionated Oils

In general, the polar antioxidants accumulate in the liquid fraction (if the fat originally contained any), and so the more saturated fraction becomes the less protected one. Trace heavy-metal salts (soaps) also accumulate in this fraction and so it is doubly endangered despite its more saturated character. Therefore it can be a good housekeeping measure to add synthetic antioxidants to the fraction(s) which do not contain any natural ones. The "solids" should preferably be kept melted (10°C above their melting point) for only limited periods, or filled in drums leaving very little headspace. Aeration of the solids (stearines) in the liquid state before filling with nitrogen gas is a sensible, but sometimes expensive, measure.

5. Hardened Oils

Hardening reduces the unsaturation of the oil and does not reduce its antioxidant content. In some cases the antioxidant level may even slightly increase, because hydrogen regenerates some antioxidant dimers. In this light, one might think that hardened oils do not need any special protective care. However, in some partially-hardened oils, like soya-bean, rape-seed and fish oils, certain iso-linoleic acids are produced during the process (see Chapter 5, Section I). On autoxidation the hydroperoxides may decompose to specific aldehydes which have offensive off-flavour character and extremely low flavour-threshold values. Therefore, protection against oxidation by keeping these very sensitive hardened oils only for short periods in well-filled closed tanks with little headspace is a sensible measure. Even the application of an iso-suppressive hydrogenation technique which reduces the level of the iso-linoleic acid precursors does not invalidate the necessity of above protective measures.

6. *Bleached oils*

Some adsorbents, like acid or steam-activated bleaching earths, decompose the residual amounts of hydroperoxides which form during alkali neutralization and are still un-decomposed at the end of it. (Oils react with dissolved oxygen radicals through the ionizing action of the alkali, probably also temporarily deactivating a part of the free tocopherols.) Despite the removal of many non-volatile decomposition products (polar glycerides and the remaining non-hydratable phospholipids) by the earth, the oil remains sensitive towards oxidation at this stage. This increase in sensitivity can probably be explained by the residual bleaching earth fines (10–100 mg kg^{-1} is quite normal, depending on the filtration efficiency) serving as (ionizing) free-radical promotors which initiate autoxidation. Bleaching earths contain quite appreciable amounts of iron, partly in ionic form, because of the ion-exchanging nature of the earth. Storing bleached, flavour-sensitive oils for longer periods (more than 6 h) in unprotected conditions with access to air invites postponed off-flavour problems with the finished, deodorized oils.

7. *Deodorized Oils*

Deodorized oils are devoid of any hydroperoxides, most of the fatty acids, phosphoacylglycerols, moisture and off-flavour volatiles, and so in order to preserve them in good (flavour) quality after cooling they should be kept in closed storage tanks, under vacuum or a high-quality inert-gas blanket. Fats or fat blends should not be left to solidify until either packed in containers or pumped into the margarine/spreads producing departments. The maximum period for which they should be kept unpacked/unprocessed seems to be 12 h, although a period of 24 h is sometimes inevitable.

F. External Transport

It is obvious that all the factors which adversely influence the quality of oils at any stage of processing and storage are also applicable to the problems involved with the transport of the different qualities of oils.

1. *Transport by Water*

The most frequently studied problems are those connected with the sea transport of palm oils. This is not only because of the volume of this commodity (the volume of soya-bean oil exported is very large), but also because of the special Malaysian regulations protecting the local refining industry. The levies on crude Malaysian palm oils mean that this oil is very often exported in the semi- or fully-refined state, with all the consequences of the long-term unprotected storage of (not even very unsaturated) oils (fats). The transfer periods are 10–50 days at sea, generally in mild-steel ship tanks with sometimes unprotected headspace. The oil is not always kept at the IASC recommended maximum loading, transit and discharge temperatures for palm oil, this being 32–35°C. For this reason the heating logs of the ship should be more thoroughly controlled. Nearly all other oils are transported by sea in the crude (pre-degummed) state, which causes less transport damage than occurs in the mainly oxidative/hydrolytic changes in the semi (fully-)refined palm oils. This damage often forces the receiving party to repeat at least the final steps of the refining operation. The only advantage of obtaining the oil at this stage of processing is the relief from handling the soapstocks or acidic deodorizer/distillation condensates, thus avoiding surface-water pollution in the receiving country.

River and other inland water transport in barges causes, because of the shorter distances, fewer, but similar problems. All the general aspects of storage, construction materials of tanks, pipelines, pumps, appliances and cleaning of installations are also applicable to all types of water transport.

2. *Transport by Rail*

All the aspects already mentioned above are relevant to the transport of oils by rail, except that heating is not required for transport times of less than 48 h in normal-size special tank cars (USA 65 ton). These transport times do not need intermittent heating, if the loading temperatures are strictly applied. Of course internal heating coils, to be connected preferentially to warm-water mains, are a necessity. Filling the tanks nearly to the top with little headspace is advantageous for decreasing damage caused by oxygen.

3. *Transport by Road*

Road transport is mainly applied for the transport of 20–30 ton amounts of fully refined, deodorized oils from the refinery site to a margarine or other fat-product manufacturing unit. The tank material of construction should be either aluminium or stainless steel. All other precautions mentioned with rail transport are relevant for road transport also. The absolute cleanliness of the tank is imperative. The truck should have its own pump installation for rapid emptying of the load. No other products (e.g. chemicals) should be transported in these tanks, except products for edible use.

The recommendations of the International Association of Seed Crushers (IASC) for the loading, transit and discharge temperatures during the bulk shipment of oils (fats) are given in Table 8.1. The oils should not be heated at a greater rate than that corresponding to a rise of 5°C day^{-1}. Heating should be executed preferably with hot water, but steam up to 1.5 bar is

Table 8.1 *Bulk shipment temperature (°C) recommendations of the IASC[a]*

Oil	Maximum loading	Transit	Discharge
Cocoa	41–46	46	27–32
Palm kernel	43–48	46	27–32
Lard, greases	51–54	54	38–41
Fish oils	30–35	35	20–25
Palm oil	49–55	55	32–35
Tallow	55–60	60	44–49
Whale	30–32	32	20–25
All liquid oils	At ambient temperatures, but not below 15		

[a] Reproduced, with permission, from IASC (1980).

acceptable. The log should include a record of temperature, how the weight of the cargo was calculated originally, and the locations of the points where samples should be drawn.

II. Energy Demands of the Oil-milling and Edible-fat Processing Operations

Mainly because of the huge volumes of seeds/fruits and oils/fats processed in the oil-milling industry and in the refining business, in the past the proportion of the labour, energy, amortization, overheads, etc. costs compared to the price of the raw materials were quite modest. This proportion was somewhat more significant in hydrogenation, fractionation, interesterification and edible-fat production (margarines, shortenings), speciality fats being an exception. Besides rising wages and social burdens the two oil crises in the recent past made the generation and use of energy an important cost factor. In any oil mill and/or edible-fat processing plant, energy management has become a major issue, covering, besides availability, safety, reliability and the economic aspects.

The two main forms of energy needed are heat energy (its carrier being mainly steam, notwithstanding the use of some synthetic heat-transfer fluids) and electrical energy. Steam can be generated in boiler houses, using liquid, solid or gaseous fossil fuels. In a few cases side-products of the oil milling, such as the husks of sunflower seeds and cotton seeds or empty bunches and nut shells of the palm and palm-kernel oil production, can be burnt. Electrical energy is generally fully (exceptionally partially) provided by the commercial (municipal) electricity networks. Internal power plants such as steam turbines or steam engines and gas-driven turbines, all coupled with generators are, in certain cases (see below), also possible methods of (smaller scale) electricity production. Diesel engines coupled with generators for raising electric power are often applied as reserve means.

A. Steam Generation

The temperature requirements of the technologies applied in a given factory determine the pressure ranges of steam to be generated in the boiler houses. Of course the maximum temperature to be achieved during a certain process and the amount of steam needed to provide it will be decisive factors. In general there are four pressure ranges of steam which might be required in different operations. These are the very low pressure (0.2–1 bar), low pressure (2–5 bar), medium pressure (up to 10–20 bar) and high pressure (up to 35–50 bar) ranges. The needs of oil mills with or without extraction are covered by the very low and low pressure ranges, edible-oil processing factories (older refineries and hydrogenation plants) will work with medium-pressure steam and only the most modern refineries require the high-pressure ranges. The common temperature ranges are thus 60–100, 120–155, up to 180–210 and up to 240–265°C. The industry generally uses saturated dry steam. Slight overheating of the steam (say by 30°C) can help to cope with losses during transport, if necessary. Injection of (condensed) water into the pipes (resaturation) just before reaching the consumer will restore the decreased specific heat of the superheated steam to that of the saturated one, this resulting in better heat transfer.

Medium- to large-sized factories like oil mills combined with refinery and hardening facilities, may consume 25–50 ton h^{-1}, or even more, steam.

There are different types of boilers which supply steam in different amounts and pressure ranges, the main difference being the method of firing (burners, nozzles, stokers) and the method of contacting the combustion gases with the water to be boiled. The furnaces serving the boilers should be able to use at least two different types of fuel. The boilers, which generate high pressure (up to 40–50 bar) and high temperature steam (up to 265°C, without superheating) are the so-called water-tube ones. Here the water circulates inside the (bent) tubes and the hot fuel gases pass along the outside tubes through baffled sections. In general not only steam turbines, but also the so-called back-pressure steam engines require a water-tube boiler, if working above 15–20 bar. Water-tube boilers are expensive, but generate the steam needed rapidly. The efficiency of water-tube boilers is about 80% (the best ones being rated at 90%). With only low and medium pressure steam needs (up to 10–20 bar) fire-tube boilers are sufficient. Fire-tube boilers burn the fuel inside or before the kettle provided with one or more internal fuel-gas tubes, the water residing around these tubes. Their efficiency is lower (70%) and the generation of steam is rather slow, but they are cheap and easy in use.

The boiler(s), for safety reasons more than one may be installed, should have overcapacity in order to be able to meet the demands during peak load. Below optimal load fire-tube boilers work with low efficiency, resulting in higher unit costs for the steam. For this reason it was proposed that 3–4 smaller fire tube boilers be installed in parallel, which can be run in an automated modular mode. At low load only one boiler is in operation with the others in stand-by mode. The necessary extra number of units are fired automatically if the load increases. This system, although requiring a lot of extra investment, should have low unit costs and maintenance.

B. Feed-water Pretreatment

The water needed for steam generation is obtained from the municipal mains, from surface waters, (after filtration/purification, etc.), or regained as condensate from heating equipment and the water cooling towers. Even at (low to) medium pressures the fresh and recovered water must be treated to render it free of hardness (softening by chemical means or by ion exchangers), oil and any other organic impurity. Freedom of dissolved gases, such as oxygen (corrosion) and nitrogen, can be achieved by, for example, cascade degasers placed on top of the pressurized feed-water tanks or on the boilers.

C. Heat-loss Prevention

The maxim of good energy management is to re-use energy instead of wasting it. To this end all heat losses to the environment (air and water), raising its temperature unneccesarily, should be avoided. The general means of reducing heat loss are: (i) to produce the highest temperature differences (pressure differentials) between the steam raised and ultimately condensed; (ii) to reduce steam pressure losses by choking through dimensioning pipes and heating surfaces appropriately; (iii) to avoid the unnecessary mixing of cooling waters and steams of different temperatures; (iv) good heat insulation of tanks and other apparatus, where needed; and (v) the rigorous prevention of the loss of steam by leakage.

Because of the relatively large area covered by an oil mill and/or edible-oil factory, the distances to be served by the steam or condensate pipelines may be long. For this reason there will be some energy losses (10%), even with pipe insulation of optimal thickness. The slight superheating of steam, as a means for reducing losses, has already been mentioned. In cold-climate regions the connection of the pipelines to low voltage (10–15 V) electrical-resistance heating in the insulation (e.g. by the use of the resistance of two parallel pipelines suspended on electrical isolators and short-circuited at one end) can also lead to steam savings. Where handling condensates, the use of steam traps (e.g. the Armstrong inverted bucket type made of stainless steel, fully welded) can help reduce steam losses. Even the latent heat of the hot condensed water leaving the steam traps of the high pressure heating coils (10–15 bar) can be utilized for low-pressure heating equipment after separation of the water from the steam.

The possible re-use of otherwise wasted heat-energy carriers, such as burnt-fuel gases, reduced-pressure steam, hot air, hot water, hot processed materials to be cooled, etc. by suitable and well-dimensioned heat exchangers, such as regenerators, recuperators, feed water economizers, etc., should be remembered. The main aim is to re-use all these heat-energy carriers at the highest possible temperatures, as this has the strongest influence on the cost savings. Some indications of how practical energy savings can be made by contacting hot/cold oily material processed at the highest possible temperature differentials by the use of appropriate heat exchangers have already been given (e.g. in Chapter 2, Section III.B; Chapter 4, Section V; Chapter 5, Sections I and II). The heat exchange of very hot oil with softened water for cooling and, thereafter, for secondary steam generation, or that of hot oil with softened water for cooling and, thereafter, use of the hot water for boiler feeding are other possibilities. An economic use of energy prescribes the thorough and regular measurement and registration of all those parameters which make better energy management possible. This goal requires not only accurate instruments mounted at the right places and in the necessary number, but also people who can interpret the data for the benefit of the total processing operation.

However, one should never forget to apply common-sense principles when attempting to save energy, as besides investments the run-on costs and operating difficulties could cancel out any benefits arising from the original idea (Marchand, 1985).

D. Internal Power Plant

The steam generated in the oil and fat industries is primarily and directly used as a transfer medium for heating materials in the different processes. However, it should not be forgotten that the thermal energy of the steam generated can be economically utilized as "useful work" for the production of mechanical/electrical energy first, and then the heat of condensation remaining in it used for heating purposes thereafter.

This option can be realized in factories where a not too high proportion of the high to very high pressure steam generated is used for high-temperature (180–265°C) heating. In other factories steam may be generated in boilers at 15–25 bar high pressure, again only a smaller part of it being used for heating at medium to high temperatures. In both cases the bulk of the steam can be directed into a non-condensing type of steam turbine with 20–50 bar initial (upstream) pressure. The steam leaving the turbine at 2–8 bar back-pressure could be used for heating at low to medium temperatures. So, for example, a lower pressure turbine, using steam of 26 bar upstream pressure, superheated to 360°C and releasing it at 4 bar back-pressure would need 7–8 kg of steam to produce 1 kW of electrical energy. In case of, for example, 25 ton h^{-1} steam production, 3000 kW electricity could be generated. Of course, depending on the size and construction of the plant, much higher or lower figures of steam consumption have also been reported. A warning may be the incomplete use of the back-pressure steam for heating, this causing a serious relapse in the calculated economy.

The price of energy will differ in different countries, and even within one country, depending on the availability of fuel and/or hydrological resources and the methods of generation used. Where commercial networks can deliver all the electrical energy required for the operation, there is little need for the internal co-generation of electrical energy with steam. In the vicinity of large power plants, even the use of externally generated steam is a realistic proposition. If the already mentioned side products of oil milling are produced in abundance, or there are local electricity distribution problems, the erection of an internal power plant should be considered more closely.

E. Cooling

The medium to cool condensed or process water is generally air, the natural circulation of it sometimes being enhanced by the use of fans in the cooling towers. The indirect cooling of barometric water via heat exchangers with water cooled in towers by air reduces the blinding, clogging and cleaning of its elements.

Artificial cooling (as required for fractionation and mainly for margarine and shortening production) is achieved primarily by the mechanical compression of ammonia gas, evaporating the liquefied ammonia subsequently in the relevant chilling tubes or elements. The re-use (heat exchange) of cooling water, warmed up in the condensers of the compressors for heating, is an ideal way of saving energy (a heat pump) if there is a use for it, e.g. room heating or cooking in the factory canteen. In tropical countries the cooling of the recirculated water used for the steam-jet ejectors by refrigeration units will lower the steam and water consumption, but will increase the electrical energy costs. If the last stage of the steam-jet unit is replaced by a water-ring vacuum pump, even the water and power consumption can be further reduced.

Other general means of energy savings will be found by avoiding unnecessary lifting of solids and liquids, by designing transfer lines of the correct dimensions, by mounting the correctly dimensioned motors for solid, liquid and gas transfer, mixing, crushing, etc., and, ideally, running the whole installation continuously and, if possible, at design capacity.

F. Some Indicative Data on the Energy Needs of Processing

The energy needs of different types of equipment or technological processes cannot be given exactly, as they differ according to the type of items used and the processes applied. However, as a guide to the level of energy requirements some figures from different sources have been compiled and are given below.

1. *Transportation*

The main source of the energy consumption of the different means of seed transportation (mechanical and pneumatic conveyors) is the power consumption of the installation itself, only a fraction of the total power being needed for the transportation of the material. The only way to use any installation efficiently is to apply a uniform and constant (not intermittent) load (continuous operation). If the load varies, overdimensioned sizes (volumes) are required, which means extra investment costs.

8. MISCELLANEOUS SUBJECTS

In the following tables (Tables 8.2–8.5) 1 ton of normal steam at 100°C and 100 kPa (1 bar) is taken as being equivalent to 555 kW h or 2000 MJ.

2. Seed-oil Milling

The energy needs of the various stages involved in seed-oil milling are given in Table 8.2.

Table 8.2 *Energy needs of seed-oil milling*

	Steam (ton ton^{-1})	Electricity kW h	(MJ ton^{-1})
Cleaning of seeds			
(sieves, sifters, aspirators, etc.)		0.4–3	(1.4–11)
Seed drying (steam of 2–3 bar)	0.06–0.1	4–6	(15–21)
Front-end dehulling of soya beans		12	(43)
Decortication of sunflower seeds			
(on non-decorticated seeds)		5–10	(18–36)
Size reduction of seeds			
Crushing of soya beans into 4–8			
parts on fluted rollers		2–3	(7.2–11)
Fine crushing on rollers		6–10	(22–36)
Flaking on smooth-surface rollers		7–12	(25–43)
Cooking (steam of 3 bar)	0.07–0.18	4–5	(15–18)
Conditioning	0.04–0.9	2–2.5	(7.5–9)
Hydraulic pressing		2–3	(7.2–11)
Expeller pre-pressing		20–35	(72–126)
Expeller full pressing		50–70	(180–252)
Cake breaking		1–3	(3.6–11)
Batch extraction	0.4–0.6	6–15	(21–54)
Compactors/extruders	0.01	8–10	(29–36)
Meal grinding		3	(11)
Meal drying by hot/cold air	0.06–0.1	4–9	(14–32)
Meal cooling in fluid bed		8–10	(29–36)
Tail-end dehulling of soya beans		2–9	(7.2–32)
Continuous extractors, total			
(steam of 2–3 bar)	0.3–0.4	25–45	(90–162)
More detailed data for continuous extractors			
Seed conditioning	0.06	0.3–0.7	(1–2.5)
Hexane preheater	0.008		
Extraction	0.02		
Desolventizing/toasting	0.15	2–3	(7.2–11)
Distillation	0.025–0.085	1.5	(5.4)
Condensers/cooling tower		2.5–6	(9–22)
Tube desolventizer	0.15	1	(3.6)
Mineral-oil system	0.02	0.4	(1.4)

Table 8.3 *Energy needs of oil processing*

	Steam (ton ton^{-1})	Electricity kW h	(MJ ton^{-1})
Pre-desliming or total desliming (mixing, filtering or centrifuging)	0.02–0.1	0.5–4.5	(1.8–17)
Batch alkali neutralization	0.15–0.25	4–8	(16–21)
Batch bleaching	0.1	4–8	(16–21)
Continuous alkali refining	0.2	10–12	(36–40)
Heat bleaching of palm oil	0.45	42	(150)
Batch deodorization, depending on final temperatures			
Live stripping steam	0.09		
Booster (vacuum set)	0.15		
Medium-pressure steam heating	0.15–0.18		
Semicontinuous deodorization			
Live stripping steam	0.03–0.05		
Booster (jet) steam	0.07–0.25		
Very high temperature steam heating			
Without heat exchange	0.2–0.3		
With heat exchange	0.1–0.15		
Physical refining/deacidification (oil with 2.5–8% free acidity)	0.5–0.6	10	(36)

3. Palm-oil Milling

Medium-sized palm-oil mills use 0.25 ton of steam (3 bar) for sterilization and for the rest of the processing use 0.25 ton steam of 3 bar and about 22 kW h (80 MJ) electrical energy for each 1 ton of oil produced. The boilers used are suitable for burning shells and fibres, the wastes of the production.

4. Animal-fat Rendering

The different continuous processes for animal-fat rendering consume about 0.1–0.17 ton of steam (3 bar) and about 16–22 kW h (48–80 MJ) of electrical energy for each 1 ton of fat produced.

5. Oil Processing

The energy needs of the various stages of oil processing are given in Table 8.3. Continuous systems may save somewhat on live and jet steam. The primary cooling water needs without heat exchange would be 10–50 tons per ton of oil.

8. MISCELLANEOUS SUBJECTS

Table 8.4 *Energy needs of palm-oil fractionation*[a]

Fractionation method	Steam (ton ton^{-1})	Electricity kW h	(MJ ton^{-1})
Dry	0.03–0.1	6–10	(21–36)
Lanza	0.23	100	(360)
Tirtiaux	0.08	22	(80)
Lipofrac (Alfa Laval)	0.1	27	(97)
Solvent aided (hexane)	0.6	100	(360)

[a] All data at 30 000–35 000 ton of annual production and 65–75% yield on oleine, the yield on oleine and its iodine value being crucial for the costs, because of the limited marketability of the stearine fraction.

Table 8.5 *Energy needs of margarine production*

	Electricity kW h	(MJ ton^{-1})
Precrystallization (e.g. in 2 A-units)	11	(40)
Crystallization (in shortening B- or C-units)	4.5	(16)
Post-cooling	11	(40)
Proportioning pump unit	4.5	(40)
Emulsion or retour pumps, each	1.5	(5.5)
Ammonia pumps (2), each	0.4	(1.5)

6. Miscellaneous

Electrolytic hydrogen production for hydrogenation needs 300–400 kW h per ton of oil, of which the starting iodine value has been reduced by approximately 70 units. The heating up of the oil batch to 150–180°C needs 0.15–0.2 ton of medium-pressure steam and the mechanical (compression, mixing, pumping) energy amounts to 3–5 kW h per ton of oil. The use of dropping tanks, coil, pipe or spiral blade heat exchangers highly reduces the heating energy needs.

The energy requirements for the fractionation and subsequent separation of palm-oil fractions are given in Table 8.4.

Interesterification, in continuous execution (6–8 ton h^{-1}) requires with, for example, metallic sodium as catalyst, 15 kW h of electricity and 0.1 ton of steam per 1 ton of input material.

Margarine (and shortening) processing requires, at 1.5–2 tons h^{-1} capacity in a Votator-type system, a total cooling surface of 1.5 m^2. The details of the energy needs of margarine processing are given in Table 8.5.

The volume of recirculated ammonia is about 2 m^3 h^{-1}. The mechanical energy for the compressors is 10–12 kW h (26–43 MJ).

III. Automation

Many process-variable data, such as temperatures, pressures, flow rates, weights, volumes and motor loads, have to be measured on a single processing unit. These variables are analogue in nature, as their values can be expressed by voltage or current changes. Data on the state of the equipment, such as the start/stop of motors, open/close of valves, can be easily registered. These variables are of discrete nature. All these data can be first collected, registered and evaluated by the operator of the unit. He must then compare all these data with the set values and states prescribed for this part of the process and also using his epxerience, he must then take any appropriate correcting action(s).

Instead of the operator many process variables/parameters can be controlled by automatic devices. A generalized automatic control sytem "in the field" consists basically of a (e.g. temperature or flow) measuring device, the sensor and a final regulating/controlling element, e.g. a valve. The analogue or digital signals emitted by the sensor are transmitted to and registered by the controller. This device compares the deviation of the process signals (the measured variables) from the set point, i.e. the required value. Based on this deviation a compensating steering signal is calculated by the device. The steering signal is fed back as the control signal via mechanical/pneumatic or electric/electronic transmitters to the field, in order to actuate the final control element. This is called a feedback control loop.

All the data collected/registered by the controller can be visualized on a control panel. The manual/automatic switching hardware for the final control element is also located on this panel.

Not only variables, but also more complex operations can be regulated automatically by electronic plugged-in programmable logic controllers (PLC) e.g.: the transfer, the weighing and/or the mixing/blending of different ingredients, a logical sequencing and product routing operation. Analogue and digital process data and parameters representing the state of a process operation (normal, steady state, abnormal), safety measures secured by built-in interlocks, limiting/restarting switches and hazard sensors (fire, explosion, overflow, etc.) can all be linked to different PLCs, which are devices half-way between a microcomputer and a simple electro/mechanical/pneumatic control.

In modern factories many of the analogue and digital control data kept on the same unit can be communicated to and collected on a control panel and visualized on mimic displays, and the necessary commands can be given from the same control panel either manually or automatically.

A dedicated process microcomputer covering some or all the basic aspects of process control can run a unit operation. It can quickly handle the input and output of data, the necessary single or sequential controls, the operator and analytical instrument interfaces as well as alarms. Thus, if so programmed, it can constantly register and update the necessary control action to be taken on a certain part of the process. Process computers should be able to communicate to other process or supervisory computers (star or bus type distributed network).

Process or supervisory computers can also be used for some management functions, such as stock control, accounting, the measurement and registration of raw material and energy inputs and product and waste outputs, and the registration of given sets of process parameters. The statistical analysis of longe-range data makes possible not only the optimization of the process control, but also that of the whole process. For example, if specific energy consumption figures, such as tons of steam per ton of product, are known, then, any short-term deviation from the optimum can be easily corrected on-line.

Of course the aspects of reliability, security, capacity, maintenance, ease of altering programmes, price and space requirements will determine the type and extent of automation of any process in the oil and fat industry, in which, until recently, the level of automation was relatively low because of the slow development of the different processes applied.

References

General References

Applewhite, T. H. (1980). *Fats and Fatty Oils*. In: *Kirk-Othmer; Encyclopedia of Technical Chemistry* Vol. 9, 3rd edn. pp. 795–831 (Grayson, M., ed.) New York: Wiley.

Baltes, J. (1975). *Gewinnung und Verarbeitung von Nahrungsfetten, Grundlagen und Fortschritte der Lebensmitteluntersuchung*, Vol. 17. Hamburg: Verlag Paul Parey.

Bernardini, E. (1985). *Oilseeds, Oils and Fats*, Vols 1 and 2, 2nd edn. Milan: Publ. House BE.- Oil.

Gunstone, F. D. and Norris, F. A. (1983). *Lipids in Foods: Chemistry, Biochemistry and Technology*. Oxford: Pergamon Press.

Hamilton, R. J. and Bhati, A. (eds) (1980). *Fats and Oils: Chemistry and Technology*. London: Applied Science Publishers.

Hamm, W. (1983). Vegetable Oils. In: *Kirk-Othmer: Encyclopedia of Technical Chemistry* Vol. 23, 3rd edn., pp. 717–741 (Grayson, M., ed.). New York: Wiley.

Laisney, J. (1984). *L'huilerie Moderne; Art et Techniques*. Paris: Comp. Français pour le Developpemant des Fibres Textiles.

Markley, K. S. (ed.) (1950, 1952). *Soybean and Soybean Products*, Vols 1 and 2. New York: Interscience.

Perry, R. H. and Green, D. (eds), (1984). *Perry's Chemical Engineers' Handbook*, 6th edn. New York: McGraw-Hill.

Swern, D. and Applewhite, T. H. (eds) (1979, 1982, 1985). *Bailey's Industrial Oil and Fat Products*, Vols 1–3, 4th edn. New York: Wiley.

Thomas, A. (1987). *Fats and Fatty Oils*. In: *Ullmann's Encyclopedia of Industrial Chemistry*, Vol. A. 10, 5th edn., pp. 173–243. Weinheim: VCH Verlagsgesellschaft.

Treybal, R. (1981). *Mass Transfer Operations*, 3rd edn. Auckland: McGraw-Hill.

Journals

Fett, Wissenschaft, Technologie [Fat, Science, Technology; formerly *Fette, Seifen, Anstrichmittel]*
Grasas y Aceites
Journal of the American Oil Chemists' Society
Journal of Lipid Research
Oleagineux
Revue Francaise des Corps Gras
Rivista Italiana della Sostanze Grasse

Special References

Introduction

Hoffmann, G. (1986). Edible Oils and Fats. In *Quality Control in the Food Industry*, (Herschdoerfer, S. M., ed.), Vol. 2, 2nd edn., pp. 407–504. London: Academic Press.

Chapter 1

Gunstone, F. D. (1967). *An Introduction to the Chemistry and Biochemistry of Fatty Acids and their Glycerides*. London: Chapman and Hall.
Gunstone, F. D., Harwood, J. L. and Padley, F. B. (1986). *The Lipid Handbook*. London: Chapman and Hall.

Chapter 2

Armstrong, R. T. and Kammermeyer, K. (1942). Ind. Eng. Chem. **34**, 1228–1231.
Bailey, A. E. (1948). *Cottonseed and Cottonseed Products*. New York: Interscience.
Bernardini, E. (1976). *J. Am. Oil Chem. Soc.* **53**, 275–278.
Bird, H. R., Boucher, R. V., Caskey, Jr., C. D., Hayward, J. W. and Hunter, J. E. (1947). *J. Assoc. Off. Agric. Chem.* **30**, 354–364.
Caskey, C. D. and Knapp Frances, C. (1944). *Ind. Eng. Chem. Anal. Ed.* **16**, 640–641.
Christensen, C. M. (ed) (1974). *Storage of Cereal Grains and Their Products*. St. Paul, Minnesota: American Association of Cereal Chemists, Inc.
Coats, H. B. and Karnofsky, G. (1950). *J. Am. Oil Chem. Soc.* **27**, 51–53.
Coats, H. B. and Wingard, M. R. (1950). *J. Am. Oil Chem. Soc.* **27**, 93–96.
Dada, S. (1983). *J. Am. Oil Chem. Soc.* **60**, 409.
Fetzer, W. (1983). *J. Am. Oil Chem. Soc.* **60**, 203–205.
Florin, G. and Bartesch, H. R. (1983). *J. Am. Oil Chem. Soc.* **60**, 193–197.
Gemeinschaftsarbeiten der DGF (Deutsche Gesellschaft fuer Fettwissenschaft; German Society of Fat Science), Publications Nos 63. and 65 (in German): The Extraction of Oils and Fats from Vegetable Plant Raw Materials, I–II, (1975, 1976). *Fette, Seifen, Anstr.* **77**, 373–382; **78**, 217–223.
Hamm, W. (1983). Extraction, liquid–solid. In *Kirk-Othmer: Encyclopedia of Technical Chemistry* (Grayson, M., ed.) Vol. 9., 3rd edn., pp. 721–739. New York: Wiley.
Homann, Th., Knuth, M., Miksche, K.-D. and Stein, W. (1978). *Fette, Seifen, Anstr.* **80**, 146–149.
Johnson, L. A. and Lusas, E. W. (1983). *J. Am. Oil Chem. Soc.* **60**, 229–242.
Juristowsky, G. (1983). *J. Am. Oil Chem. Soc.* **60**, 226–228.
Karnofsky, G. (1986). *J. Am. Oil Chem. Soc.* **63**, 1011–1014.
Kehse, W. (1974). *Fette, Seifen, Anstr.* **76**, 15–18.
Kock, M. (1983). *J. Am. Oil Chem. Soc.* **60**, 198–202.
Lebrun, A., Knott, M., Beheray, J. and Poschelle, G. L. (1985). *J. Am. Oil Chem. Soc.* **62**, 793.
Mangold, H. K. (1983). *J. Am. Oil Chem. Soc.* **60**, 226–228.
Milligan, E. D. (1976). *J. Am. Oil Chem. Soc.* **53**, 286.
Nieuwenhuyzen, W. van (1976). *J. Am. Oil Chem. Soc.* **53**, 425–427.
Olomucki, E. and Bornstein, S. (1960). *J. Assoc. Off. Agric. Chem.* **43**, 440–442.
Othmer, D. F. and Jaatinen, W. A. (1959). *Ind. Eng. Chem.* **51**, 543–546.
Pardun, H. (1983). *Fette, Seifen, Anstr.* **85**, 1.

REFERENCES

Ringers, H. J. and Segers, J. C. (1977). U.S. Patent 4.049.686.
Schoenemann, K. and Voeste, Th. (1952). *Fette, Seifen, Anstr.* **54**, 385–393.
Spencer, M. R. (1976). *J. Am. Oil Chem. Soc.* **53**, 238.
Szuhaj, B. F. (1983). *J. Am. Oil Chem. Soc.* **60**, 306–309.
Tindale, L. H. and Hill-Haas, S. R. (1976). *J. Am. Oil Chem. Soc.* **53**, 265–270.
Ward, J. A. (1976). *J. Am. Oil Chem. Soc.* **53**, 261–264.
Zajic, J., Volhejn, E. and Jirousek, A. (1986). *Fette, Seifen, Anstr.* **88**, 231.

Chapter 3

Barlow, S. M. and Stansby, M. E. (eds) (1982). *Nutritional Evaluation of Long Chain Fatty Acids in Fish Oils.* New York: Academic Press.
Duppjohann, J. and Hemfort, Jr. H. (1975) *Fette, Seifen, Anstr.* **88**, 231.
Heublum, R. and Japhe, H. (1936). *Moderne Oelgewinnung und ihre Grundlagen*, a shortened version of the original Russian book: Goldowskij, A., *Theoretical Questions of Vegetable Oil Processing* Berlin: Allgemeiner Industrie-Verlag GmbH.
Rehbinder, P. A. (1933). *Kolloid-Zeitschrift* **65**: 268–283.
Reuter, H. (1982). *Fette, Seifen, Anstr.* **84**, 567.

Chapter 4

Duppjohann, J. and Hemfort, Jr. H. (1975) see Chapter 3.
Forster, A. and Harper, A. J. (1983). *J. Am. Oil Chem. Soc.* **60**, 265–271.
Gavin, A. M. (1983). *J. Am. Oil Chem. Soc.*, **60**, 420.
King, R. R. and Wharton, F. W. (1949). *J. Am. Oil Chem. Soc.* **26**, 389–392.
Kramer, J. K. G., Sauer, F. D. and Pigden, W. J. (eds.) (1983) *High and Low Erucic Rapeseed Oils, Production, Usage, Chemistry and Toxological Evaluation.* London: Academic Press.
Mag, T. K. (1973). *J. Am. Oil. Chem. Soc.* **50**, 251–254.
Segers, J. C. (1983). *J. Am. Oil Chem. Soc.* **60**, 262–264.
Singleton, W. A. and McMichael, C. E. (1955). *J. Am. Oil Chem. Soc.* **32**, 1–6.
Watson, K. S., Meierhofer, C. H. (1976). *J. Am. Oil Chem. Soc.*, **53**, 437.
Woerfel, J. B. (1983). *J. Am. Oil Chem. Soc.* **60**, 310–313.

Chapter 5

Albright, L. F. (1965). *J. Am. Oil Chem. Soc.* **42**, 250–253.
Albright, L. F. (1973). *J. Am. Oil Chem. Soc.* **50**, 255–259.
Bailey, A. E. (1949). *J. Am. Oil Chem. Soc.* **26**, 644–648.
Bailey, A. E. (1950). *The Melting and Solidification of Fats.* London: Interscience.
Boekenoogen, H. A. (ed.) (1964). *Analysis and Characterization of Oils, Fats and Fat Products.* New York: Interscience.
Van den Borre, H. (1983). Proceedings of the Third American Soybean Association Symposium on Soybean Processing: Hydrogenation of Soy Oil, 7–9 June, 1983, Antwerp, pp. 22–30.
Coenen, J. W. E. (1976). *J. Am. Oil Chem. Soc.* **53**, 382–389.
Coleman, M. H. and Fulton, W. C. (1961). In: *Enzymes of Lipid Metabolism* (Desnuelle, P. ed.), pp. 127–137. Oxford: Pergamon Press.
Eckey, E. W. (1956). *J. Am. Oil Chem. Soc.* **33**, 175–179.

Gemeinschaftsarbeiten der DGF, Publication Nos 57 and 58 (in German): The Interesterification of Edible Fats, I–II, (1974). *Fette, Seifen, Anstr.* **75**, 467–474 and 587–593.

Gemeinschaftsarbeiten der DGF, Publication Nos 67 and 69 (in German): The Hydrogenation of Fats I–II, (1976, 1977, 1978). *Fette, Seifen, Anstr.* **77**, 385–394; **79**, 181–195; **80**, 1–9.

Hamilton, R. J. and Bhati, A. (eds.) *Fats and Oils: Chemistry and Technology*, London: Applied Science Publishers.

Hannewijk, J., Haighton, A. J. and Hendrikse, P. W. (1964). *Dilatometry of Fats*. In: *Analysis and Characterization of Oils, Fats and Fat Products* (Boekenoogen, H. A. (ed.), Vol. I, pp. 119–182. London: Interscience.

Haraldson, G. (1983). *J. Am. Oil Chem. Soc.* **60**, 251–256.

Keppler, J. G., Schols, J. A., Feenstra, W. A. and Meijboom, P. W. (1965). *J. Am. Oil Chem. Soc.* **42**, 246–249.

Koslowsky, L. (1974). *Oleagineux*, **30**, 221–227

Latondress, E. G. (1983). *J. Am. Oil Chem. Soc.* **60**, 257–261.

Lush, E. J. (1923). *J. Soc. Chem. Ind. (London)* 219–225.

Patterson, H. B. W. (1983). *Hydrogenation of Fats and Oils*. London: Applied Science.

Patterson, W. J. (1938). U.S. Patent 2.123.811.

Proceedings of the Third American Soybean Association Symposium of Soybean Processing: *Hydrogenation of Soy Oil*, 7–9 June, 1983, Antwerp.

Puri, P. S. (1980). *J. Am. Oil Chem. Soc.* **57**, 850–854A.

Rakowski, A. (1906). *Z. Phys. Chem.* **51**, 321–340.

Rossell, J. B. (1973). *Chem. Ind.* **1st Sept**: 832–935.

Rozendaal, A. (1973). *Chem. Weekblad.* **28**, K7.

Sieck, W. (1936). U.S. Patent 2.054.889.

Van der Wal, R. J. (1960). *J. Am. Oil Chem. Soc.* **37**, 18–20.

Chapter 6

Boldingh, J. and Taylor, R. J. (1962). *Nature* **194**, 909–913.

Enden, J. C. van den, Haighton, A. J., Putte, K. van, Vermas, L. F. and Waddington, D. (1978). *Fette, Seifen, Astr.* **80**, 180–196.

FDA Standard, *Federal Reporter* **38**, 25672.

Haighton, A. J. (1959). *J. Am. Oil Chem. Soc.* **36**, 345–348.

Haighton, A. J. (1976). *J. Am. Oil Chem. Soc.* **53**, 397–399.

Hoffmann (1986). See Introduction.

Mattson, F. H. and Grundy, S. M. (1985). *J. Lipid Res.* **26**, 194–202.

Stuyenberg, J. H. van (1969). *Margarine*. Liverpool: Liverpool University Press.

Trenka, E. (1987). *Olaj, Szappan, Kosmetica.* **36**, 81.

USDA Standard (1983), *Federal Reporter* **48**, 52692.

Wiedermann, L. H. (1978). *J. Am. Oil Chem. Soc.* **55**, 823–829.

Chapter 7

Asahi Denko Kogyo, K. K. (1984). Japanese Patent 28482.

Macrae, A. R. (1983). *J. Am. Oil Chem. Soc.* **60**, 291–294.

Chapter 8

Hoffmann, G. (1961). *J. Am. Oil Chem. Soc.* **38**, 1–3.
International Association of Seed Crushers (1980). *Oilseeds, Oils and Fats*, 3rd edn. London: IASC.
Luyben, W. L. (1974). *Process Modeling, Simulation, and Control for Chemical Engineers*. Auckland: McGraw-Hill.
Marchand, D. E. (1986). Paper no 27 of "World Conference on Emerging Technologies in the Fats and Oils Industry", Cannes, France 3–8 November, 1985. *J. Am. Oil Chem. Soc.* **63**, 139.

Index

Acetone
  process, fractionation, 259–260
  removal of gossypol, 89
Acetune, cocoa butter extender, 340
Acetyl coenzyme A, activity, 9, 19–20
Acid oils (split soap stocks)
  for soap production, 160
  treatment, during refining, 158–160
Acid-resistant tanks, 159
Active oxygen method (AOM), fat stability, 332
Acyl carrier protein, 9
Acylglycerols (glycerides)
  density, 12
  fractionation retardation, 261
  interesterification, 264–277
  iodine value, 12
  nomenclature, 10
  physical and chemical properties, 10–12
  rate of crystallization, 252
  removal on stripping, 187
  structures, 2
  symmetric/asymmetric, 11
  terminology, 2
  see also Diacylglycerols; Monoacylglycerols; Triacylglycerols
Additives *see* Margarine, additives
Adipose tissue, structure, 123
Adsorbents, 164–167
  activated carbons, 167
  bleaching earths, 166
Adsorption
  chemisorption, 164
  physical, 164

Adsorptive bleaching *see* Bleaching
Aflatoxins, destruction, 114
Aglucons, 20
Air pollution
  avoiding, deodorization, 198
  in fat processing, 138
Alcohols
  components of lipids, 14
  fatty, 14
Alcon process
  degumming, 118
  soft oil desliming, 146
Aldehydes, off flavour, 216
Aldrin, removal, 187
Alkali-metal soaps, catalyst, poisons, 231
Alkali neutralization (deacidification)
  caustic-based, batch, 151–154
  fatty acid factor, 154
  general principles, 149–150
  refining factor, formula, 154
  re-refining, 153, 156–158
  semi-continuous, 154–155
  strong caustic-based continuous, 155–158
Ammonia, removal of aflatoxin, 114
Amygdalin, 20
Amyrins, 16
Anatto colouring, 309
Animal fats
  adipose tissues, 123
  back, leaf, 124
  bone fats, 130
  cracklings, 131
  killing, 124

372 INDEX

Animal fats (*cont.*)
  marine animals, 131–133
  pollutants produced, 138
  quality control, 137–138
  raw materials, pretreatment, 124–126
  rendering processes, 126–130, 360
  tallows, 130–131
Antinutrients, 113–114
Antioxidants
  activity, 25–26
  in esterification, 275
  in margarine, 320
  natural, 26
  synergists, 26–27
Antispattering agents, 308
Aspirators, separation technology, 49
Automation, energy needs, 362–363
Autoxidation, 22–27
  conjugation, 23–24
  primary oxidation products, 23
  in refining, 140
  retardation of autoxidation, 25–27
  secondary oxidation products, 24

Bakery creams, 330
Baking, fundamental principles, 330–332
Bancroft's empirical rule, 149
Batch deodorizers, 189–191
Batch extractors
  leaching equipment, 91–92
  multiple stage, 90
  single stage, 89–90
Batch processes
  hydrogenation, 231–238, 239–243
  margarine production, 312–314
Bentonites
  bleaching earths, 166
  discarded, animal feeds, contra-indications, 172
Berl's saddles, 233
BHA (Butylated hydroxy anisol), 26
Bingham liquids, 291
Bleached oils, storage, 352
Bleaching (decolorization)
  adsorptive bleaching, 168–169
  batch-wise operations, 169–170
  bleaching earths, 166–167
  cake filtration, 178–179
  calculations, 168–169
  chemical aspects, 167–168
  chemical bleaching, 182

  chemical, edible oils, 182
  filtration/separation, 172–178
  heat bleaching, 179–182
  hydro bleaching, 182
  physical aspects, 164–165
  polishing filtration, 179
  removal of oxidized glycerides, 118
  semi- and full-continuous bleaching, 170–172
  Thomson process, 172
Bones, production of fats, 130–131
Box presses, 71
Brassicasterol, 16
Butter
  flavour components, 310–311
  production, 281
Butylated hydroxy anisol, 26
Butyrospermol, 16

Cage presses, 70
  vs screw expellers, 75
Cake filtration, 178–179
Calciferol, 21
Campesterol, 16
Candelilla wax, 17
Candida, microbial lipase, 341
Carbons, activated, 167
Carbon monoxide, action on catalyst, 231
Cardiolipins, structure, 13
Carnauba wax, 17
β-Carotene, 309
Carotenoids
  acid sensitivity, 179
  as antioxidants, 25–76
  chemistry, 17–19
  decomposition in bleaching, 168
  heat sensitivity, 179–180
  structure, 18
Catalysts
  hydrogenation, 221–225
    commercial catalysts, 225
    dry-reduced, 223–224
    electrolyte precipitation, 224
    selectivity and structure, 208–209
    theoretical background, 204–211
    wet-reduced, 222
  interesterification, 264–265, 272–273
  isomerizations, 27, 212–213
  maintaining quality, 241
  poisoning by fish oils, 219
  poisons, 230–231

INDEX 373

precursors, interesterification, 264
spent, extraction, 242
types
    copper, 213
    nickel, 213
Caustic-based batch neutralization, 151–154
Celite, enzymatic interesterification, 341
Centrifugal separators, refining, 157
Chemical bleaching, edible oils, 182
   see also Bleaching
Chocolate, composition, 339
Chlorophylls, 19
Cholesterol
   in margarines, 301
   in palm oil, 15
   structure, 15
Cholesterol benzoate, double refraction, 148
Chromium carbonyl catalyst, 222
5,6-Chroman quinon, 219
Citric acid, addition to degumming process, 118
cis/trans configuration, 6–7
Clays, as bleaching earths, 167
Cleaning of seeds, 45–50
Cloud point, 246, 257
Cobalt carbonyl catalyst, 222
Cocoa butter
   chemistry, 339
   equivalents, 339
   extenders, 340
   melting point, SFC curve, 215
   replacers, 340
Coconut oil
   hydrolysis, 22
   low temperature crystallization, 258
Colours, permitted, 309–310
Computers, in processing, 362–363
Compactors, 79–80
Conditioning of bulk seed, 63–68
   before oil extraction, 64–65
   equipment, 65–68
Cone penetrometers, 292
Confectioners' baked goods, 331
Conjugation, in refining, 140
Contact systems of extraction, 89–90
Continuous process *see* specific processes
Convenience foods, 330
Conveyors
   bucket, 41

continuous flow, 42
flight, 42
pneumatic, 43–44
screw, 41–42
troughed belt, 41
Cooking, functions, 64
Cooking/conditioning of bulk seed, 63–68
Cooking, energy management, 358
Copper catalyst, 221–222
   see also Catalysts
Copra
   analysis, 34
   final moisture content, 64
Cotton seed
   dehulling, 62
   final moisture content, 64
Cotton seed oil
   analysis, 34
   cocoa butter extender, 340
   fractionation, 253
Cracklings
   fat separation, 131
   formation, 126
Creams, filled, 327, 330
Crude oils
   acetone—insoluble phosphatides, 161
   products of milling, 111–112
   refining losses, 161
   storage tanks, 345, 350
Crushing processes, 50–55
Crystal agglomerators, 260
Crystallization of fats
   demixing, 284–285
   mechanism, 250–252, 284
   polymorphism, 285–286
   retarders, 287
   supersaturation, 250
Crystallization vessels, 260
Cyanohydrin, 20
Cycloartenol, 16
Cyclopentanophenanthrene (sterol), structure, 15

D'Arcy equation, 173
DDT, DDE removal, 187
Deacidification, in refining
   deodorization by distillation, 182–198
   dry methods, 162–163
   by neutralization, 146–162

Deacidification, in refining (*cont.*)
    automation, 160–162
    process, 149–162
    theoretical background, 147–149
Decortication/dehulling, 55
  cotton seed, 62
  groundnuts and rape seed, 63
  soya beans, 57–59
  sunflower seed, 59–62
(Post)degumming, 143–146
  dry, 145
  intermediate, 146
  wet, 144
(Pre)degumming, 143
(Super)degumming reagents, addition to refining mixtures, 144
Dehydration, 140
Deodorization/deacidification
  chemical changes, 187–189
  effluent treatment, 198–200
  energy savings, 195–196
  equipment
    batch, 189–191
    continuous, 191–193
    continuous-film, 195
    cross-current continuous, 193–194
    semicontinuous, 194
    spherical liquid film, 195
    steam-jet ejectors, 196–197
  fat recovery, 198, 200
  losses, 197–198
  process conditions, 185–187
  steam stripping, 183, 185
  theoretical background, 183–184
Deodorized oils, storage, 352
Desliming, continuous post desliming sequence, 146
Desolventizer/toaster/driers, 103–105
  cooler, 106
  drier/cooler, 107–108
Dewaxing of oils, 246, 257–258
Diacylglycerols
  as basic catalysts, 264
  glucosides, 20–21
  structure, 2
Diatomaceous earth, 179, 341
  added to catalyst, 223
Dieldrin, removal, 187
Dimethyl sterols, 16
Disc dehuller, 60
Disc mills, 55, 56
Disintegrators, 55, 56

Distillation, chemistry of fatty acids, 4–5
Drying
  driers, 36–39
  rate of drying, curve, 36
  recirculation, 37–38
  theoretical background, 35–36
Dust, critical explosion limits, 110

Earths, bleaching
  adsorbents, 166–167
  de-oiling, 171–172
  spent, hazards, 172
ECC (end-of-carbon chain) system, 6
EDTA, sequestering agent, 27
EFA *see* Fatty acids, essential, 9
Eggs, use in mayonnaise, 325–326
Elaidinic acid, hydrogenation, 215–216
Elastic shearing modules, 291
Elastic solids, rheology, 290
Elevators *see* Conveyors; storage and transport
Emulsions and emulsifiers
  definition, 147
  filled creams, 327
  margarine processing
    addition, 308
    theoretical background, 283–284
  in mayonnaise, 325
  mesomorphic behaviour, 148
  pasteurization, 326
  theoretical background, 147–149
End-of-carbon chain (ECC) system, 6
Energy demands, 354
  cooling, 358
  data on the energy needs of processing, 358
  feed water pretreatment, 356
  heat loss prevention, 356
  internal power plant, 357
  steam generation, 355
Energy saving
  deodorization, 195
  internal power plant, 357
Equilibrium relative humidity, 32
Ergosterol, 16
Ester interchange *see* Interesterification
Essential fatty acids *see* Fatty acids
Ethenes, 5
Ethanol, as solvent in separation, 89
Ethylene diamine tetra-acetate, sequestering agent, 27

Ethylhydroxyethyl cellulose emulsifier, 160
Ethynoic acids, 5
Expellers
  L-form connection, 77
  screw presses, 71–79
Expression, mechanical, 68–80
Extraction processes
  continuous, 92–99
  definition, 81
  mechanical, 68–80
  solvent-aided, 80–111
  systems, 89–90

Fat modification processes
  historical notes, 201–203
  see also Fractional crystallization; Hydrogenation; Interesterification
Fat modifying reactions, 22–28
Fat scrubbers, 197–198
Fat separation processes
  principal methods, 122
    animal tissues, 123, 124–126
    marine animals, 131–133
    plant tissues, 123–124, 133–138
  rendering processes, 126–131
  summary, 46
Fat settling, grease trap, 199
Fats and oils
  classification, 1
  fate, during refining, 154
  formulation of blends, 293–298
  general characteristics, 1–3
  hardness, measurement, 292
  interesterification
    heating and drying, 272
    adding catalyst precursors, 272–274
    pretreatment, 272
    reaction, 274–275
    side reactions, 275–276
  major components, 3–14
  minor components, 14–22
  modification, reactions, 22–28
  pretreatment, 272
  quality control, 262–263
  reactions, 22–28
  rheology, 289–292
  separation *see* Fat separation processes
  solid fat content (SFC), 214, 297–298
  solid fat index (SFI), 12

  thixotropic character, 293
  work softening, 292
Fatty acids
  biosynthesis, 8–10
  classification
    saturated, 3–5
    unsaturated, 5–8
    other types, 8
  essential, biosynthesis, 9
  extraction
    distillation of free fatty acids, 162
    dry deacidification, 162
    re-esterification, 163
  factor, formula, 154
  free fatty acids
    and alkali neutralization, 153
    in sampling, 3
    removal on stripping, 187, 188
  interesterification, 264–267
  isomerism, 6
  nomenclature, 3–4, 5–6
  in natural fats, 266–267
  polyunsaturated (PUFA), 7, 9
    isomerization, 27, 205
    labelling, 282
    margarines, 301
  random distribution, 267
  removal during refining by deacidification, 146
  SAFA, in margarines, 301
  shorthand notations, 6–7
  table, 4
  in triacylglycerols, 265–266
Fatty alcohols, sources, 14
FFA *see* Fatty acids, free
Filtration
  adsorptive bleaching, 168–169, 172–178
  cake, 178–179
  depth filtration, 178
  disc-filter element, 175–176
  filtering devices, 173–178
  filter presses, 174
  fixed bed, 170
  Florentine filter, 254
  Kelly filters, 99
  media, 172–173
  of miscella, 99
  polishing, 179
  rate, formula, 173
  vertical disc filter, 177–178
  vertical leaf filter, 175, 178

Fish oils
  bulk shipment temperature, 354
  off-flavours, 217
  production, 132–133
Flaking, 50–55
Flash desolventizers, 105–106
Flocculation
  in emulsions, 148
  with lecithin, 117
  waste water disposal, 199
Fluidized-bed contactors, 67–68
Fluidized-bed driers/coolers, 107–108
Foreign material, separation, 45–50
Fractionation
  cost, 262–263
  dry, 253–255
  process steps, 252
  quality control, 263–264
  raw materials, 261–262
  solvent-aided, 259
  storage of oils, 351
  surrounding equipment, 260–261
  wetting agent mediated, 255–257
  winterization, 257–259
Free fatty acids (FFA) see Fatty acids
Freezing point, eutectic/monotectic, 247–248
Freundlich equation, 164–165
Frying oils, 332
Fuller's earth, 166

Gases, as solvents, 89
Ghees, 329–330
Gliding angles, 31–32
Glucosides, in fats, 20–21
Glue solution, 130
Gluten formation, 330
Glycerides see Acylglycerols
Glycerol, chemistry, 14
Glycerol-lactylpalmitate/stearate, 308
Gossypol, 19, 39
  removal, 89
'Green bases' (nickel silicate precipitates), 223
Grinding, 50–55
Ground-nut oils
  analysis, 34
  replacing, cocoa butter, 340
Groundnuts
  dehulling, 63
  moisture content, final, 64
  storage, 114
Gums, as catalyst poisons, 230

Haemagglutinin, 102
Haemin, 19
Haighton's relation, hardness, 292
Hammer mill, 55, 56
Hardened fats
  post-hydrogenation treatment, 243
  post-treatment, 243
Hardened oils, storage, 351
Hardness, measurement, 292
Heat exchangers, 195–196
Heat loss prevention, energy management, 356
Heat treatment (conditioning), 63–68
Henry's law, 212
Hexachlorhexane, removal, 187
Hexane
  amount used, 87
  composition, commercial, 88
  flash point, 100
  in fractionation, 259
  heat of vaporization, 100
  safety measures, 110
  solvent extraction, 171
  superheated to 160°C, 105
  and synthetic fibres, 172
  water treatment, 109
  see also Miscella
Horiuti–Polanyi mechanism, 212
Horizontal-basket extractor, 95–96
Horizontal rotary extractor, 97
Horizontal single-belt type extractor, 98
Humidity see Moisture content
Hydraulic presses, 68–71
Hydrobleaching, palm oil, 182
Hydrocarbon reforming process, 229–230
Hydrocarbons, in fats, 19–20
Hydrogen gas
  catalyst poisons present, 230–231
  iodine value, and calculation of amount required, 240
  quality specification, 226
  safety measures, 244
Hydrogen peroxide, treatment of lecithin, 119
Hydrogen sulphide, formation during extraction, 108–109

Hydrogenation
  activated adsorption, 204–206
  activity and selectivity, 206–209
  catalysts, 221–225
  cis/trans isomerization
    substrate/catalyst, 205
    trans selectivity ratio, 208
  concentration in oil, 210–211
  cyclic and polymeric side products, exclusion, 218–219
  'dead-end' systems 238
  dispersion of gas bubbles, 234–235
  economy, different processes, 242–243
  electrolysis of water, 226–229
  electrolytic hydrogen, energy needs, 361
  endpoint calculation, 240
  external gas circulation, 232–235
  external liquid circulation, 235–238
  'flash', 218
  general survey, 202–204
  hardened fats, post-treatment, 243
  heat of reaction, 206
  hydrobleaching, 219–220
  hydrogenation equipment, 231–238
  hydrogenation processes, 238–243
  and iodine value, 240
  isomerizations with different catalysts, 212–213
  isosuppressive hydrogenation, 217
  loop system, 235–238
  low-temperatures, 216–217
  mass-transfer effects, 209–211
  medium-temperature, 220
  Normann system, 232–235
  off-flavour reducing techniques, 216
  oil quality, 230–231
  pressure-swing adsorption, 230
  quality, 230–231
  quality control, 243–244
  reaction rate, effect of process variables, 211–212
  selectivity differences, effects, 214–216
  side products, 218
  solid fat content, 214–216
  steam reforming process, 229
  theoretical background, 204–211
  total, 220–221
  trans-promoting, 220
  two-stage, 220
  Wilbushevits (loop) system, 235–238
  see also Hydrogen gas
Hydrolysis
  principles, 22
  in refining, 140
Hydroperoxides
  acid-catalysed dehydration, 24
  decomposition, 167–168

Iljin oil-separation method, 124
Illipe
  cocoa butter extender, 340
  deacidification, 262
Immersion extractor, 93
Impact breakers, 55, 56
Impact dehuller, 60–61
Interesterification: ester interchange
  applications, 267–271
  'directed', 270–271
  energy needs, 361
  enzymatic method, 340–341
  hazards, 276
  mixtures, 269
  oils, storage, 351
  processing methods, 271–275
  quality control, 276–277
  side reactions, 275–276
  theoretical background, 264–267
Inventory control, 343–352
Iodine value (IV) and refractive index
  calculation of hydrogen, 240, 243–244, 293
  high IV margarine, production, 221
  low IV margarine, production, 220
  in measuring degree of unsaturation, 7–8, 12
Ion-exchange resins, 341
Isomerization
  fats, 27
  fatty acids, 6
  positional, 205
  in refining, 140
Iso-propyl alcohol (isopropanol), in fractionation, 259–260
  as solvent, 89
Iso-suppression, 220
Isothio-oxazolidones, 65
IUPAC rule system, nomenclature, 3–4
IV see Iodine value

Jojoba, fatty alcohols, 14

INDEX

Karitene, 20
Katharometer, use in quality control, 245
Kelly-type filters, 99
Kieselguhr *see* Diatomaceous earth
Knife bars, 72
Knife dehuller, 60
Kokum fat, cocoa butter extender, 340

Lanza fractionation, 255–257
Lard
  bulk shipment temperature, 354
  composition, 269
  interesterification, 269, 270
  palmitic acid, 266
  rendering process, 126–130
Lauric fats
  hydrogenation, 340
  interesterification, 268
  see also Coconut oil; Palm kernel oils
Leaching (solvent-aided extraction), 80–111
  calculations, 82–85
  definition, 81
  equipment, 91–99
  extraction systems, 89–90
  miscella filtration, 99
  safety, 110–111
  solvent recovery, 100–102
  solvents, 87–89
  special aspects, 85–87
  waste recovery, 108–111
Lecithins
  lysolecithin, 119
  production
    commercial production, 119
    critical temperature, 118
    predegumming of oils, 115–119
    uses, 115
  'sludges', 21
  structure, 13, 14
Legislation, margarine, 281–282
  fat/water content, 281
  terminology, 282
Leuconostoc, milk souring, 306–307
Licanic acid, 8
Lindane, removal, 187
α-Linoleic acid
  chemistry, 10
  frying oil, 332
  in health margarines, 301
  importance, 9–10
  notation, 7
Linolenic acid
  α and γ forms, 10
  hardening flavour, 216
  importance, 9–10
  α-methylenic structure, 23
  methyl esters
    hydrogenation, rates, 206
    saturation selectivity, 207–208
Liquid crystals, cholesterol benzoate, 148
Lipids, definition, 1
Lotox, 180
Lupenol, 16
Lycopene, structure, 17
Lye
  automatic ratio control, 161
  concentration
    and FFA content, 153
    and oil type, 153
  re-refining, 156–158
  strong continuous caustic process, 155–158
Lyotropic mesophase, 116
  definition, 160

Magnets, seed cleaning, 49–50
Maillard reactions, 113, 306
Maize germ, moisture content, final, 64
Malaxation, 136
Margarines
  additives, 304–311
    antioxidants, 310
    antispattering agents, 308–309
    citric acid, 307
    colours, 309–310
    cows' milk, 305
    emulsifiers, 308
    lactic acid, 307
    preservatives, 307
    tracers, 310
    salt, 309
    vitamins, 309
    vegetable proteins, 305
    water, 305
  bacterial souring, 306
  chemical souring, 307
  confectioner's, 303
  consistency of edible-fat products, 288–292

INDEX

creaming power, 300
'diet' margarines, 301–341
double-emulsion, 325
energy needs of production, 361
filling, of packaging, 334–336
flavours, additions, 310–311
formulation of fat blends, 293–298
general survey of processes, 280
hardness of fats and its measurement, 292
hardness and solid fat content, correlation, 298–299
'health', 282, 301, 302
historical background, 279–281
legal aspects, 281–282
iodine value (IV), 240, 243–244, 293
  high IV production, 221
  low IV production, 220
low calory spread (LOCAS), 282
low fat, production, 324
maturing, 323–324
microbiological quality, 337–338
oleomargarine, production, 130
packaging, 332–336
pasteurization, 306
production line
  cleaning, 338
  energy needs, 361
quality assessment, 336–338
rheology, 289–292
soft, 298–301
solid fat content (SFC)
  and hardness, 298–299
  product properties, 286–288
  table 302–303
  vs temperature, 300
spreading and creaming, 301
storage life, 336
technology of processing, 311–325
  Complector plant, 314
  continuous, 315–323
  discontinuous, 312–314
  hardness after storage, 323
  minimum-calory production, 324
  packaging, 323
terminology, 282
theoretical background
  crystallization of fats, 284
  demixing, 284
  emulsions and emulsifiers, 283–286
  polymorphism, 285–286
types produced, 299–303

vitaminization, 21
warehousing, 336
work softening, 292–293
Marine animals
  fat processing, 131–132
  source of fatty alcohols, 14
Mass-transfer effects, 209–211
Mass-transfer operations, 80–81
  extraction pseudophases, 82
Mayonnaise
  emulsions and emulsifiers, 283–284
  production, 325–326
Meals
  and cakes, 112–113
    antinutrients, 113–114
    quality indices, 113
    storage, 114–115
  drying and cooling, 105
Metal scavengers, 26–27
Micelles
  composition, 13
  critical micelle concentration, 149
  definition, 116
  differentiation from *miscella*, 82
Microbial lipases, 265, 341
Microwave drying, 38
Microwave heating, in vacuum, 68
Miers theory, 250–251
Milk fat, composition, 281
Milling processes
  main products, 111–115
  mechanical extraction, 68–80
  side products, 115–119
  solvent-aided extraction, 80–111
Miscella
  in batch extractors, 92
  calculations, 82–83
  definition (footnote), 82
  distillation, 100–101
  filtration, 99
  refining, 158
Mixes, convenience products, 330
Moisture content
  during milling, 78
  regulation
    drying, 35–39
    equilibrium moisture content, 32–34
    of oil seeds, *table*, 34
    storage defects, 39–45, 346–347
    types of moisture, 33
  urease activity *vs* residence time, 103

Monoacylglycerols
  emulsifiers, 308
  in margarine, 283–284
  structure, 2
Montmorillonites, soda-boiled, 172
Mucilage, as catalyst poison, 231
Myrosinase, action, 65

Naphthas, 88
Natural fats, 249–250
  see also specific names
Neutralized oils, storage, 352
Newtonian liquids, 290, 291
Nickel catalyst
  activity, 221
  commercial types, 225
  production, 222–225
  Raney nickel, 222
  waste control, 245
  see also Catalysts
Nickel/silver catalysts, 222
Nitrogen solubility index (NSI), 102, 113
2-Nitropropane, in fractionation, 259
Nomenclature, fatty acids, 3–4, 5–6, 10
Normann hydrogenation systems, 232–235

Oenothera, γ-linolenic acid, 10
Off-flavours, 24–25, 216–217
Oil processing, energy needs, 360
  see also Fats and oils: and specific headings
Oleic acid, cetyl alcohol, 14
Oleic acid methylesters, hydrogenation, rates, 206–207
Oleine
  definition, 246
  fractionation, 258
  partial hydrogenation, 340
  production, 245
  quality criteria, 262–263
  use, 259
Olive oil, production, 136–138
Organochlorine pesticides, removal, 187
Orthophosphoric acid, 3
Ostwald's empirical rule, 150

Packaging, 332–336

PAH see Polycyclic aromatic hydrocarbons
Pall rings, 108
Palm kernel oil
  analysis, 34
  bulk shipment temperature, 354
  crystallization, 287
  fractionation, 253
  hydrolysis, 22
  processing, 135
  quick processing, 29–30
Palm oil
  bulk shipment temperature, 354
  carotenoids, 179–180
  in chocolate manufacture, 340
  colour quality, 180
  earth and heat bleaching, 181–182
  enzymatic interesterification, 341
  fatty acid distribution, 268
  fractionation, 253
    energy needs, 361
  heat bleaching, 181
  hydrobleaching, 182, 219
  interesterification, 268
  milling, energy needs, 360
  mixture with lauric fats, 269–270
  prebleaching, 180
  pretreatment, 134
  processing, 135
  pyrolysis, 28
  separation, 134
  standardization, 180
Palm oleine see Oleine
Pan-releasing agents, 327
Pastries, 330–331
Perchloroethylene, de-oiling solvent, 171
Percolation extractors, 93–99
Penetrometers, 292
Phosphatides see Phosphoacylglycerols
Phosphatidyl choline (lecithin), structure, 13, 14
  see also Lecithin
Phosphatidyl ethanolamine, structure, 13
Phosphatidyl inositol, structure, 13
Phosphatidyl serine, structure, 13
Phosphoacylglycerols
  characteristics, 13–14
  definition, 3
  effect of degumming, 116, 117–118
  non-hydratable, 118

INDEX

structure, 13
see also Lecithin
Phospholipase inactivation, 119
Phospholipids
   solubility in hexane, 86
   see also Phosphoacylglycerols
Phosphorus, as catalyst poison, 230–231
Photo-oxidation, 25
Phytomenadion, 22
Pigments *see* Carotenoids; Chlorophylls; Gossypol
Plant fats, separation methods, 123
Plastic solids, rheology, 291
Plate heat exchanger, 101
Plate presses, 70–71
Polishing filtration of fines, 179
Pollutants, from processing methods treatment, 138
   see also Air pollution; Water
Polycyclic aromatic hydrocarbons
   health risks, 38–39
   removal, 157
Polymerization of fat
   chemistry, 28
   in refining, 140
Pot presses, 70
'Premier jus', 130
Pressure-swing adsorption (PSA) process, 230
Pristane, 20
Programmable logic controllers, 362
Propionic acid, 8
Prostaglandins, 10
Protein
   changes during extraction, 112–113
   dispersibility index (PDI), 102, 113
   meals and cakes, content, 112–114
PUFA *see* Fatty acids, polyunsaturated
Pyrolysis
   of carotenoids, 28
   of fats, 28
   on refining, 140
   triacylglycerols, 188–189

Rail and road transport *see* Storage and transport
Rape seed
   analysis, 34
   dehulling, 63
   formation of H_2S, 108
   pre-degumming, 117
   storage, 114
   oil, interesterification, 269
Raschig rings, 100, 108
Raw materials, vegetable oil sources, 29–45, 340
Refining
   automation, 161–162
   bleaching, 163–182
   chemical *vs* physical, 140–146
   deacidification by neutralization, 146–162
   deodorization/deacidification by distillation, 182–198
   dry deacidification methods, 162–163
   effluent treatment, 198–200
   fate of fat classes, 142
   general survey, diagram, 141
   quality control, 200
   terminology, 139
Refining factor, formula, 154
Refractive index, and iodine value, 7–8, 12, 243–244
Rendering processes
   continuous, 128–130
   large scale batch, 126–128
   'premier jus', 130
   small scale batch, 126
   tallows, 130
   Titan wet-rendering, 127
Residual oil, extraction, *see* Leaching
Retinol, 21
Rheology of fats, 289–292
Rhizopus, microbial lipase, 341
Ricinoleic acid, 8
Roll(er) mills, 51–55
   capacity, 52
   disc mills, 55
   fluted surfaces, 51–53
   impact breakers, 55, 56
   smooth surfaces, 54–55
Rotary driers, 37–39
Rotational viscometers, 292
Ruminants, decomposition of fatty acids, 8
Russian dehuller, 60

SAFA labelling, 282
Safety measures, oil separation, 110–111

Safflower seed
  analysis, 34
  dehulling, 62
Sal fats, 340
Salt, added to margarine, 307
Sapogenins, 21
Saponification value, 12
Saponins, 20–21
Scales, bulk, 44–45
Schnecken type meal desolventizer, 41–42
Schnecken heaters, 103
Screening, 47–49
Screw press (expeller), 71–79
  de-oiling techniques, 76
    maintenance, 73
      working, 73–76
      vs conventional, 75
Seed-oil milling, energy needs, 359
Seeds
  botanical structure, 55–56
  cleaning, 45–50
  storage, 31–32
  see also Vegetable oil; and specific plants
Sedimentation tanks, prerefining, 143
'Selective wetting' of surfaces, 121, 123
Separation technology *see* Vegetable oil separation
Sequestering agents, 26
Sesamolin, 20
Shea fat, 20
  cocoa butter extender, 340
  deacidification, 262
Shear stress, and deformation rate, 290
'Shift' reaction, hydrogen production, 229
Shortenings, fluid, plastic, solid, 327–329, 331
Silos, 31–34, 39–45
Simmondsia, fatty alcohols, 14
β-Sitosterol, 16, 20
Size reduction, 50–55
Skipin oil separation method, 124
Soaps
  formation, 5, 149
  salting-out, 149
Soapstock phase
  collection, 158
  miscella refining, 158
  neutralizations, 160

as poultry feed, 160
removal, 156
splitting, 159
treatment, 158–160
Sodium alcoholate catalyst precursor, 272, 276
Sodium hydroxide/glycerol catalyst precursor, 273
Sodium, metallic catalyst precursor, 273
  hazards, 276
Sodium/potassium catalyst precursor, 270, 274
Sodium xylene sulphate, 160
Solid fat content (SFC)
  statistical equivalents, 297–298
  vs temperature, 214
Solid fat index (SFI), 12
Solvent(s)
  desolventization, 101–102
  ideal, 87–88
  recovery from fractionation, 261
  safety, 110–111
  types, 87–89
  waste recovery, 100–102, 108–111
  see also Leaching
Soya beans (oil)
  analysis, 34
  degumming, 117
  dehulling, 57–59
  extraction, H_2S formation, 108–109
  fractionation, 253
  interesterification, 269
  moisture content, final, 64
  sapogenins, 102
  solid fat content *vs* temperature, 214
  storage, 114
  storage time, allowable, 39–40
  toasting, 102–104
  trypsin inhibition, 113–114
'Speciality fats', 341
Spreads, 326–327
Squalene, structure, 20
Squirrel-cage treadmill dehuller, 60
Staley process, 118
  super degumming progress, 118
Steam generation, energy demands, 355–356
Steam-jet ejectors, 196–197
Steam/natural gas reforming process, 229–230
Steam stripping, 185–186

Steam tube rotary drier, 37
Stearic acid, desaturation, 9–10
Stearines
  biosynthesis, 9
  definition, 245–246
  fractionation, 258
  uses, 259, 262
Steroids, properties, 15
Sterols
  components of lipids, 13–16
  dehydration, 168
  glucosides, 20–21
  losses on deodorization, 187
Stick water, 130
Stigmasterol, 16
Storage and transport
  air access, 348–349
  bleached oils, 352
  bulk storage, 31–34, 39–45
  bulk transport
    external, 29–34
    internal, 40–44
  contaminants, 349–350
  crude oils, 350
  deodorized oils, 352
  external transport
    rail, 353
    road, 353–354
    water, 352
  fractionated oils, 351
  hardened oils, 351
  heat, 347
  interesterified oils, 351
  light, 347
  internal transport, 344
  inventory control (weighing), 343–352
  moisture *see* Moisture content
  neutralized oils, 350
  special cases, 350–352
  storage conditions, 345–348
  storage tanks, crude oils, 345
  vegetable oil sources, 29–34, 39–45
Streptococcus, milk souring, 306–307
Sucrose polyesters, 'synthetic fats', 301
Sulphur, as catalyst poison, 230
Sunflower seed
  analysis, 34
  dehulling, 59–62
  husks, fuel energy, 57
  moisture content, final, 64

Tallow
  bulk shipment temperature, 354
  fractionation, 253
  rendering, 130
Thioglucosinolates, 20
Tirtiaux process, 254
'Toasting', 102–108
Tocol, tocotrienols, 16
α-Tocopherols
  as antioxidants, 187
  chemistry, 21–22
  components of lipids, 16–17
  in interesterification, 275–276
  properties, 16–17
  structure, 15, 16
Tocotrienols, 16
Totox, 180
Transesterification, 271
  see also Interesterification
Transport *see* Storage and transport
Triacylglycerols
  biosynthesis, 10–11
  classification, 265–266, 294
  crystallization, 247–248
  fatty acid distribution, 265–266
  in formulation of blends, 293
  MP, 246
  optical activity, 12
  phase diagrams, 248
  physical characteristics, 11
  polymorphism, 285–286, 295–296
  positional isomerism, 11
  pyrolysis, 188–189
  random fatty acid distribution, 267
  SSS, SSU, SUU, UUU types, 246
  selectivity, 208
  structure, 2
Trichloroethylene, toxicity, 88
Trielaidinglycerol, MP, 215
Triglycerides *see* Acylglycerols;
  Triacylglycerols
1,2,3-Trioxypropane, 14
Trypsin inhibiting factor, 102
  deactivation, 133–144

Urease activity *vs* residence time, 103

Vanaspati, 329–330
Vapour contractor, 101
Vegetable butters, 340

Vegetable oil separation, technology
  drying, 35–39
  general survey, 122–124
  milling, 68–111
  preparatory treatments, 45–68
  products, 111–119
  raw material handling, 29–45
  storage and transport, 29–34
Vent-gas treatment, 108–109
Vernolic acid, 8
Vertical-basket extractor, 94–95
Villavecchia–Baudouin test, 20
Viscometers, 292
Viscosity, rheological constant, 290
Vitamin A, 21
  carotenoids, 17–18
Vitamin D, 21
Vitamin E, 21–22
  activity, 16
Vitamin K, 22

Warehousing *see* Storage and transport
Waste recovery, 108–111
Water
  feed water, pretreatment, energy
    demands, 356
  pollution
    acid oil formation and removal, 160
    from fat processing, 138
  and storage *see* Moisture content
  transport, 352
    see also Storage and transport
  treatment, 109
  waste, disposal, 199–200
  see also Drying; Moisture content
Water activity, of seed, 32
Water-cooling, 198
Waxes
  characteristics, 17
  definition, 17
Weighing, bulk, 44–45
Whale oil
  bulk shipment temperature, 354
  processing, 131
  rendering, 132
  source of fatty alcohols, 14
Wilbushevits hydrogenation system, 235–238
Winterization, 246, 257–258

Young modulus, 290

Zenith semi-continuous refining line, 145
Zeta potential, 148